Numerical methods in offshore piling

Numerical methods in offshore piling

Proceedings of a conference organized by
The Institution of Civil Engineers and
held in London, 22–23 May 1979

Institution of Civil Engineers
London, 1980

ORGANIZING COMMITTEE
I. M. Smith (Chairman)
P. George
W. J. Rigden

PRODUCTION EDITOR
Thelma Darwent

Published by The Institution of Civil Engineers, PO Box 101, 26–34 Old Street, London EC1P 1JH, and produced and distributed by Thomas Telford Ltd at the same address

First published 1980

ISBN: 0 7277 0086 3

Typeset and printed by Henry Ling Ltd at The Dorset Press, Dorchester, and bound by Leighton Straker, London

Contents

Conversion factors

Length	1 in = 0.0254 m 1 ft = 0.3048 m
Second moment of area	1 in^4 = 41.62 × 10^{-8} m^4
Mass	1 lb = 0.4536 kg
Force	1 lbf = 4.448 N 1 tonf = 9.964 kN 1 kip = 4.448 kN
Moment of force	1 lbf ft = 1.356 N m 1 lbf in = 0.1130 kN m
Pressure	1 lbf/in^2 = 6.895 kN/m^2 1 UStonf/ft^2 = 95.76 kN/m^2

1. A survey of numerical methods in offshore piling

I. M. SMITH, BSc, MS, PhD, MICE (University of Manchester)

A review of numerical methods used in the offshore piling industry is given. The purpose of these is to compute the static and dynamic behaviour of single piles and of pile groups at working load and at 'failure' for various types of loading, principally axial and lateral. Using discrete element methods (the 't–z' and 'p–y' types of calculation) it is possible to follow accurately the load transfer mechanism operating between single piles and the ground up to collapse, for monotonic and cyclic loading. The calculation is well within the scope of mini-computers, subject to adequate discretization of the ground's resistance. Back-analyses are essential. Using finite element methods, load transfer can be predicted, for single piles at working loads, using ground resistance parameters which are in everyday use in soil mechanics. It is within the scope of faster computers to continue such computations up to collapse, and this is a likely development. A particularly interesting case of 'failure' of single piles occurs during installation. Attempts to link driving resistance with ultimate static capacity still depend on the use of parameters which are not measured every day in soil mechanics laboratories. Drivability itself is reasonably predicted, particularly at soft sites, and back-analyses are encouraging. Flutter has been identified as an instability mechanism worthy of consideration. Analysis of pile group behaviour still rests heavily on the assumption of linear ground properties and on the principle of superposition. Errors of 20% or more are not uncommon in the computation of influence factors for single piles if the pile is inadequately represented. Effective stress analyses depend on a better knowledge of excess pore pressures due to driving, and on more realistic ground stresses after installation than can be measured or computed at present. Dynamic analyses of piles and groups in situ (as distinct from during driving) subject to wave and earthquake loading are at an early stage of development, but will clearly be pursued intensively in the future.

INTRODUCTION

Numerical methods have been widely used in the offshore piling industry for the past 25 years or so. This Paper cannot attempt to be a literature review, but merely sets out the current state of achievement and tries to point the way to future developments. In addition it draws together the various strands of work presented at this Conference. These fall naturally into four main subject areas; namely, quasi-static behaviour of piles and groups, drivability, pore pressure considerations and dynamics.

QUASI-STATIC BEHAVIOUR OF PILES AND PILE GROUPS

Single axially loaded pile

2. *Deflexion: the t–z method.* A basic problem is the calculation of the deflexion of a single axially loaded pile subjected to prescribed load or displacement at the head. Even this problem is quite intractable without resort to numerical methods. For these purposes the pile can be discretized as a series of finite difference stations or as a series of line 'finite elements', and the ground as a series of discrete axial 'springs' or as some continuously distributed axial spring stiffness (Fig. 1). Although called springs, the ground resistance–displacement relationships can be as complicated as is necessary. These are usually called t–z curves and can be introduced into the computer program either as mathematical functions or, more usually, as a series of points between which linear interpolation is assumed (Fig. 2).

3. Various methods can be used in the computation to follow the prescribed t–z curves. The most modern and efficient are borrowed from genuine finite element analysis and work with a constant stiffness in each ground spring[1] (e.g., the slope of the first segment of the t–z plot (Fig. 3)). As loading proceeds, any excess force in a spring over and above the t which ought to be carried for that value of z is redistributed to the other springs by processes called 'initial stress' or 'viscoplastic strain' in finite element work. In these methods, the simultaneous equations have constant coefficients and are merely re-solved for varying loads. Displacements at the pile head rather than forces should be prescribed for two reasons. First, it is more efficient since fewer iterations are required in the numerical process (typically two before failure is approached) and secondly, displacement control is the only way of continuing the analysis beyond peak load on the pile. Some of the older finite difference algorithms are rather cumbersome by modern standards.

4. This type of calculation is well within the scope of mini-computers. Of course, the difficulty lies in selecting the t–z curves appropriate to various soil types and conditions. The suggestion has been made[2] of the dimensionless relationship

$$\frac{t}{t_{max}} = 2\left\{\frac{z}{z_c}\right\}^{1/2} - \frac{z}{z_c}$$

for the side springs, where t_{max} is the maximum soil resistance which is mobilized at a critical displacement z_c. For the end bearing spring, the corresponding suggestion is

$$\frac{t}{t_{max}} = \left\{\frac{z}{z_c}\right\}^{1/3}$$

5. Figure 4 shows how field data from test piles in sand and clay can be back-analysed using this approach.[3] A

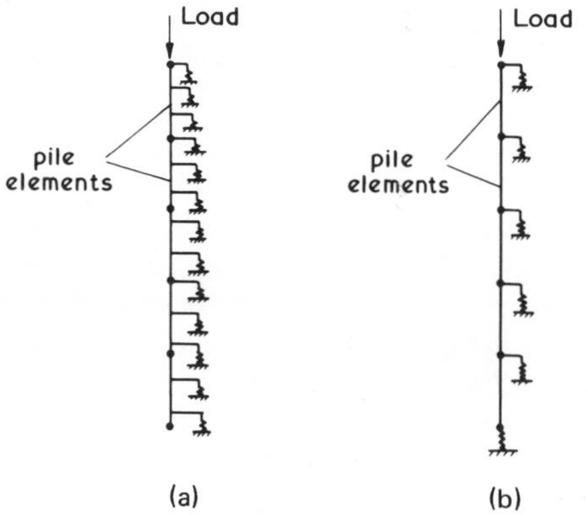

Fig. 1. Discretization of pile and ground: (a) continuous axial ground resistance; (b) discrete axial ground resistance

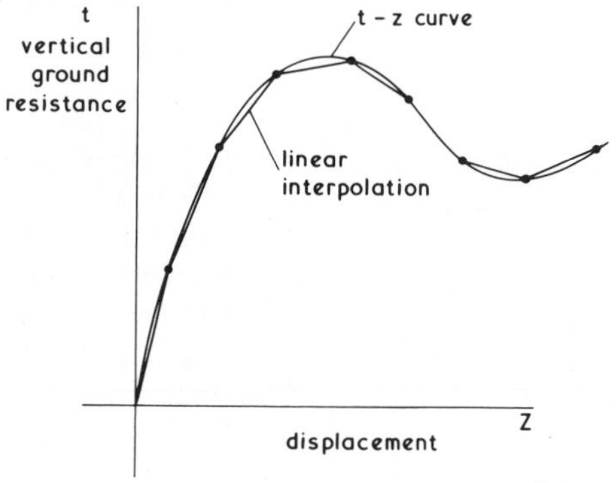

Fig. 2. The t–z approximation

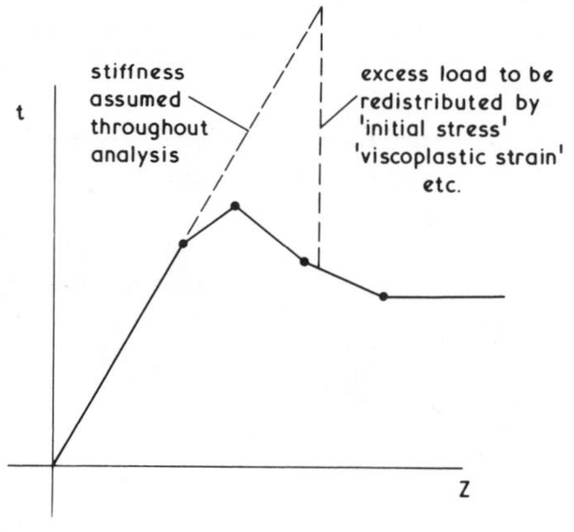

Fig. 3. Load redistribution process

large number of such fits would be necessary to build up confidence in the use of the method in new situations.

6. *Deflexion: finite element methods (Paper 11).* An improved representation of the ground is as a solid, in the simplest case an elastic solid bonded to the pile. However, mesh design problems arise when modelling pipe piles with open or closed ends. It is difficult to achieve both the right end bearing area and the right pile stiffness at that diameter when analysing equivalent solid piles.

7. The real benefits come when non-linear, stress-dependent properties are taken for the soil, together with slippage allowance between pile shaft and soil. For example, Desai[4] originally showed that the load transfer in a pile in sand is quite non-uniform with depth, as shown in confirmatory calculations in Fig. 5. When a field test was back-analysed by this method, the load–displacement curve and load transfer profile could be rather well reproduced in such calculations.[3] However, the non-linear elastic assumption for the soil and interfaces means calculation becomes unreliable when a large number of elements 'fail', so ultimate loads are best not computed in this way.

8. *Deflexion: boundary element methods (Paper 14).* Particularly when ground conditions are uniform and linear stress–strain properties can be assumed to prevail, boundary element methods can be superior to finite elements because fewer equations have to be solved. Poulos and Davis[5] have provided widely used charts based on a simplified form of this method, subsequently somewhat refined by Butterfield and Banerjee.[6] For layering and other forms of non-homogeneity, or when non-linear soil properties have to be considered, the method is less attractive.

9. *Failure.* The t–z computations for load–deflexion can be continued to collapse, and by means of displacement control can take residual conditions into account. The non-linear elastic type of finite element calculation is not recommended for computing collapse. Instead, initial stress or viscoplastic strain algorithms[7] should be used. Examples of displacement fields at collapse of deep foundations in cohesive and cohesive–frictional materials are shown in Fig. 6, together with load–displacement graphs for base pressure. By these means, the bearing capacity factors N_c, N_q and N_γ can be obtained numerically and the load transfer mechanism at failure identified.

10. Boundary element methods can of course in principle be used in this area, but have not so far found practical application.

11. *Cyclic loading (Paper 16).* An important feature of offshore loading conditions is their cyclic nature. It is well known that under (slow) cyclic loading, engineering materials degrade and become softer and weaker. Because of their particulate nature, clays are prone to the formation of low strength, slickensided rupture surfaces under large and repeated alternating displacements. The t–z and finite element methods can cope with cyclic loading, given that the material behaviour can be defined.

12. For example, Fig. 7 shows a possible t–z behaviour for side springs under cyclic loading.[1] Peak t is a function of N, the number of cyclic load applications, as is the ratio of peak t to displacement z at which it is attained. Fig. 8 shows typical results of this kind of computation for varying cyclic load (displacement) amplitude. At lower levels stabilization takes place but as the level increases, the pile fails in cyclic loading. Tip resistance can be ignored in tension and so on.

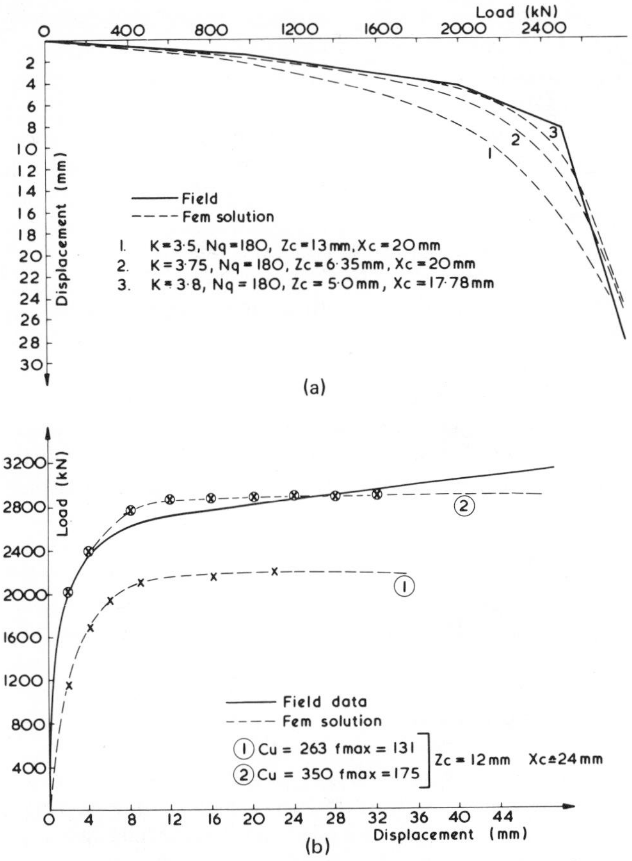

Fig. 4. Back-analysis of field results by t–z method: (a) closed-ended pile in sand; (b) pile with shoe in clay

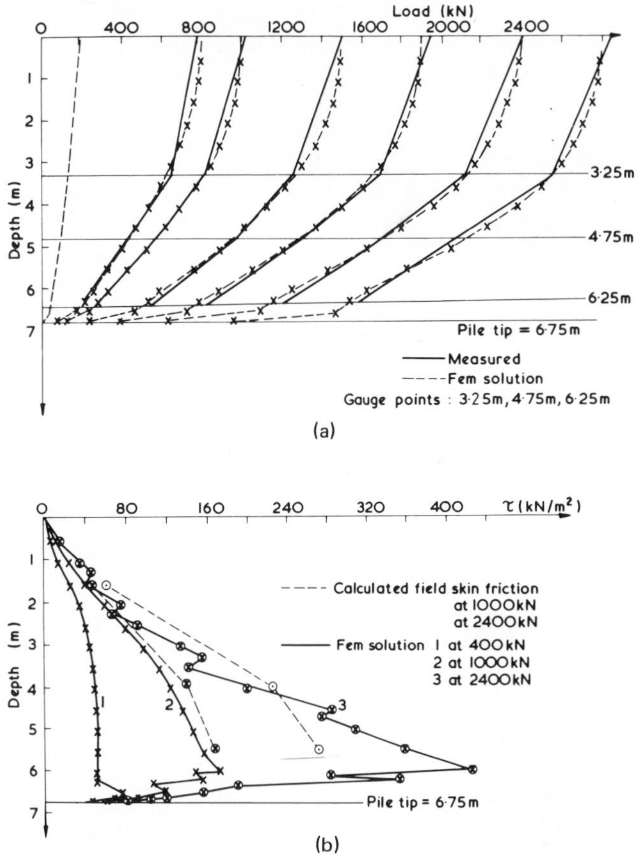

Fig. 5. Back-analysis of field results by finite element method: (a) load distribution of closed-ended pile; (b) graph of τ with depth for various load increments

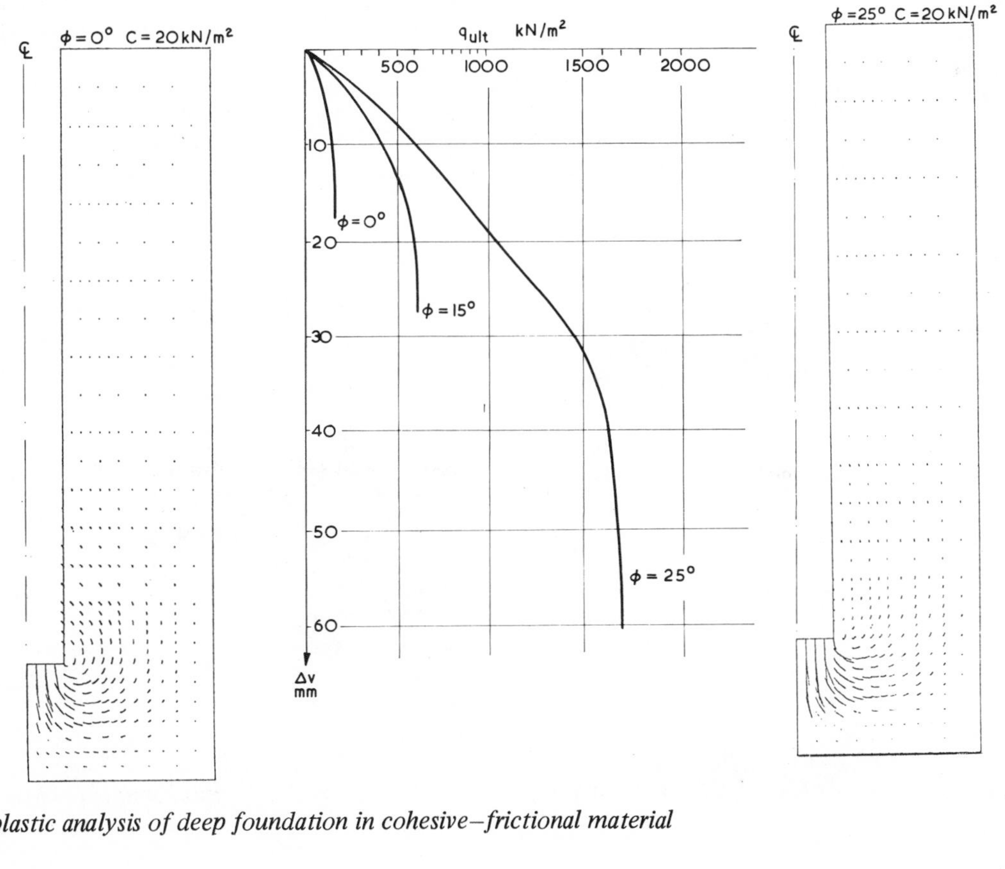

Fig. 6. Viscoplastic analysis of deep foundation in cohesive–frictional material

13. Another feature is the generation and dissipation of pore pressures during cycling. Finite element approaches can deal with this, but practical cases do not seem to have been solved yet.

Groups of axially loaded piles

14. *Deflexion (Papers 14 and 15).* For deflexion of groups of axially loaded piles, the boundary element methods come into their own. The t–z approach ignores interaction completely and the three-dimensional nature of the problem makes finite element computations expensive, even for linear soils.[8] Therefore linearity of the soil's stress–strain response is usually assumed for the purposes of interaction computations, and so boundary elements are attractive. Charts have been produced by Poulos and by Butterfield and Banerjee which enable group behaviour to be computed for various typical pile geometries. Sometimes the interaction factors from such a linear analysis are combined with t–z curves for single piles to yield an empirical non-linear group behaviour.

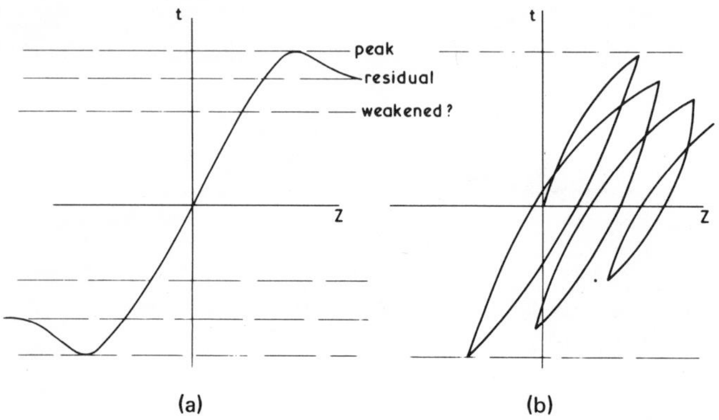

Fig. 7. Degradation of ground: (a) static loading; (b) cyclic loading

15. *Failure.* Three-dimensional finite element computations can be done, but are rather expensive. If an equivalent axisymmetric pier can be assumed, stability parameters follow.

16. *Cyclic loading.* The writer is not aware of work of this nature.

Single laterally loaded pile

17. *Deflexion: the p–y method (Paper 17).* The computations in the p–y method are entirely analogous to those of the t–z method, with p replacing t and y replacing z. As long as satisfactory curves can be specified this calculation is very quick and cheap.

18. *Deflexion: finite element methods (Paper 12).* For axisymmetric pipe piles the commonly used simplifications for axisymmetric structures under non-axisymmetric loads can be used. Displacements and so on are expanded in Fourier series so that the analysis becomes quasi-two-dimensional. As long as the geometry remains axisymmetric there is no difficulty in incorporating layered soils. Typical results for the deflexion of a pipe pile under lateral load[9] are shown in Fig. 9, together with those originally published by Poulos,[10] who assumed that the pile was a thin rectangular strip, and used a boundary element method. The latter can overestimate deflexions by 25% or more. The power of the finite element method is shown in Fig. 10, where deflexion profiles in layered soils are given. The radius of influence is computed to be typically 10 pile diameters for homogeneous soil and 6 pile diameters for soils increasing in stiffness with depth. The depth of influence is never greater than about 6 pile diameters. In the boundary element method (e.g., Banerjee and Davies[11]), rather radical simplifications have to be made to cope effectively with arbitrary inhomogeneity. The finite element technique has been extended to consider non-linear soils.[12] The difficulties here are merely in storing enough information about the circumferential variations in properties.

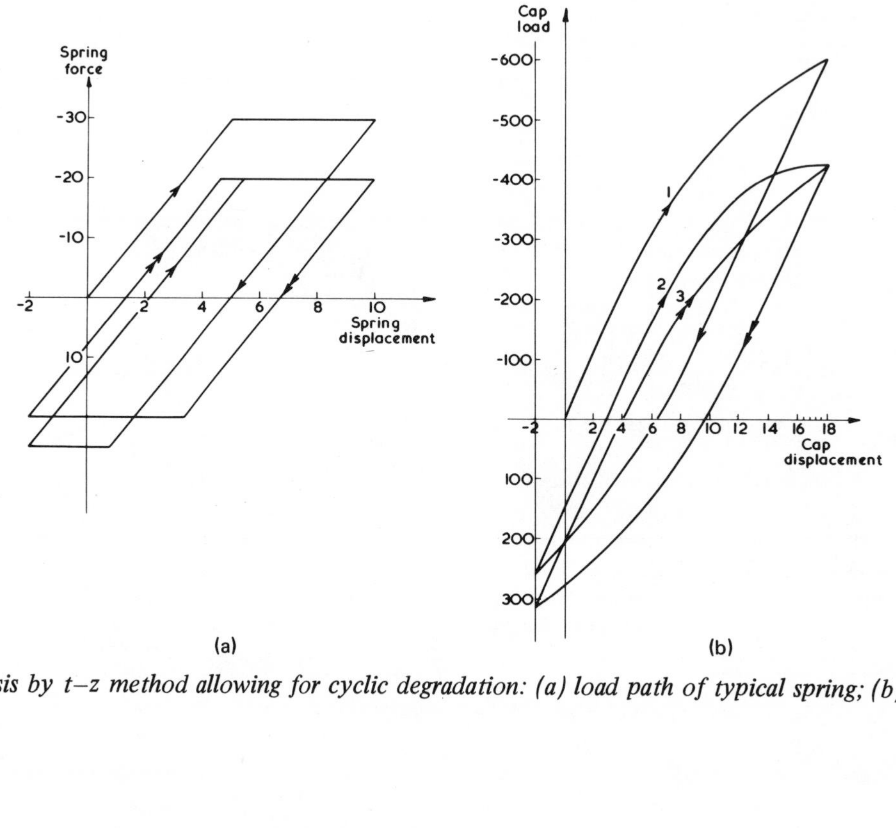

Fig. 8. Pile analysis by t–z method allowing for cyclic degradation: (a) load path of typical spring; (b) load–displacement response of pile

19. *Deflexion: boundary element methods (Paper 13).* With the provisos given above, the boundary element technique is quite suitable for analysis of linear problems, especially in homogeneous soils.

20. *Failure.* Because of the very high bending moments at the mudline carried by piles when laterally loaded, material failure in the piling is much more likely than soil failure for deep-driven offshore piles. Stiff, stubby piles could fail by a rotating mechanism but this does not seem to be of practical interest.

21. *Cyclic loading.* Comments are entirely analogous to those made about axial loading.

Groups of laterally loaded piles (Papers 13 and 18).

22. For groups of laterally loaded piles, the boundary element method has again seemed the natural choice. As was the case with axially loaded piles, empirical marriage of linear interaction factors with non-linear $p-y$ response is often attempted to obtain practical solutions.

DRIVABILITY

The one-dimensional wave equation

23. Because of the difficulties of ordering equipment ahead of time, fleeting weather windows and so on, predictions of pile drivability have assumed great importance in offshore operations. The techniques are not widely used in on-land situations, in the UK at least. The original recommendations of E. A. L. Smith with respect to material parameters such as elastic compressions appear to be adhered to. Probably the major difficulty attaches to the estimate of the viscous component of resistance (Smith's parameter *J*). It has been pointed out that the success of the method in estimating pile set may well be due to the insensitivity of this quantity to the method of computation.[13] Other factors, which are often measured in instrumented pile tests, are much more sensitive to the method of computation employed. It is also fair to say that predictive capacity has turned out to be much better in soft as opposed to hard sites, especially for clays and clay–sand mixtures.

Special problems of offshore piling (Papers 4–7)

24. Apart from the large scale of the operations, involving very massive piling and novel capacities of driving equipment, special problems concern, for example, gravity connectors. Offshore piles tend to be driven through a long follower at the end of which is a heavy connector. Thus in the wave propagation procedure there is a significant reflection back up the pile from the connector, and moreover a separation (no tensile stress transfer) occurs. Nevertheless, impressive back-analyses of driving records on some sites have been achieved.[14] If this can be done, an obvious extension is to analyse the driving record in situ and hence to predict the ultimate static capacity of the pile, thus preventing costly over-driving. Considerable experience of these techniques has been built up on land sites in certain areas of the world[15] and it remains to be seen how general the extrapolation procedure is. Again the difficulty in the drivability phase is the viscosity effect, and one would expect 'sands' to be more amenable to prediction than 'clays'. In addition the phenomenon of set-up due to pore pressure effects is clearly an additional difficulty in 'clays'.

Effect of curvature and/or kinks

25. Conductor piles can be deliberately driven with a curvature, the better to exploit the resources in a reservoir. Alternatively piles can be imperfectly welded so that there is an induced curvature or even a sharp kink between adjacent sections. Because of the great length of conductors particularly, concern has been expressed as to the effects of such disturbances on the drivability predictions, and on the stresses in the piling and forces on the guides.

26. The problem has been tackled in a rather mathematical way by Fischer,[16] in the form of finite difference

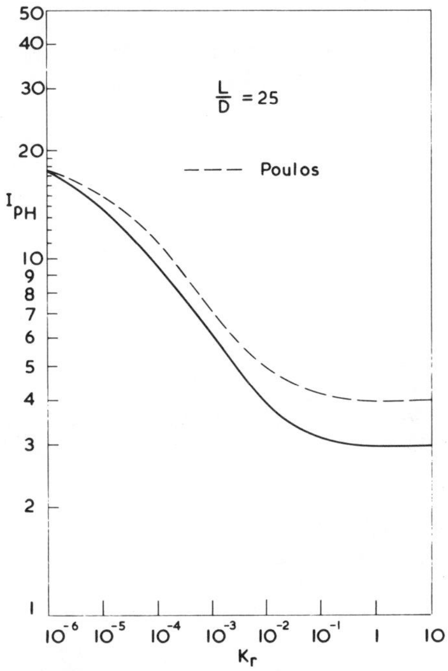

Fig. 9. Finite element analysis of laterally loaded pile in uniform ground; L/D = 25

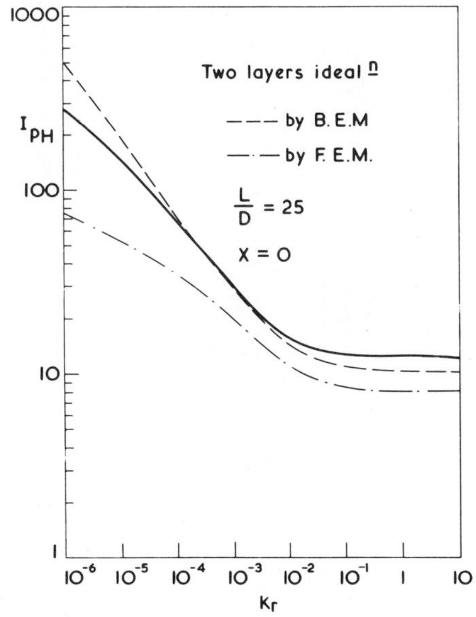

Fig. 10. Laterally loaded pile in layered ground; L/D = 25, x = 0

approximation to a rather obscure pair of coupled differential equations formulated by Isakovich and Komarova.[17] A simpler method of attack, which shows immediately whether curvature and/or kinks have any great significance, involves finite element approximations of typical piles.[18] As shown in Fig. 11, the pile elements can be genuinely curved, or curves can be approximated by a series of kinks. In either case there will obviously arise coupling between the compressional wave and the transverse motions of the pile. If the kinked pile is a reasonable approximation to the truly curved one, this representation is clearly preferable, since arbitrary kinks can readily be treated.

27. If curvature is to have any effect on the driving process it must be manifested in a shift of the eigenvalues of the curved pile relative to those of the straight pile. Table 1 shows the eigenvalues for three cases, namely straight, truly curved, and kinked piles. It can be seen that the differences between non-straight and straight piles are quite small, as are those between truly curved and kinked. On this basis one would expect the effects of curvature on drivability to be small for typical curvatures.

28. A second analysis involves the stresses in the piling (compressional plus flexural) during driving. Fig. 12 shows a representation of a pile which failed due to overstressing during driving. The computed stress in the pile at a point close to the failure position is shown in Fig. 13, from which it can be seen that the additional stress due to flexure was a second-order effect and could not really have contributed much to the failure. Material imperfection is a more likely cause.

Flutter (Papers 2 and 3)

29. Recently, attention has been drawn to the nature of the soil forces which resist the penetration of piles during installation. It has been pointed out that these forces may be of the 'follower' type (i.e., they remain tangential to the pile rather than taking up some fixed (usually vertical) direction). This being so, instabilities of a type frequently encountered in aerodynamics can be met at load levels far short of those required to cause instability in the classical buckling sense. Fig. 14 shows typical results in terms of load combinations at which various types of instability occur.[19] The finite element method proves to be a particularly simple means of solving these problems. So far, publications have merely indicated the possibility of instability arising; they have not shown the effects on drivability of a tendency towards instability. This tendency again manifests itself as a shift in the eigenvalues of the pile–soil system and can readily be incorporated in drivability programs.

Three-dimensional effects

30. In hard driving, it seems quite likely that significant energy is expended in deforming the pile laterally against the sides of the hole. In addition, the presence or absence of a soil 'plug' inside the pipe can clearly affect the mechanisms of wave transmission. It is perfectly possible to analyse the influence of these factors using axisymmetric finite element representations of pile and soil in a dynamic calculation but this does not yet seem to have been achieved.

PORE PRESSURE EFFECTS (Papers 19 and 20)

31. So far, soil resistance and strength have been represented exclusively in terms of total stresses. However, it is well known that, particularly for piles driven into normally consolidated, impermeable clays, large excess pore water pressures can be generated, the dissipation of which governs the pile's ability to resist loads applied at various times after driving.

32. Numerical methods have recently been applied in this area,[20–22] but the problem is a difficult one, and the writer doubts whether the state of excess pore pressure existing adjacent to piles driven into normally consolidated or overconsolidated clays and sand–clay mixtures, such as those in the North Sea, can be predicted with much confidence analytically. Field observation seems to be necessary here. However, rates of dissipation of the generated excess pore pressures should be perfectly adequately computed by present analytical techniques, given adequate values of permeability coefficients, obtained from tests in the field or on large specimens.

DYNAMICS (Papers 8–10)

33. In a cyclic loading environment the frequencies of the alternating forces and their relationship to the critical frequencies of the structure–soil system assume a decisive importance. The main types of dynamic loading experienced by offshore structures appear in the form of sea waves and/

Table 1. Natural frequencies of simply supported curved beam represented by straight and curved finite elements; radius of curvature 50 m, subtended angle 30°

Mode	Frequency, mHz	Percentage error for straight elements with kinks			Percentage error for curved elements		
		4	12	20	4	12	20
1	0.6345	+ 5.67	+5.48	+5.73	+ 1.50	1.4	1.46
2	2.6190	+ 2.4	+1.9	+2.10	+ 0.76	0.36	0.35
3	5.9269	+ 2.3	+0.57	+0.57	+ 2.01	0.20	0.16
4	10.5581	+11.6	+0.52	+0.50	+11.1	0.20	0.01
5	16.5125	+10.8	–0.11	–0.37	+13.0	0.80	0.01
6	19.362	+ 3.9	+0.14	+0.22	< 0.1	<0.1	<0.1
7	23.790	+23.0	+0.50	+0.15	24.1	0.5	0.09
8	32.391	+31.0	+0.89	+0.30	18.3	1.0	0.13
9	38.330	+15.6	+0.51	–0.30	15.5	<0.1	<0.1
10	42.315	+27.8	+1.63	+0.70	27.4	2.0	0.19

or earthquakes. These two forms of excitation are radically different, as has been pointed out in the context of gravity structures.[23] For piled structures, many of the same considerations apply.

Sea waves and earthquakes

34. Ocean wave loading is of low frequency and long duration, whereas earthquake loading is of high frequency and short duration. The magnitudes of total load imposed on the structure–soil system by the two excitations are, however, of the same order. In the case of excitation by earthquakes, it is usually assumed that shear waves propagate vertically through the soil from bedrock, and of course all of the soil and any deep-driven piles embedded in it are subject to the earthquake motions. These lead to rather large shear strains everywhere in the soil mass and a great deal of 'primary' interaction between the soil and the piles. This interaction causes shaking of the structure which in turn exerts a 'secondary' interaction back through the piling into the soil. The descriptions primary and secondary dynamic interaction are borrowed from usage in the nuclear power plant industry and may not realistically express the relative importances of the two effects. When the excitation is by sea waves, the interaction is by the above definition totally secondary. This type of interaction is much more localized to the immediate vicinity of the structure, and in the case of typical deep-piled offshore structures, will not be felt at all by the ground below quite a shallow depth of a few metres below mudline.

Methods of analysis

35. Idealizations of piled foundations for dynamic analysis follow closely their static counterparts. Often the

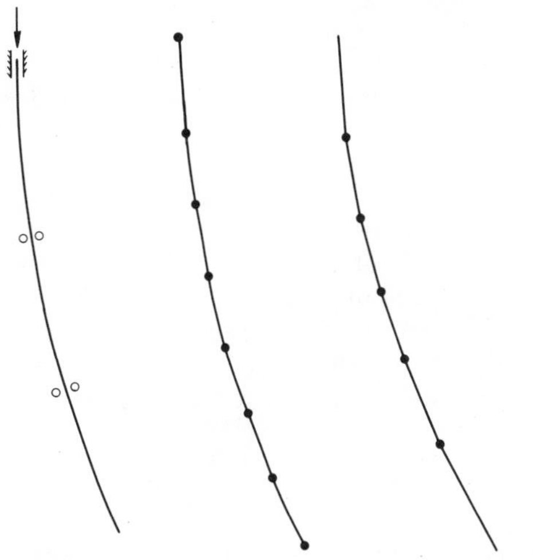

Fig. 11. Representation of curved conductors by finite elements: (a) curved pile; (b) curved elements; (c) straight elements

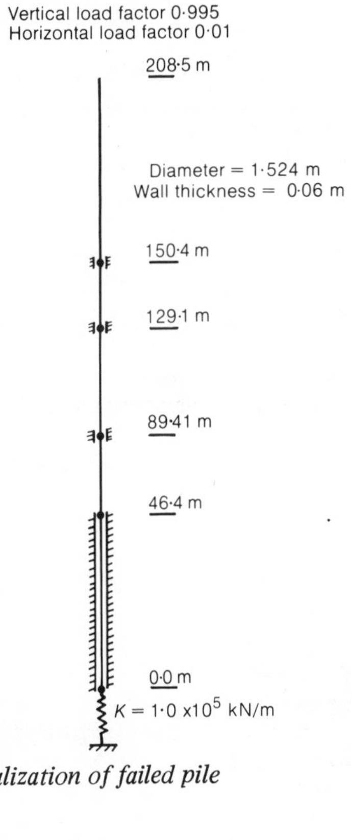

Fig. 12. Idealization of failed pile

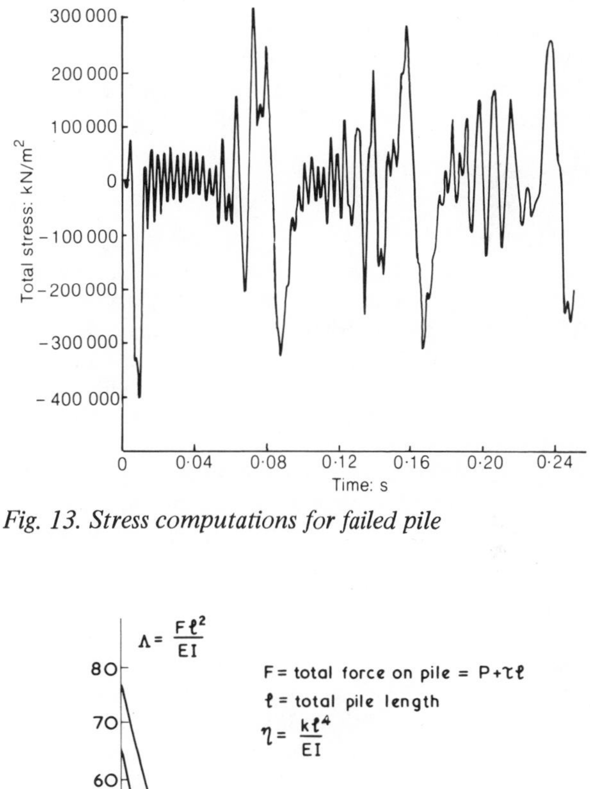

Fig. 13. Stress computations for failed pile

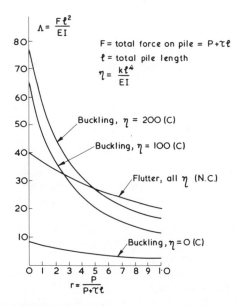

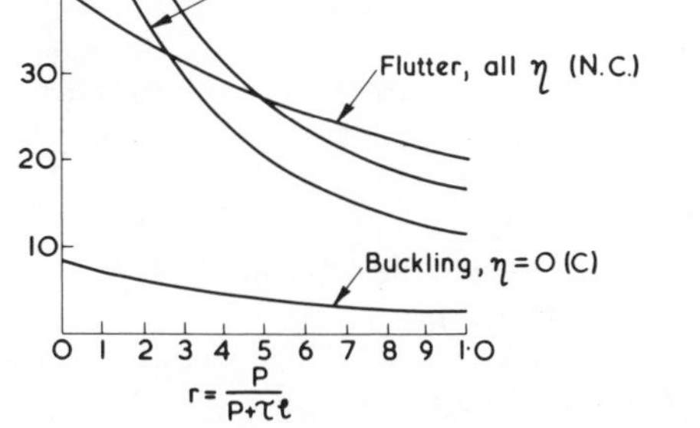

Fig. 14. Flutter analysis of piles by finite elements; F = total force on pile = $P + \tau l$, l = total pile length, $\eta = kl^4/EI$

ground resistance is approximated by a set of springs, but this method is more questionable than in the static case because of the inability of the springs to cope effectively with a major source of energy dissipation, namely by geometric or radiation damping. This is particularly important in earthquakes where there is a lot of energy present throughout the excited ground. In order to include a measure of geometric damping, finite element or boundary element approximations can again be used,[24] and a few solutions for simple cases can even be produced analytically.[25]

36. Among the various computational tactics which can be adopted, the most widely used are integration of the equations of motion in the time domain, which allows rather general energy dissipation mechanisms to be included. These tend to comprise the non-linear effects of friction, plasticity, viscosity and so on. Alternatively, linearized calculations can be made by the 'complex response' or 'impedance' methods, taking account of hysteretic damping only. Both methods have their advantages, and experience with the analysis of gravity platforms indicates that lateral stiffnesses and motions of the structure–soil system will probably be similarly predicted by both methods, given similar initial assumptions. However, the truly non-linear calculations produce some results, such as permanent displacements and sub-resonances, which cannot be present in any linearized analysis.[23] This is a fruitful field for further study.

CONCLUSIONS

37. Numerical methods in offshore piling are in many cases more sophisticated than the physical data which is input to the programs. What is often required is back-analysis from the field since full-scale tests are prohibitively expensive. Modelling, especially true scale modelling using centrifuges, can also contribute valuable data for calibration of the numerical results.

38. Areas where further numerical developments can be made include constitutive relationships for soil, especially during cyclic loading and penetrating of piles, three-dimensional effects and dynamics.[26]

REFERENCES

1. SMITH I. M. (BELL F. G. (ed.)). Computer predictions in difficult ground conditions. In: Foundation engineering in difficult ground. Newnes–Butterworth, London, 1978, 143–160, chapter 4.

2. VIJAYVERGIYA V. N. Load–movement characteristics of piles. Ports 77 Conf., Long Beach, California, 1977.

3. CHIN Y. K. The finite element analysis of axially loaded piles. Report to Lloyd's Register of Shipping, London, 1978.

4. DESAI C. S. Numerical design–analysis for piles in sands. J. Geotech. Div. Am. Soc. Civ. Engrs, 1974, 100, June, GT6, 613–635.

5. POULOS H. G. and DAVIS E. H. Elastic solutions for soil and rock mechanics. Wiley, London, 1974.

6. BUTTERFIELD R. and BANERJEE P. K. The elastic analysis of compressible piles and pile groups. Géotechnique, 1971, 21, Mar., No. 1, 43–60.

7. ZIENKIEWICZ O. C. et al. Associated and non-associated viscoplasticity and plasticity in soil mechanics. Géotechnique, 1975, 25, Dec., No. 4, 671–689.

8. OTTAVIANI M. Three-dimensional finite element analysis of vertically loaded pile groups. Géotechnique, 1975, 25, June, No. 2, 159–174.

9. SUEN P. C. Axisymmetric piles under non-axisymmetric loads. Report to Fugro Ltd, Ruislip, 1978.

10. POULOS H. G. Behaviour of laterally loaded piles—single piles. J. Soil Mech. Fdns Div. Am. Soc. Civ. Engrs, 1971, 97, May, SM5, 711–731.

11. BANERJEE P. K. and DAVIES T. G. The behaviour of axially and laterally loaded single piles embedded in non-homogeneous soils. Géotechnique, 1978, 28, Sept., No. 3, 309–326.

12. MEISSNER H. E. (DESAI C. S. (ed.)). Laterally loaded pipe pile in cohesionless soil. In: Numerical methods in geomechanics. American Society of Civil Engineers, New York, 1976, Vol. 3, 1353–1365.

13. SMITH I. M. (ZIENKIEWICZ O. C. et al. (eds)). Transient phenomena of offshore foundations. In: Numerical methods in offshore engineering. Wiley, London, 1977, 483–513, chapter 15.

14. SUTTON V. et al. A full scale instrumented pile test in the North Sea. Proc. 11th Offshore Technology Conf., Houston, Texas, 1979.

15. GOBLE G. G. et al. Bearing capacity of piles from dynamic measurements. Case Western Reserve University, Cleveland, 1975, final report to Ohio Dept of Transportation.

16. FISCHER F. J. Driving analysis for initially curved marine conductors. Proc. 7th Offshore Technology Conf., Houston, Texas, 1975, paper 2309.

17. ISAKOVICH M. A. and KOMAROVA L. N. Longitudinal-flexure modes in a slender rod. Soviet Physics-Acoustics, 1968, 13, 491–494.

18. YANDZIO E. Effect of curvature and kinks on driveability of piles. MEng thesis, University of Liverpool, 1979.

19. SMITH I. M. Discrete element analysis of pile instability. Int. J. Num. Meth. Engng, 1979, 3, No. 2, 205–211.

20. DESAI C. S. Effect of driving and subsequent consolidation on behaviour of driven piles. Int. J. Num. Meth. Engng, 1978, 2, No. 3, 283–301.

21. CARTER J. P. et al. Stress and porepressure changes in clay during and after the expansion of a cylindrical cavity. Int. J. Num. Anal. Meth. Geomech., 1979, 3, No. 4, 305–322.

22. RANDOLPH M. F. and WROTH C. P. An analytical solution for the consolidation around a driven pile. Int. J. Num. Anal. Meth. Geomech., 1979, 3, No. 3, 217–229.

23. SMITH I. M. and MOLENKAMP F. (GUDEHUS G. (ed.)). Linearised and truly nonlinear dynamic response of offshore structure–foundation systems. In: Plastic and long term effects in soils (Dynamical methods in soil and rock mechanics, Vol. 2). Balkema, Rotterdam, 1978, 299–320, chapter 11.

24. PENZIEN J. and TSENG W. S. (ZIENKIEWICZ O. C. et al. (eds)). Three dimensional dynamic analysis of fixed offshore platforms. In: Numerical methods in offshore engineering. Wiley, London, 1977, 221–243, chapter 7.

25. NOVAK M. (PRANGE B. (ed.)). Effects of piles on dynamic response of footings and structures. In: Dynamic response and wave propagation in soils (Dynamical methods in soil and rock mechanics, Vol. 1). Balkema, Rotterdam, 1978, 185–200.

26. SMITH I. M. Installation and performance of piled foundations. Proc. 3rd Int. Conf. Numer. Meth. Geomechanics, Aachen, 1979.

2. Effect of driving support conditions on pile wandering

I. W. BURGESS, BA, PhD (Lecturer in Civil and Structural Engineering) and
C. A. TANG, BEngSc, MSc(Eng) (Research Student; University of Sheffield)

It has recently been demonstrated in theoretical terms that driven or jacked piles may undergo a directional instability in placement which could lead to large final deflexions from their assumed directions. This is supported by field evidence that certain slender piles in soft soils do acquire large curvatures in their lower regions. The problem is rather complex, there being complete interaction between the pile and the soil continuum, and the usual form of instability is flutter during the pile's downward motion rather than buckling under load. The analytical approaches so far have therefore been rather tentative, with a severe restriction on the numbers of interactive parameters taken into account, and have attempted progressively to increase their degree of realism. The cases treated so far have covered piles guided at the soil surface, driven into soils of constant or linearly increasing shear strength and reaction modulus, and have shown depths at which instability occurs for practical examples which compare well with the small amount of numerical site evidence which is available. The theoretical study is here extended to cover a range of driving support conditions which may be more relevant to piling practice than the guided conditions hitherto assumed. The effects of changing the driving support conditions are illustrated with reference to typical steel pile sections in soft soils of different profiles with depth of shear strength and lateral reaction modulus.

INTRODUCTION

The directional stability of driven piles during installation has recently been studied[1–4] as an inherently different phenomenon from the piles' buckling stability under their in-service loads. The qualitative difference between these superficially similar cases lies in the nature of the forces which act on piles when they are moving downwards through the ground and when they are essentially static. In the latter case, so long as all elements of the pile–soil system remain elastic during an infinitesimally small displacement, the forces which act on the pile are totally conservative, being derivable in the normal way from the potential of the applied load. In the installation case, however, while the pile is moving downwards the forces acting on it remain tangential to any infinitesimally displaced profile and are thus of the non-conservative type known as 'follower' forces. Whereas conservative forces can cause instability of the usual structural buckling type, follower forces are capable of causing two types of structural instability. These are buckling (also known as divergence), which is similar in nature to that caused by conservative forces; and oscillations of exponentially increasing amplitude, known as flutter. Analysis of the stability of such systems is necessarily dynamic rather than quasi-static, and studies the changes in fundamental frequencies as loads are increased. Buckling occurs when a fundamental frequency changes from a real to an imaginary quantity, and flutter occurs when two adjacent real frequencies coalesce at a certain load, producing a pair of complex conjugate roots.

2. If flutter or buckling instability can occur during the very brief periods of pile motion in a driving process the question of interest becomes, 'Does the instability actually cause any significant effects on the pile?' The answer to this would appear to be twofold. It is probable that in general neither type of instability would cause collapse of, or damage to, the pile during the very brief time period within which it is required to respond, although this can be said with less confidence of buckling instability. However, since the pile is moving downwards as its lateral deflexion during instability increases, and since its final profile is largely determined by the path taken by its tip, it is distinctly probable that either type of instability would cause a sudden divergence in the pile's path below the critical depth at which instability initially occurs. Although few piles are in practice instrumented for inclinometer measurements and the open literature on the results of such instrumentation is extremely scanty, evidence does exist[5, 6] that slender piles in soft soils do experience a wandering phenomenon below certain depths. This does not, in general, cause any considerable loss of load capacity in isolated piles, but could obviously cause damage to neighbouring piles if there is close spacing within a group. There are, in any case, many occasions on which a reasonable guarantee of straight driving is required by the foundation engineer and this, rather than any effect on load capacity, must remain as the justification for the work.

3. The main problem with a mathematical investigation of the phenomenon, even in the simplified form attempted so far, is the very large number of parameters which can affect the equations of motion. For a pile driven into a soil of shear strength increasing linearly with depth, with a significant tip force, there are three load parameters and two lateral reaction modulus parameters. In the previous studies, which have assumed that a pile is guided into

the soil at surface level and that tip force constitutes only a minor effect, the load parameters were reduced to two and the reaction modulus parameters effectively to one. Any greater number of active parameters makes a parametric study of any real meaning very difficult to achieve, and means that the designer may be forced to carry out particular studies related to individual cases.

4. The case examined in the previous studies, with the piles being guided at the soil surface, could represent fairly well the installation of piles guided into the soil along the legs of an offshore oil platform but is inadequate to represent most installation conditions of more traditional piledriver set-ups. In these the head of a continuous pile length would be held at the leaders, some distance above the soil surface, but no other support would be provided. If the leaders are rigidly supported this could be modelled as support against lateral deflexion but not rotation at the pile head. Obviously the design of piledrivers is by no means uniform and other cases might provide effectively no lateral restraint at the head. Allowance would need to be made in this case for the concentrated mass of the driving gear attached to the pile head. In the present study the work is extended to cover the former case, with effective lateral restraint at a certain distance above the soil surface. This brings into the analysis one new and major independent parameter, the pile's stick-up above ground, and makes presentation of comprehensive general characteristics even more difficult. An extension to the other case, without lateral support, has no difficulties in principle but the inclusion of a concentrated mass would again increase the scope of the parametric study.

5. There is some evidence to suggest that computer programs for general solution of such problems might best be based on a finite element procedure. Smith[7] has already shown that identical results can be achieved by this method, which can trade off a lack of algebraic complexity for larger numbers of elements. The method would also be more amenable to the use of continuum soil elasticity rather than the Winkler approach which has so far been used, although this is still a fairly complex problem in finite element terms. The Winkler reaction modulus idealization seems an unsatisfactory one for a proper stability analysis, although its convenience does make it attractive in the present stages of the study.

NOTATION

A_i, a_i, b_I	deflexion mode amplitudes
B	pile breadth
c	apparent soil cohesion at pile tip
EI	pile bending stiffness
F	driving force
h	pile height above ground
i, j, m, n, s, I, J, N	integers
k_0	soil Winkler stiffness (constant)
l	pile length
m_k	slope of soil Winkler stiffness
m_τ	slope of soil's apparent shear traction
P	tip force
p	pile section perimeter length
Q_i, q_i, r_I	generalized co-ordinates
S, X	parts of soil strain energy
T	kinetic energy of pile
$U_s = 0$	discrete equation of motion
V	strain energy of pile
v	pile deflexion co-ordinate
$\overline{W}$, $\overline{Y}$, $\overline{Z}$	parts of instantaneous potential of loads
x	pile length co-ordinate
α	dimensionless tip force
β, β_0, μ_τ	dimensionless shear force components
δ_{ij}	Kronecker delta
η, η_0, μ_k	dimensionless soil stiffness components
Λ	dimensionless driving force
ξ	dimensionless frequency
ρ	pile mass per unit length
σ	stress in pile
τ	shear per unit length on pile
ω	pile frequency

ANALYSIS

6. Consider the pile installation conditions shown schematically in Fig. 1. A pile of initially straight uniform section is pushed 'slowly' into the soil. Its head, which protrudes by some distance out of the soil, is held against lateral deflexion but is free to rotate. The shear strength of the soil which is transmitted to the pile walls (which may be only a proportion of the actual shear strength) is assumed to vary in a linear manner with depth, as also is its lateral reaction modulus. The pile is, therefore, subject to a distributed shear along its walls and a tip force, both of which follow the deflected shape of the pile when it undergoes a displacement from its assumed straight-line driving path. Since these forces are non-conservative a stability analysis must be dynamic rather than static, and takes the form of a study of the pile's fundamental vibration frequencies and their changes as placement proceeds. The method which has been used so far by the present Authors, and which is again used here, is to set up the Lagrange equations of motion for periodic motion of the system in terms of a discrete set of trigonometric Rayleigh–Ritz displacement functions. For previous analysis it has been assumed that the pile is guided at its head, and the displacement functions have thus been restricted to cosine forms. Here the additional freedom to rotate at the head means that sines also need to be included, and the displacement takes the general form

$$V = \sum_{i=1,3,\ldots}^{(2n-1)} l\,q_i\left(1-\cos\frac{i\pi x}{2l}\right) + \sum_{I=1}^{m} l r_I \sin\frac{I\pi x}{l} \qquad (1)$$

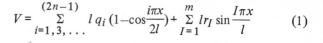

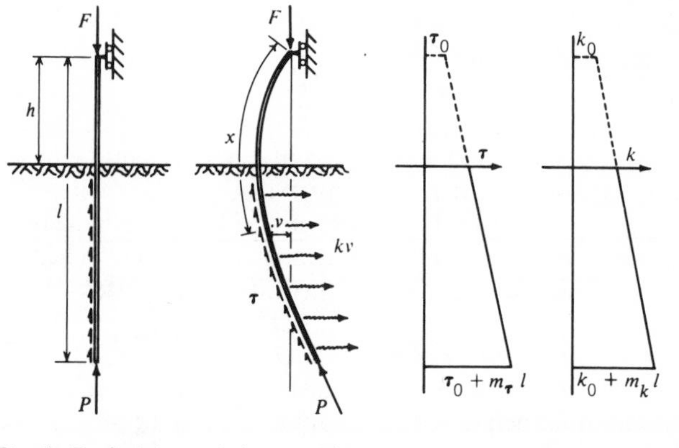

Fig. 1. Definition of the problem

The n mode amplitudes q_i and the m mode amplitudes r_I are assumed to change harmonically with time as

$$\left. \begin{aligned} q_i &= a_i \cos \omega t \\ r_I &= b_I \cos \omega t \end{aligned} \right\} \qquad (2)$$

It is convenient to define a unitary set of $N = n + m$ mode amplitudes

$$\left. \begin{aligned} Q_j &= q_{(2j-1)}, \quad (j = 1,2, \ldots, n) \\ Q_j &= r_{(j-n)}, \quad (j = (n+1), (n+2), \ldots, N) \end{aligned} \right\} \qquad (3)$$

and to re-state (2) as

$$Q_j = A_j \cos \omega t \qquad (4)$$

The Q_j and A_j co-ordinates are used in the aspects of the analysis concerned with setting up the equations of motion in general form. Obviously when the exact nature of a particular equation or term is required it is necessary to return to its original co-ordinates q_i, a_i, r_I and b_I.

7. The Lagrange equations of motion at any instant of time require the kinetic and elastic strain energies of the pile and foundation, and the generalized forces derived from the instantaneous potentials of the shear and tip forces, to be found in terms of the system's discrete generalized co-ordinates. Since new mode forms are introduced in the present study, and since part of the pile's length is not buried, it is necessary to re-derive these potentials. In continuum terms (and with the notation of Burgess[2]) they are given by

$$T = \tfrac{1}{2}\rho \int_0^l \dot{v}^2 \, dx \qquad (5)$$

$$V = \tfrac{1}{2} EI \int_0^l v''^2 \, dx \qquad (6)$$

$$X = \tfrac{1}{2} \int_h^l (k_0 + m_k x) \, v^2 \, dx \qquad (7)$$

$$\overline{Y} = \tfrac{1}{2} \int_h^l (\tau_0 + m_\tau x) \int_0^x v'^2 \, dx \, dx - \int_h^l (\tau_0 + m_\tau x) \, v' | v \, dx \qquad (8)$$

$$\overline{W} = \tfrac{1}{2} P \int_0^l v'^2 \, dx - P v'(l) | v(l) \qquad (9)$$

Where these potentials are found in terms of the Q_j generalized co-ordinates their lowest order terms can all be expressed as quadratic forms as follows:

$$T = \tfrac{1}{2} \rho l^3 \, T_{ij} \, \dot{Q}_i \, \dot{Q}_j \qquad (10)$$

$$V = \tfrac{1}{2} \frac{EI}{l} \, V_{ij} \, Q_i \, Q_j \qquad (11)$$

$$X = \tfrac{1}{2} (k_0 + m_k l) l^3 \, X_{ij} \, Q_i \, Q_j - \tfrac{1}{2} m_k l^4 \, S_{ij} \, Q_i \, Q_j \qquad (12)$$

$$\overline{Y} = \tfrac{1}{2} (\tau_0 + m_\tau l) l^2 \, (Y_{ij} + 2\overline{Y}_{i|j}) \, Q_i \, Q_j - \tfrac{1}{2} m_\tau l^3 \, (Z_{ij} + 2\overline{Z}_{i|j}) \, Q_i \, Q_j \qquad (13)$$

$$\overline{W} = \tfrac{1}{2} Pl \, (W_{ij} + 2\overline{W}_{i|j}) \, Q_i \, Q_j \qquad (14)$$

The well known form of the Lagrange equations of motion used here is

$$U_s = \frac{d}{dt}\left(\frac{\partial T}{\partial \dot{Q}_s}\right) - \frac{\partial T}{\partial Q_s} + \frac{\partial V}{\partial Q_s} - \overline{P}_s = 0 \qquad (15)$$

Using the quadratic forms (10) – (14) the general form for one of the N equations of motion then becomes

$$\begin{aligned} U_s &= \rho l^3 \, T_{sj} \, \ddot{Q}_j + \frac{EI}{l} V_{sj} \, Q_j + (k_0 + m_k l) l^3 \, X_{sj} \, Q_j \\ &\quad - m_k l^4 \, S_{sj} \, Q_j - (\tau_0 + m_\tau l) l^2 \, (Y_{sj} + \overline{Y}_{j|s}) Q_j \\ &\quad + m_\tau l^3 \, (Z_{sj} + Z_{j|s}) Q_j - Pl(W_{sj} + W_{j|s}) Q_j \\ &= 0 \end{aligned} \qquad (16)$$

If the assumption of periodic motion is made then the equations of motion become completely algebraic. It is convenient at the same time to change the coefficients of the various terms to the dimensionless forms

$$\xi^2 = \omega^2 \rho l^4 / EI, \quad \alpha = Pl^2/EI, \quad \beta_0 = \tau_0 l^3 / EI,$$

$$\mu_T = m_\tau l^4 / EI, \quad \eta_0 = k_0 l^4 / EI, \quad u_k = m_k l^5 / EI$$

and since the tip values of shear and reaction modulus appear in (16) it is also found convenient to define

$$\beta = \beta_0 + \mu_T$$

and

$$\eta = \eta_0 + \mu_k$$

In this form the typical algebraic equation of motion becomes

$$\begin{aligned} U_s &= [-\xi^2 \, T_{sj} + V_{sj} + \eta X_{sj} - \mu_k \, S_{sj} - \beta \, (Y_{sj} + \overline{Y}_{j|s}) \\ &\quad + \mu_T \, (Z_{sj} + \overline{Z}_{j|s}) - \alpha (W_{sj} + \overline{W}_{j|s})] \, A_j = 0, \\ &\quad (s = 1,2, \ldots, N) \end{aligned} \qquad (17)$$

The fundamental frequencies are then given by the solutions of the eigenvalue problem $[U_{s|j}] \, A_j = 0$, in which the matrix $[U_{s|j}]$ is asymmetric, due to the non-conservative nature of the loads. Recalling also that the set of mode amplitudes A_j is actually composed of two sets a_i and b_I it is obvious that the matrix $[U_{s|j}]$ is composed of four quadrants whose elements are differently derived. For these derivations it is necessary to return to the original a_i and b_I nomenclature, and they are performed simply by doing the required algebra on the continuum forms (5)–(9) with the discretization (1) inserted. This algebra, although fairly trivial in nature, is lengthy in presentation and only the results for each term are shown in Appendix 1. The terms are very easily programmed for the computer, and once written the program can deal with as many degrees of freedom as the required accuracy dictates. In fact very good accuracy can be obtained with a few degrees of freedom, the present results having been obtained with four sine and four cosine modes. As usual, loads converge much more rapidly than deflected shapes with numbers of degrees of freedom.

RESULTS

8. Since a complete parametric study covering the range of distributions of soil properties with depth is not really feasible here, the results presented are for the two

traditional idealized cases of purely cohesive and purely frictional soil behaviour. It may then be possible to infer the behaviour of intermediate cases at least qualitatively. The numbers of parameters involved in any case encountered in practice suggest that a specific analysis should be performed for the case. A further restriction on the results presented has been that the tip force has been neglected. This is based on the facts that the tip force forms a very small part of the total driving force for a slender pile section, and that trial analyses show that it has only a minor effect on the instability behaviour.

9. For an analysis of the purely cohesive case the dimensionless parameters under consideration are then the driving force Λ, the reaction modulus η, and the exposed length ratio h/l. It is inevitable that the analysis essentially considers the pile as a structural element of constant length subjected over a constant part of this length to a load whose intensity can increase until some form of instability is experienced. In practice, of course, an identical end-result is produced by the reverse of this process, with the embedded length increasing progressively through a soil imposing a load of constant intensity. In either case the load parameter Λ increases with penetration, but in the latter case this is accompanied by increasing η and decreasing h/l.

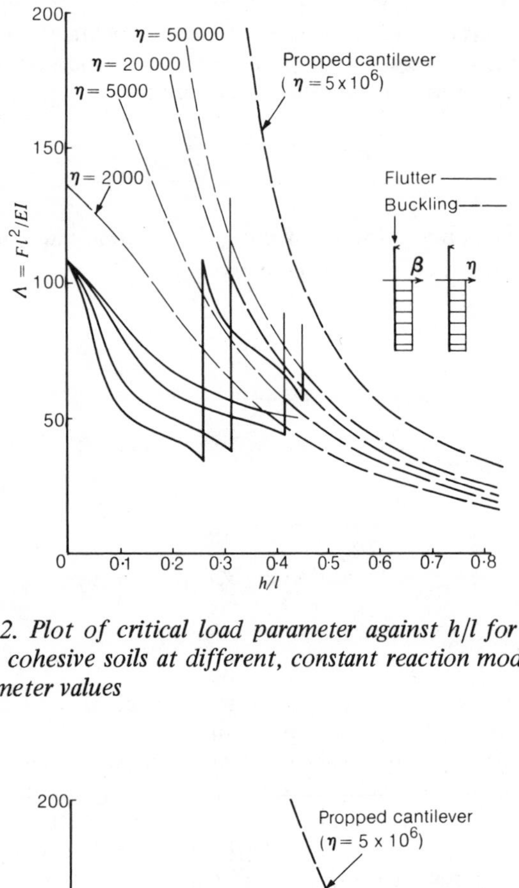

Fig. 2. Plot of critical load parameter against h/l for uniform cohesive soils at different, constant reaction modulus parameter values

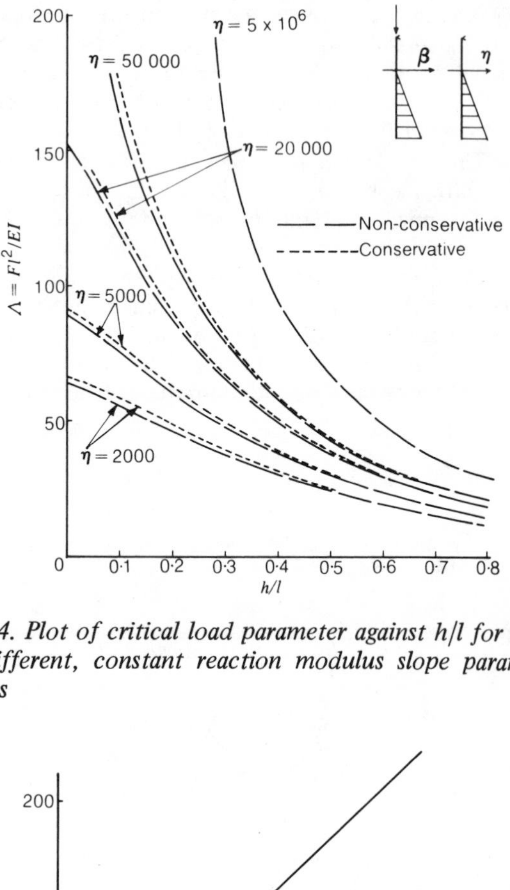

Fig. 4. Plot of critical load parameter against h/l for sands at different, constant reaction modulus slope parameter values

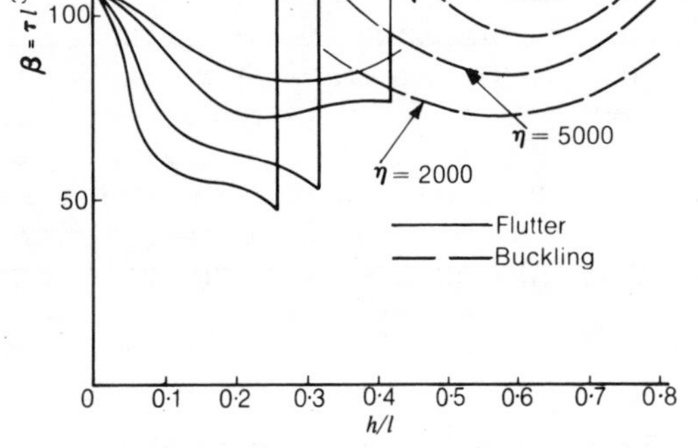

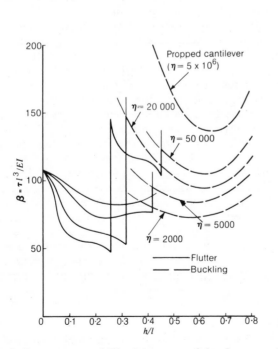

Fig. 3. Restatement of Fig. 2 in terms of the shear traction parameter β rather than the overall load parameter Λ

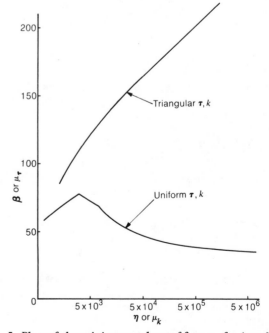

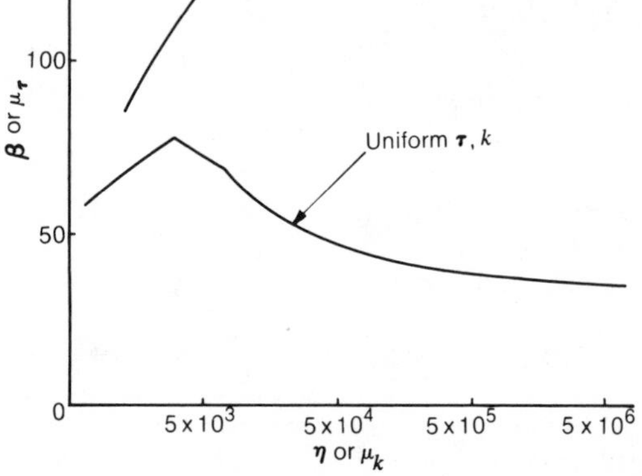

Fig. 5. Plot of the minimum values of β or μ_τ for instability against the corresponding η or μ_k values

This means that if an analysis is to represent this case identically then it can only consider a particular combination of pile section and soil properties, and thus loses all generality. A plot of critical load parameter Λ against h/l is shown in Fig. 2 for different constant reaction modulus parameters η. It can be seen from this figure that for each η value there exists a range of h/l for more than 50% penetration within which flutter instability limits stable behaviour. Beyond this range buckling instability is the limit to stable behaviour. It is notable that the flutter characteristics are not completely continuous, but undergo sharp step increases of Λ at certain values of h/l. This is caused by a splitting of the two load–frequency curves which have been coalescing to form the flutter load, and by a simultaneous coalescence of the higher of these with the next higher load–frequency curve at a higher Λ value. For large η values this may happen more than once before the flutter loads are superseded by buckling loads. For those parts of the flutter curves which are directly comparable it is noticeable that increasing η reduces the critical value of Λ, while the effect of increasing η in the buckling regions is to increase Λ values in the normal way. The four curves which are plotted for η between 2000 and 50 000 cover a fairly practical range of η values for slender piles in soft clay. As a check to the validity of the analysis it is interesting to note that for very high values of η (for example, the 5×10^6 plotted) the buckling curve coincides with the well known solution for a propped cantilever of length h, which is in this case

$$\Lambda = 20.2/(h/l)^2 \tag{18}$$

10. An alternative presentation of these curves is given in Fig. 3. Instead the dimensionless form of the total driving force, which may be expressed as $\Lambda = \tau(l-h)l^2/EI$, being plotted, the dimensionless form of the shear traction $\beta = \tau l^3/EI$ is plotted against h/l. On this plot a pile of a particular length and section, driven into a soil of constant shear strength and reaction modulus, has its stability limits marked at any depth of penetration by the curve of the correct η value. This η value will not change during driving for the particular length of pile considered. So, for example, if $\eta = 20\,000$, the lowest soil shear strength which will cause instability does so in the flutter mode and at $h/l = 0.32$. For any soil shear strength above this value the type of instability is that given by the highest h/l value at which the shear strength parameter value intercepts the appropriate stability limit curve.

11. If slopes of shear traction and reaction modulus with depth are superposed on the constant values of the purely cohesive case the load–frequency curves progressively change, with consequent effect on the stability limit curves. The slope of shear traction with depth is the minor parameter of these two in its effect on the stability limits. If this is increased without a simultaneous increase of the reaction modulus slope then only fairly minor changes take place in the curves comparable to those of Fig. 2, these changes having the effect of decreasing the Λ values in both the flutter and buckling ranges. The major effect comes from changes in the slope of reaction modulus with depth. Increases in this parameter tend to cause the load–frequency curves to separate, destroying flutter instabilities. For the case of an ideally frictional soil, with triangular distributions of shear traction and reaction modulus, the flutter range has been eliminated altogether, as is shown in Fig. 4. It must be emphasized in this context that the buckling curves which form the stability limits during driving are not equivalent to those which would be generated if the loading on the pile were purely conservative. These can easily be found by removing the non-conservative potential terms from the equations of motion, and are shown for comparison with the corresponding non-conservative cases in Fig. 4. The conservative buckling loads are seen to lie slightly above, but very close to, the corresponding non-conservative buckling loads in this case, although this is by no means general. The plots of Fig. 4 could be converted to a form similar to that of Fig. 3, with the dimensionless rate of increase of shear traction $\mu_T = m_T l^4/EI$ plotted against h/l. As in Fig. 3, the most important feature of such a plot for any given value of η is the minimum value of μ_T for which instability occurs. In Fig. 5 these minimum values of μ_T are plotted against the corresponding η values, along with the more complex characteristic of the minimum β values taken from Fig. 3.

DESIGN CONSIDERATIONS

12. If constant lengths of pile section are driven into clay soil transmitting an effective cohesion c to the pile walls, then the minimum depths at which instability can be encountered are found by using Fig. 5. The dimensionless shear traction parameter is given by

$$\beta = \rho c l^3/EI \tag{19}$$

While it is far from easy to ascribe a firm numerical value to the lateral reaction modulus, it is at least possible to gain an approximate value from Terzaghi,[8] who suggests that for clay soils it is proportional to shear strength and decreases somewhat with the width of the pile. Converted into SI units his postulated values give

$$\eta = \frac{437cBl^4}{EI}\left\{\frac{0.3048}{B}\right\} \tag{20}$$

If c is eliminated between equations (19) and (20) then

$$\frac{\beta}{\eta} = \frac{P}{133l} \tag{21}$$

For a given length of steel section this enables the appropriate values of β and η to be identified on Fig. 5, and the corresponding values of c and, if required, h/l can then be evaluated. Figs 6 and 7 show plots of critical length against apparent cohesion for examples of H section and tubular section steel piles. For each example critical lengths are shown for the pile guided into the soil, the maximum length which can be completely embedded without instability under the present support conditions, and the embedded length at which this case would become unstable with an infinitesimal increase of overall length or soil cohesion (the minimum depth for stable driving). The maximum pile lengths capable of being completely installed are very close to those for the guided case but in general are slightly greater. The minimum depths for stable driving are rather less than the guided critical lengths. Figs 8 and 9 show the maximum pile stress during placement for each of the example cases of Figs 6 and 7. From these it is possible to define the useful range of each section, as limited by the

Table 1. Specimen values of μ_k and μ_T for loose, medium and dense sands, and example critical pile lengths

	Loose	Medium	Dense
μ_k	$\frac{2512l^5}{EI}$	$\frac{8800l^5}{EI}$	$\frac{21353l^5}{EI}$
μ_T	$\frac{1.275l^4}{EI}p$	$\frac{1.982l^4}{EI}p$	$\frac{2.871l^4}{EI}p$
Critical length for 305 X 305 X 79 kg/m H section, m	33	29.5	26

allowable stress limits in common use for driven and jacked piles.

13. In the case of sands Terzaghi suggests that subgrade reaction modulus increases with depth, as has been assumed in this analysis. His suggested values for submerged sands of loose, medium and dense compositions give the values of μ_k shown in Table 1. Making the assumptions that the friction angles are 18°, 22° and 26° respectively and that the horizontal stress is half the vertical, Terzaghi's suggested densities give also the μ_T values shown in the table. For each of these sands there is a unique critical length for each pile section, thus eliminating curves of the type derived for clays. As an example the approximate critical lengths for buckling of the 305 X 305 X 79 kg/m H section are also shown for the three sand types quoted. These, basically, are associated with buckling of the exposed part of the pile at $h/l \approx 0.7$ and should not be of any importance in practice in the context of the wandering phenomenon.

14. It seems, therefore, that soils which are mainly frictional in nature do not give rise to either flutter or buckling instabilities at any considerable embedded lengths with the present support conditions. This has been shown not to be true for guided support conditions which do give rise to flutter instabilities. Soils which are mainly cohesive would seem to cause both flutter and buckling instabilities, and accurate calculation of driving length for particular soils is evidently important if the designer is to take advantage of the maximum stable lengths predicted. It might be possible, by welding on successive short lengths after the original length is almost completely embedded, further to increase the stable embedded length to some extent.

15. The highly idealized nature of the analysis has been pointed out before and this is obviously still the case. It is difficult even to see how a very sophisticated numerical analysis using a continuum approximation for the soil can be more than a rather crude representation of reality. The most pressing need in developing workable design criteria from the theoretical basis is a careful experimental programme to highlight any deficiencies in the theory and perhaps to point out areas of importance for further theoretical study.

APPENDIX 1. TERMS OF THE MATRIX $[U_{s|j}]$

16. The matrix $[U_{s|j}]$ is composed of four separate quadrants

$$\left[\begin{array}{c|c} \text{I} & \text{II} \\ \hline \text{III} & \text{IV} \end{array}\right]$$

at least some of whose terms are separately derived. It is necessary to derive these in terms of the co-ordinates a_i and b_I, where the integers i and I are those appearing in equation (1).

Quadrant I (n X n)

$$T_{si} = 1 + \tfrac{1}{2}\delta_{si} + \frac{2}{\pi}\left[\frac{(-1)^{(s+1)/2}}{s} + \frac{(-1)^{(i+1)/2}}{i}\right] \tag{22}$$

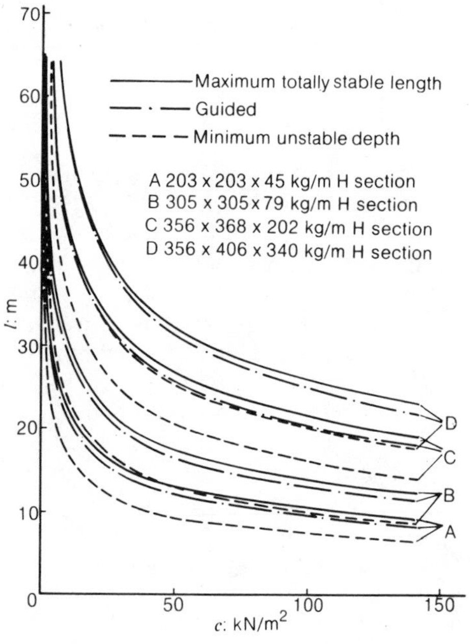

Fig. 6. Plot of critical length (in flutter) against corresponding soil cohesion value for typical H section piles

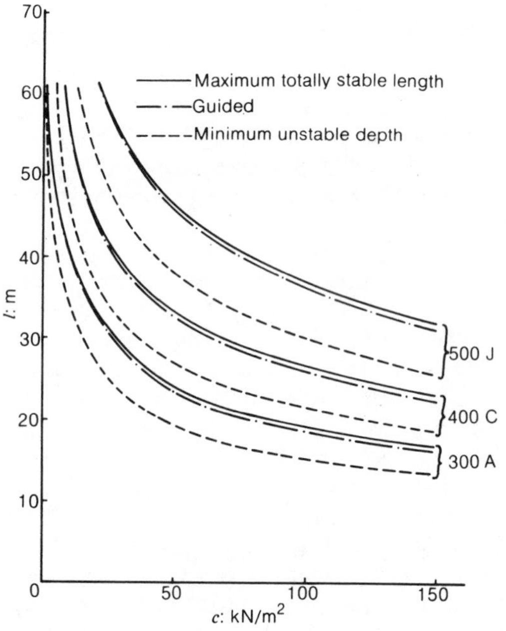

Fig. 7. Plot of critical length (in flutter) against corresponding soil cohesion value for typical tubular piles

$$V_{si} = \frac{\pi^4 s^4}{32}\delta_{si} \tag{23}$$

$$X_{si} = \left[\frac{x}{l}(1 + \tfrac{1}{2}\delta_{si}) - \frac{2}{\pi}\left(\frac{1}{i}\sin\frac{i\pi x}{2l} + \frac{1}{s}\sin\frac{s\pi x}{2l} - \frac{1}{2(s+i)}\sin\frac{(s+i)\pi x}{2l} - \frac{(1-\delta_{si})}{2(s-i)}\sin\frac{(s-i)\pi x}{2l}\right)\right]_h^l \tag{24}$$

$$S_{si} = X_{si} - \left[\tfrac{1}{2}\frac{x^2}{l^2}(1 + \tfrac{1}{2}\delta_{si}) - \frac{2x}{l\pi}\left(\frac{1}{i}\sin\frac{i\pi x}{2l} + \frac{1}{s}\sin\frac{s\pi x}{2l} - \frac{1}{2(s+i)}\sin\frac{(s+i)\pi x}{2l} - \frac{(1-\delta_{si})}{2(s-i)}\sin\frac{(s-i)\pi x}{2l}\right) - \frac{4}{\pi^2}\left(\frac{1}{i^2}\cos\frac{i\pi x}{2l} + \frac{1}{s^2}\cos\frac{s\pi x}{2l} - \frac{1}{2(s+i)^2}\cos\frac{(s+i)\pi x}{2l} - \frac{(1-\delta_{si})}{2(s-i)^2}\cos\frac{(s-i)\pi x}{2l}\right)\right]_h^l \tag{25}$$

$$Y_{si} = \frac{\pi^2 si}{8}\left[\tfrac{1}{2}\frac{x^2}{l^2}\delta_{si} + \frac{4}{\pi^2}\left(\frac{1}{(s+i)^2}\cos\frac{(s+i)\pi x}{2l} - \frac{(1-\delta_{si})}{(s-i)^2}\cos\frac{(s-i)\pi x}{2l}\right)\right]_h^l \tag{26}$$

$$\overline{Y}_{i|s} = \left[\cos\frac{i\pi x}{2l} - \frac{i}{2(s+i)}\cos\frac{(s+i)\pi x}{2l} + \frac{i(1-\delta_{si})}{2(s-i)}\cos\frac{(s-i)\pi x}{2l}\right]_h^l \tag{27}$$

$$Z_{si} = Y_{si} - \frac{si}{8}\left[\frac{\pi^2 x^3}{3l^3}\delta_{si} + \frac{4x}{l}\left(\frac{1}{(s+i)^2}\cos\frac{(s+i)\pi x}{2l} - \frac{(1-\delta_{si})}{(s-i)^2}\cos\frac{(s-i)\pi x}{2l}\right) - \frac{8}{\pi}\left(\frac{1}{(s+i)^3}\sin\frac{(s+i)\pi x}{2l} - \frac{(1-\delta_{si})}{(s-i)^3}\sin\frac{(s-i)\pi x}{2l}\right)\right]_h^l \tag{28}$$

$$\overline{Z}_{i|s} = \overline{Y}_{i|s} - \left[\frac{x}{l}\left(\cos\frac{i\pi x}{2l} - \frac{i}{2(s+i)}\cos\frac{(s+i)\pi x}{2l} + \frac{(1-\delta_{si})}{2(s-i)}\cos\frac{(s-i)\pi x}{2l}\right) + \frac{2}{\pi}\left(\frac{1}{i}\sin\frac{i\pi x}{2l} - \frac{i}{2(s+i)^2}\sin\frac{(s+i)\pi x}{2l} + \frac{i(1-\delta_{si})}{2(s-i)^2}\sin\frac{(s-i)\pi x}{2l}\right)\right]_h^l \tag{29}$$

$$W_{si} = \frac{\pi^2 s^2}{8}\delta_{si} \tag{30}$$

$$\overline{W}_{i|s} = \frac{i\pi}{2}(-1)^{(i+1)/2} \tag{31}$$

Quadrant IV (m × m)

$$T_{SI} = \tfrac{1}{2}\delta_{SI} \tag{32}$$

$$V_{SI} = \tfrac{1}{2}S^4\pi^4\delta_{SI} \tag{33}$$

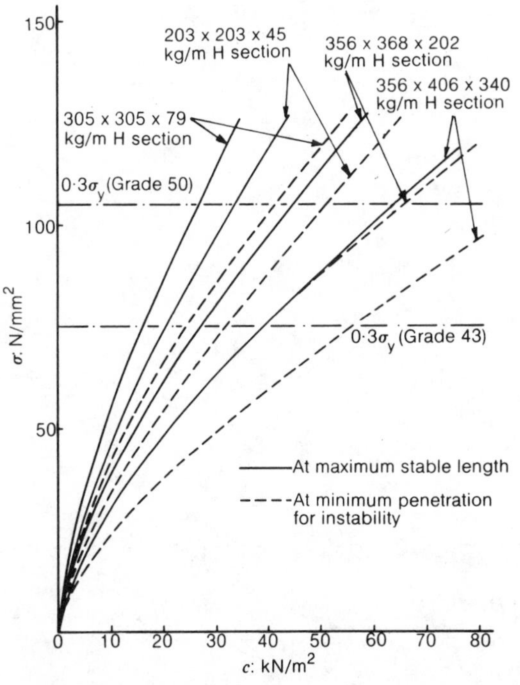

Fig. 8. Plot of axial stresses in H piles at the critical lengths shown in Fig. 6

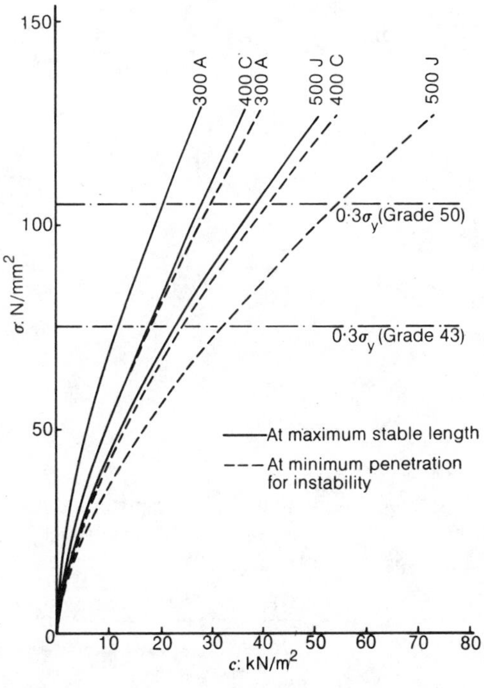

Fig. 9. Plot of axial stresses in tubular piles at the critical lengths shown in Fig. 7

$$X_{SI} = \tfrac{1}{2}\left[\frac{x}{l}\,\delta_{SI} + \frac{(1-\delta_{SI})}{\pi(S-I)}\sin\frac{(S-I)\pi x}{l} - \frac{1}{\pi(S+I)}\sin\frac{(S+I)\pi x}{l}\right]_h^l \tag{34}$$

$$S_{SI} = X_{SI} - \tfrac{1}{2}\left[\tfrac{1}{2}\frac{x^2}{l^2}\,\delta_{SI} + \frac{x}{l\pi}\left(\frac{(1-\delta_{SI})}{(S-I)}\sin\frac{(S-I)\pi x}{l} - \frac{1}{(S+I)}\sin\frac{(S+I)\pi x}{l}\right) + \frac{1}{\pi^2}\left(\frac{(1-\delta_{SI})}{(S-I)^2}\cos\frac{(S-I)\pi x}{l} - \frac{1}{(S+I)^2}\cos\frac{(S+I)\pi x}{l}\right)\right]_h^l \tag{35}$$

$$Y_{SI} = \tfrac{1}{2}SI\left[\tfrac{1}{2}\frac{\pi^2 x^2}{l^2}\,\delta_{SI} - \frac{1}{(S+I)^2}\cos\frac{(S+I)\pi x}{l} - \frac{(1-\delta_{SI})}{(S-I)^2}\cos\frac{(S-I)\pi x}{l}\right]_h^l \tag{36}$$

$$\overline{Y}_{I|S} = \tfrac{1}{2}\left[\frac{I}{(S+I)}\cos\frac{(S+I)\pi x}{l} + \frac{I(1-\delta_{SI})}{(S-I)}\cos\frac{(S-I)\pi x}{l}\right]_h^l \tag{37}$$

$$Z_{SI} = Y_{SI} - \tfrac{1}{2}SI\left[\frac{\pi^2 x^3}{3l^3}\,\delta_{SI} - \frac{x}{l}\left(\frac{1}{(S+I)^2}\cos\frac{(S+I)\pi x}{l} + \frac{(1-\delta_{SI})}{(S-I)^2}\cos\frac{(S-I)\pi x}{l}\right) + \frac{1}{\pi(S+I)^3}\sin\frac{(S+I)\pi x}{l} + \frac{1}{\pi(S-I)^3}\sin\frac{(S-I)\pi x}{l}\right]_h^l \tag{38}$$

$$\overline{Z}_{I|S} = \overline{Y}_{I|S} - \tfrac{1}{2}\left[\frac{Ix}{l}\left(\frac{1}{(S+I)}\cos\frac{(S+I)\pi x}{l} + \frac{(1-\delta_{SI})}{(S-I)}\cos\frac{(S-I)\pi x}{l}\right) - \frac{I}{\pi}\left(\frac{1}{(S+I)^2}\sin\frac{(S+I)\pi x}{l} + \frac{(1-\delta_{SI})}{(S-I)^2}\sin\frac{(S-I)\pi x}{l}\right)\right]_h^l \tag{39}$$

$$W_{SI} = \tfrac{1}{2}\pi^2 S^2\,\delta_{SI} \tag{40}$$

$$W_{I|S} = 0 \tag{41}$$

Quadrants II and III

All terms except the non-conservative potentials are transposable between quadrants II and III. These are presented in the form appropriate to quadrant II.

$$T_{sI} = \frac{1}{\pi}\left[-\frac{1}{I}\cos\frac{I\pi x}{l} + \frac{1}{(s+2I)}\cos\frac{(s+2I)\pi x}{2l} - \frac{1}{(s-2I)}\cos\frac{(s-2I)\pi x}{2l}\right]_0^l \tag{42}$$

$$V_{sI} = \tfrac{1}{4}\pi^3 s^2 I^2\left[\frac{1}{(s+2I)}\cos\frac{(s+2I)\pi x}{2l} - \frac{1}{(s-2I)}\cos\frac{(s-2I)\pi x}{2l}\right]_0^l \tag{43}$$

$$X_{sI} = \frac{1}{\pi}\left[-\frac{1}{I}\cos\frac{I\pi x}{l} + \frac{1}{(s+2I)}\cos\frac{(s+2I)\pi x}{2l} - \frac{1}{(s-2I)}\cos\frac{(s-2I)\pi x}{2l}\right]_h^l \tag{44}$$

$$S_{sI} = X_{sI} - \left[\frac{x}{\pi l}\left(-\frac{1}{I}\cos\frac{I\pi x}{l} + \frac{1}{(s+2I)}\cos\frac{(s+2I)\pi x}{2l} - \frac{1}{(s-2I)}\cos\frac{(s-2I)\pi x}{2l}\right) + \frac{1}{\pi^2}\left(\frac{1}{I^2}\sin\frac{I\pi x}{l} - \frac{2}{(s+2I)^2}\sin\frac{(s+2I)\pi x}{2l} + \frac{2}{(s-2I)^2}\sin\frac{(s-2I)\pi x}{2l}\right)\right]_h^l \tag{45}$$

$$Y_{sI} = sI\left[\frac{\pi s x}{(s^2-4I^2)l} - \frac{1}{(s+2I)^2}\sin\frac{(s+2I)\pi x}{2l} - \frac{1}{(s-2I)^2}\sin\frac{(s-2I)\pi x}{2l}\right]_h^l \tag{46}$$

$$\overline{Y}_{I|s}\ (\text{quadrant II}) = -I\left[\frac{1}{I}\sin\frac{I\pi x}{l} - \frac{1}{(s+2I)}\sin\frac{(s+2I)\pi x}{2l} - \frac{1}{(s-2I)}\sin\frac{(s-2I)\pi x}{2l}\right]_h^l \tag{47}$$

$$\overline{Y}_{i|S}\ (\text{quadrant III}) = \tfrac{1}{2}i\left[\frac{1}{(i+2S)}\sin\frac{(i+2S)\pi x}{2l} - \frac{1}{(i-2S)}\sin\frac{(i-2S)\pi x}{2l}\right]_h^l \tag{48}$$

$$Z_{sI} = Y_{sI} - sI\left[\frac{x}{l}\left(\frac{\pi x s}{2(s^2-4I^2)l} - \frac{1}{(s+2I)^2}\sin\frac{(s+2I)\pi x}{2l} - \frac{1}{(s-2I)^2}\sin\frac{(s-2I)\pi x}{2l}\right) - \frac{2}{\pi}\left(\frac{1}{(s+2I)^3}\cos\frac{(s+2I)\pi x}{2l} + \frac{1}{(s-2I)^3}\cos\frac{(s-2I)\pi x}{2l}\right)\right]_h^l \tag{49}$$

$$\overline{Z}_{I|s}\ (\text{quadrant II}) = \overline{Y}_{I|s} - I\left[\frac{x}{l}\left(-\frac{1}{I}\sin\frac{I\pi x}{l} + \frac{1}{(s+2I)}\sin\frac{(s+2I)\pi x}{2l} + \frac{1}{(s-2I)}\sin\frac{(s-2I)\pi x}{2l}\right) - \frac{1}{I^2\pi}\cos\frac{I\pi x}{l} + \frac{2}{\pi(s+2I)^2}\cos\frac{(s+2I)\pi x}{2l} + \frac{2}{\pi(s-2I)^2}\cos\frac{(s-2I)\pi x}{2l}\right]_h^l \tag{50}$$

$$\overline{Z}_{i|S}(\text{quadrant III}) = \overline{Y}_{i|S} - \tfrac{1}{2}i\left[\frac{x}{l}\left(\frac{1}{(i+2S)}\sin\frac{(i+2S)\pi x}{2l} - \frac{1}{(i-2S)}\sin\frac{(i-2S)\pi x}{2l}\right) + \frac{2}{\pi}\left(\frac{1}{(i+2S)^2}\cos\frac{(i+2S)\pi x}{2l} - \frac{1}{(i-2S)^2}\cos\frac{(i-2S)\pi x}{2l}\right)\right]_h^l \quad (51)$$

$$W_{sI} = \frac{\pi s^2 I}{s^2 - 4I^2} \quad (52)$$

$$\overline{W}_{I|s}\,(\text{quadrant II}) = -\pi I(-1)^I \quad (53)$$

$$\overline{W}_{i|S}\,(\text{quadrant III}) = 0 \quad (54)$$

REFERENCES

1. BURGESS I. W. A note on the directional stability of driven piles. Géotechnique, 1975, 25, No. 4, 413–416.
2. BURGESS I. W. The stability of slender piles during driving. Géotechnique, 1976, 26, no. 2, 281–292.
3. BURGESS I. W. Analytical studies of pile wandering during installation. Int. J. Num. Anal. Meth. Geomechanics, 1979, 3, No. 1, 49–62.
4. OMAR R. M. Instability of piles during driving. PhD thesis, Queen Mary College, London, 1977.
5. HANNA T. H. The bending of long H-section piles. Can. Geotech. J., 1968, 5, No. 3, 150–172.
6. FELLENIUS B. H. Bending of piles determined by inclinometer measurements. Can. Geotech. J., 1972, 9, No. 1, 25–32.
7. SMITH I. M. Discrete element analysis of pile instability. Int. J. Num. Anal. Meth. Geomechanics, 1979, 3, No. 2, 205–213.
8. TERZAGHI K. Evaluation of coefficients of subgrade reaction. Géotechnique, 1955, 5, 297–326.

3. Directional stability of piles during driving

R. M. OMAR, BSc, PhD (formerly at Queen Mary College, University of London; now with State Oil Consulting Services, Bagdad) and T. J. POSKITT, DSc, FICE, FIStructE (Professor of Civil Engineering and Head of Department, Queen Mary College, University of London)

A theoretical and experimental investigation of the directional stability of driven piles is described. Deviation is shown to depend on initial imperfections in the pile and on the soil characteristics. The deformation of the pile is flexural and for thin-walled open cross-sections (e.g., H sections) this is accompanied by torsion. The stability of the path followed by the pile is investigated using the methods of Routh and Lyapunov. These give combinations of the physical parameters for which the path is unstable. Skin friction, horizontal subgrade modulus, the distribution of these with depth, and the point resistance are found to influence the behaviour. Case studies reported in the literature are discussed and the results of a laboratory investigation using small piles is given. The laboratory piles were observed by means of a radiographic technique. Bending and twisting is clearly visible.

INTRODUCTION

It is well known that long driven or jacked piles may deviate from a straight course during insertion. This has important practical implications if

(a) piles are members of a group and a pile which has deviated affects the driving of other members of the group;
(b) the structural integrity of the pile is impaired due to large deviations during insertion;
(c) piles designed to go through upper layers of soft soil and bear on firm underlying strata fail to reach the underlying strata due to large deviations;
(d) due to large deviations the application of the working load produces additional deflexions of the pile which are sufficient to cause yield of the soil; the carrying capacity of the pile might then be impaired.

If one or more of the above factors is suspected it may be necessary to establish the integrity of the pile by load-testing.

2. Awareness of the problem of piles not driving straight arose with the extensive use of steel piles in Norway in the 1930s.[1] Since then many cases have been reported.[2-7] With the exception of the work described by Hanna,[5] all the above investigations are concerned with the bearing capacity of bent piles rather than the cause of bending.

3. In practice the reasons for piles not driving straight may be one or more of the following: lack of straightness in the pile; eccentric driving forces; or an asymmetric failure pattern in the soil at the pile tip. Other considerations such as slackness in the guides or leaders are equally important since these determine the direction of entry of the pile into the ground.

4. Due to the compressive nature of the forces in the pile during installation the tendency to deviate may under some circumstances be greatly increased. This will occur at a critical penetration length when the pile will begin to suffer large curvatures and twists. These may be sufficient to cause yield or fracture. The phenomenon, which is analogous with the buckling of a strut, is in the nature of a classical instability in the pile–soil system. It is of practical importance to know the circumstances which produce this.

5. To understand the behaviour of a pile as it penetrates the ground it is useful to consider an analogy with a strut. This enables two important aspects of the problem, namely the equilibrium state and the stability or instability of this state, to be focused on.

6. When an axial load is applied to a strut which is initially slightly curved it will induce further curvature. Generally for a given axial load the induced curvature will depend on the initial lack of straightness. This is also the case with a pile where the deviation is found to depend on the magnitude of the initial imperfections. With a strut as the axial load approaches the Euler critical load the deflexions become greatly magnified and the strut fails. The curved equilibrium state is then unstable. For a pin-ended strut this condition is approached when the parameter $PL^2/EI \rightarrow \pi^2$, where P is the axial load, L is the length, E is Young's modulus and I is the second moment of area. An analogous situation arises with a pile as it penetrates the ground. As the critical penetration length is approached the tendency of the pile to deviate becomes greatly magnified and the path followed by the pile becomes unstable.

7. Before analytical methods of studying pile deviation

can be developed there are some important differences which must be noted between a simple strut and a pile. A strut has a linear stress–strain law and the forces do not change direction as it buckles (i.e., they are conservative). For a pile the non-linear behaviour of the soil and the non-conservative nature of the forces which the soil exerts on the pile must be taken into account. The problem of specifying stress–strain laws for soils are well known and are not elaborated on in detail here. However, the non-conservative nature of the forces exerted by the soil on the pile is not generally recognized.[8]

8. The consequence of non-linearity and non-conservative forces is that when analysing a pile it is not possible to use the methods which find a universal application in the study of the stability of conservative linear structural systems.[9] Instead it is necessary to use the more general dynamical theory of stability as developed by Routh and Lyapunov. This, as the name suggests, requires the pile to be considered in motion. Therefore equations of motion have to be established for the pile and the stability of these is investigated by considering the motion to be slightly perturbed. For very small perturbations the equations for the perturbed component of motion are linear. If the solution of these equations is a small bounded oscillation the system is considered stable. The combination of physical parameters which give this provides the stability criteria for the system. Other combinations of physical parameters not satisfying these criteria result in an unstable system. In these cases the perturbation equations have solutions which either diverge uniformly or oscillate with an ever increasing amplitude. This latter condition is often referred to as flutter. The methods of Routh and Lyapunov are not well known in civil engineering and this is probably the reason why pile deviation has not received the attention it warrants.

9. However, before formal consideration can be given to these methods it is first necessary to establish the mode of deformation of a pile as it enters the ground and determine the forces that the soil exerts on it.

NOTATION

a	half the pile width
A_p	pile point area
C	torsional rigidity (= GJ)
$\bar{c}$	adhesion factor
c_u	undrained shear strength
d	thickness of pile
E_s	modulus of elasticity of soil
EI	flexural rigidity per unit width of pile
G	modulus of shear of pile
g	gravity
i	$(-1)^{1/2}$
I_0	polar moment of inertia of pile cross-section
J	torsion constant of pile
K_δ	coefficient of earth pressure on pile wall
K_2	$K_\delta \tan \delta$
L	embedded length of pile
L_c	critical length for flexural instability (tests in clay)
L_{cr}	critical length for flexural instability (tests in sand)
$L_{c\phi}$	critical length for torsional instability
p	pile perimeter
P_0	pile driving force
P_1	pile point resistance
t	time
w	weight per unit length for unit width of pile
α_2	soil stiffness coefficient
β_2	soil stiffness coefficient
δ	angle of friction between pile and soil
η	increase in subgrade reaction per unit depth
$\sigma_{1,2}$	normal earth pressures on the pile wall
σ_x	axial stress in pile
$\tau_{1,2}$	shear stresses on pile wall
Ω	angular velocity
ρ	$(IA_p)^{1/2}$

PILE DEFORMATION AND SOIL REACTION

10. When a pile is subjected to an axial force it will shorten in length and bend in either one or two mutually perpendicular planes. For a pile of thin-walled open cross-section such as an H section it will also tend to twist.[10] After loading the deformed shape will be as shown in Fig. 1. Cases of piles bending about two axes have been reported[5] and radiographs showing torsional/flexural behaviour are presented later in this Paper.

11. One of the difficulties with H piles is that little knowledge is available on the effects of plugging between web and flanges. This makes it difficult to determine the skin friction and normal pressure which the soil exerts on the different faces of the pile (Fig. 1). Nevertheless these forces are of sufficient magnitude to maintain the pile bent about two axes and twisted about the shear centre. To demonstrate how the necessary bending and twisting

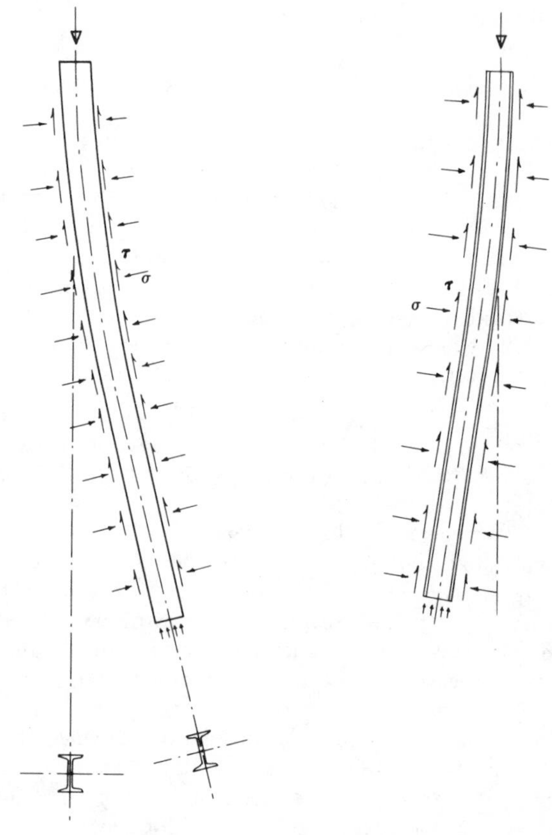

Fig. 1

moments are generated by the soil it is simplest to consider each separately.

12. Referring to Fig. 2, consider the forces acting on unit width of the pile. Resolving tangentially at D:

$$-(\tau_1 + \tau_2) - \frac{dP}{dx} + w = 0$$

Resolving normally:

$$\sigma_1 - \sigma_2 + \frac{dF}{dx} - \frac{P}{R} = 0$$

where w is the weight per unit length of a pile of unit width, and R is the radius of curvature of the element. Taking moments:

$$F + \frac{dM}{dx} = 0$$

where M is the moment induced by the soil forces and the pile self weight.

13. From these equations it may be shown that

$$\frac{d^2M}{dx^2} = -\bar{\sigma} - [P_0 + wx - \int_0^x \bar{\tau}\, dx] \frac{d^2y}{dx^2} \qquad (1)$$

where

$$\bar{\tau} = \tau_1 + \tau_2$$

$$\bar{\sigma} = \sigma_2 - \sigma_1$$

14. It is less easy to demonstrate how the soil forces generate significant torsional moments on a pile section. For simplicity consider a rectangular cross-section as shown in Fig. 3 which represents a web or flange. This shows an element of twisted pile of length dx. Due to the symmetry of the cross-section the element twists about the centroidal axis Ox. The angle of twist is ϕ. Consider a longitudinal strip at a distance r from the x axis. The surface area is dA (= $dx\, dr$) as shown in part (a) of Fig. 3. The inclination of the strip with the x–y plane is θ and is given by

$$\theta = r\frac{d\phi}{dx} \qquad (2)$$

15. Taking the moment about the x axis of the soil forces acting on the strip gives for the skin friction

$$-(\tau_1 + \tau_2)dA \sin\theta\, r$$

and for the normal pressure

$$-(\sigma_2 - \sigma_1)dA \cos\theta\, r$$

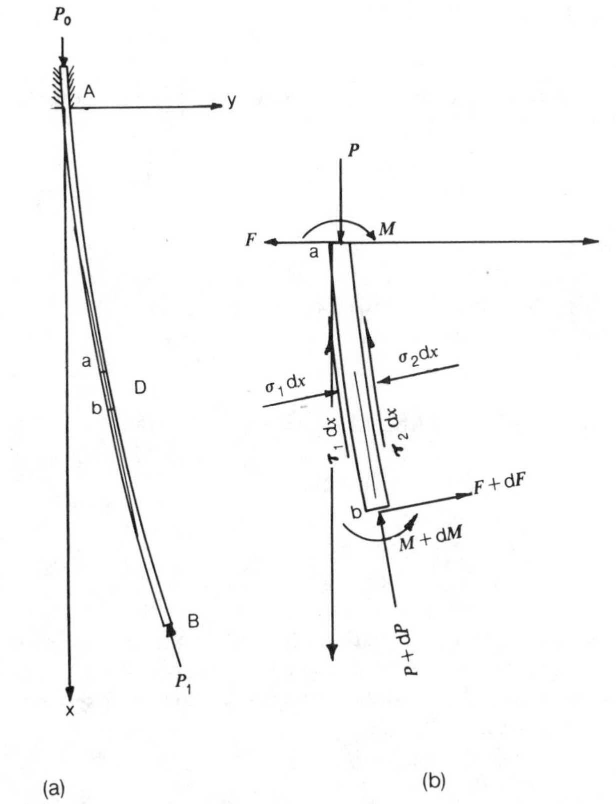

Fig. 2

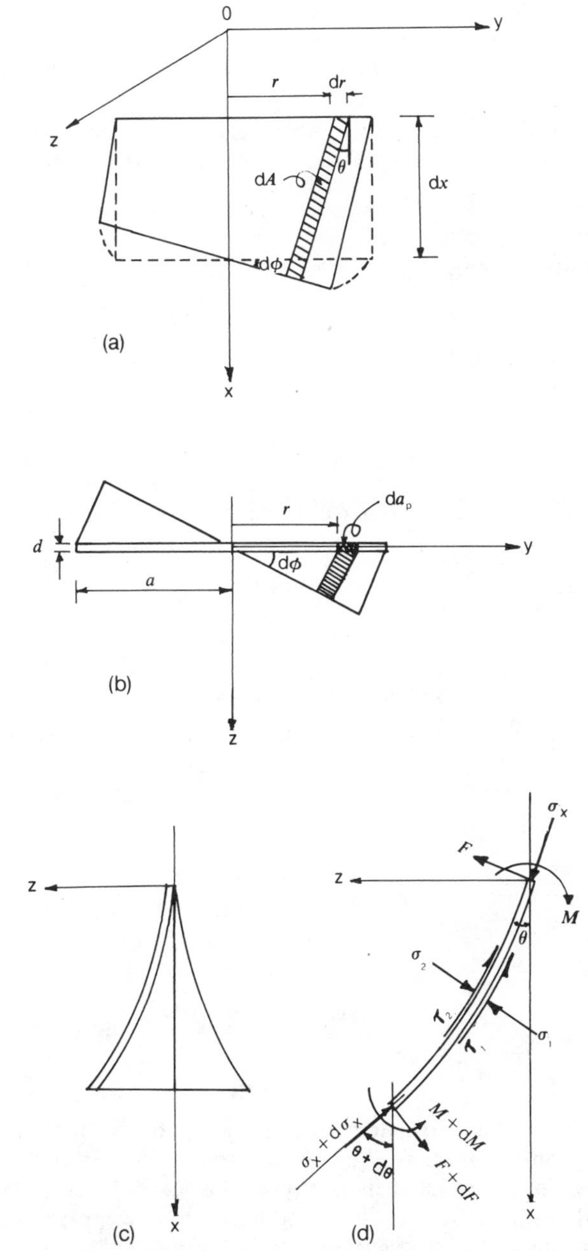

Fig. 3

Adding these two components and integrating them over the element gives

$$-2\int_0^a r(\bar{\tau}\sin\theta+\bar{\sigma}\cos\theta)\,\mathrm{d}r=\frac{\mathrm{d}T}{\mathrm{d}x} \qquad (3)$$

where T is the torque due to the soil forces. Since the angle θ is small equation (3) becomes

$$-2\int_0^a \left(\bar{\tau}r\frac{\mathrm{d}\phi}{\mathrm{d}x}+\bar{\sigma}\right)r\,\mathrm{d}r=\frac{\mathrm{d}T}{\mathrm{d}x} \qquad (4)$$

16. Equations (1) and (4) show how the soil forces produce bending and twisting in the pile. It is next necessary to consider how small changes in these forces control stability as the pile is inserted.

17. The changes in soil forces are the consequence of an assumed small deflexion, i.e., perturbation of the pile: If they are such as to cause the pile to depart further from its undisturbed state the system is unstable, while if they tend to make it return the system is stable.

18. The admissible modes in which the pile may be perturbed are those of flexure and torsion. For pure flexure y will become $y+h$ and for pure torsion ϕ will become $\phi+\psi$. The effect of these small variations on equations (1) and (4) are found by Taylor's theorem; thus for equation (1)

$$\delta\left(\frac{\mathrm{d}^2M}{\mathrm{d}x^2}\right)=-\beta_2h+y''\int_0^x\alpha_2h\,\mathrm{d}x-\left[P_0+wx-\int_0^x\bar{\tau}\,\mathrm{d}x\right]h'' \qquad (5)$$

and for equation (4)

$$\delta\left(\frac{\mathrm{d}T}{\mathrm{d}x}\right)=-2\int_0^a\left[\alpha_2r^2\psi\phi'+\bar{\tau}r\psi'+\beta_2\psi r\right]r\,\mathrm{d}r \qquad (6)$$

where

$$\left.\begin{aligned}\alpha_2&=\frac{\partial\bar{\tau}}{\partial y}\\ \beta_2&=\frac{\partial\bar{\sigma}}{\partial y}\end{aligned}\right\} \qquad (7)$$

Note:

$$\frac{\partial\bar{\tau}}{\partial\phi}=\frac{\partial\bar{\tau}}{r\partial\phi}r=\alpha_2r$$

and

$$\frac{\partial\bar{\sigma}}{\partial\phi}=\frac{\partial\bar{\sigma}}{r\partial\phi}r=\beta_2r$$

19. The parameters β_2 and α_2 are the stiffness coefficients of the soil. They are analogous to p–y parameters used in the lateral analysis of piles. For normal pile design only β_2 would be required. The reason for α_2 appearing is that coupling has been assumed between shear stress and normal strain. If as a first approximation the soil is regarded as an anisotropic elastic material then $\alpha_2=0$. For soil in the plastic range this assumption does not appear valid and further investigation is required.

20. β_2 is related to the horizontal subgrade modulus of the soil. Since soil is essentially a non-linear material, which depending on the state of stress may be either elastic or plastic (including work-softening), the horizontal subgrade moduli, which are tangents to the stress–strain curve of the current stress level, may therefore be either positive or negative.

21. The problems of determining stiffness properties of soil in the vicinity of the pile wall and of determining the stresses σ and τ exerted on the pile by the soil constitute two of the major problems in pile design. There is no general agreement on their determination and so for purposes of this work the simplest and most widely accepted published results have been taken.

FLEXURAL INSTABILITY DURING DRIVING

22. Figure 2 shows the pile at any instant of time during driving. Consider a small element of pile ab of unit width; if y_0 is the undeflected shape of the pile then the equation of motion is

$$M''-EI(y''''-y_0'''')=\frac{w}{g}\ddot{y} \qquad (8)$$

where M'' is given by equation (1). Primes indicate $\partial/\partial x$ and dots $\partial/\partial t$.

23. As mentioned earlier the flexural stability can be considered by imagining y to be perturbed a small amount δy. The effect of this perturbation on equation (8) may be found by Taylor's theorem thus

$$\delta M''-EI(\delta y''''-\delta y_0'''')=\frac{w}{g}\delta\ddot{y} \qquad (9)$$

Substituting from equation (5) and writing $\delta y\equiv h$, equation (9) becomes

$$-\beta_2h-\left[P_0+wx-\int_0^x\bar{\tau}\,\mathrm{d}x\right]h''-EIh''''=\frac{w}{g}\ddot{h} \qquad (10)$$

where $\alpha_2=0$ for reasons given in paragraph 18.

24. Stability boundaries for equation (10) corresponding to various soil types are given by Omar.[11]

TORSIONAL INSTABILITY DURING DRIVING

25. Figure 3 shows an element of pile dx at any instant of time during driving. The equation of motion is

$$T'+C\phi''-I_0(\sigma_x\phi')'=\frac{w\rho^2}{g}\ddot{\phi} \qquad (11)$$

where T' is given by equation (3), C is the torsional rigidity of the element, I_0 the polar moment of inertia of the cross-section and ρ the radius of gyration of the cross-section. The axial stress in the pile σ_x is

$$\sigma_x=\frac{2}{A_p}\int_x^L\mathrm{d}x\int_0^a\bar{\tau}\,\mathrm{d}r \qquad (12)$$

26. Substituting from equations (4) and (12) into equation (11) gives

$$(C - I_0\sigma_x)\phi'' + 2\phi' \int_0^a \bar{\tau}(\rho^2 - r^2)\mathrm{d}r - 2\int_0^a \bar{\sigma} r\,\mathrm{d}r = \frac{w}{g}\rho^2\ddot{\phi} \qquad (13)$$

The boundary conditions are

$$\left.\begin{array}{ll} x = 0, & \phi = 0 \\ x = L, & \phi' = 0 \text{ (zero torque at the pile point)} \end{array}\right\} \qquad (14)$$

(primes indicating $\partial/\partial x$ and dots $\partial/\partial t$).

27. If $(C - I_0\ \sigma_x) = 0$ equation (13) has a singularity. This is analogous to torsional buckling.[12] The buckle is located at the top of the pile where σ_x is greatest. In the absence of destabilizing influences from the soil this provides the criterion for instability.

28. As mentioned earlier the torsional stability can be considered by imagining ϕ to be perturbed a small amount $\delta\phi$. The effect of this perturbation on equation (11) may be found by Taylor's theorem thus

$$\delta T' + C\delta\phi'' - I_0\ \delta(\sigma_x\phi')' = \frac{w\rho^2}{g}\delta\ddot{\phi} \qquad (15)$$

Taking the variation of equation (15) gives

$$(C - I_0\sigma_x)\psi'' + \left[\frac{2I_0}{A_p}\int_0^a \bar{\tau}\,\mathrm{d}r - 2\int_0^a r^2\bar{\tau}\,\mathrm{d}r\right]\psi' - 2\psi\int_0^a \beta_2 r^2\,\mathrm{d}r = \frac{w\rho^2}{g}\ddot{\psi} \qquad (16)$$

where $a_2 = 0$ (§18) and $\psi \equiv \delta\phi$.

29. It is reasonable to assume that β_2 is dependent on depth and independent of r. The same can be said of $\bar{\tau}$. Equation (16) then becomes

$$(C - I_0\ \sigma_x)\psi'' + \frac{ad^2}{6}\bar{\tau}\,\psi' - \frac{2\beta_2 a^3}{3}\psi = \frac{w\rho^2}{g}\ddot{\psi} \qquad (17)$$

30. Equation (17) is linear and homogeneous and oscillatory solutions may be found by writing

$$\psi(x,t) = \psi(x)\,\mathrm{e}^{\mathrm{i}\Omega t} \qquad (18)$$

31. Substituting from equation (18) into equation (17) gives

$$(C - I_0\ \sigma_x)\frac{\mathrm{d}^2\psi}{\mathrm{d}x^2} + \frac{ad^2}{6}\bar{\tau}\frac{\mathrm{d}\psi}{\mathrm{d}x} + \left(\frac{w\rho^2}{g}\Omega^2 - \frac{2\beta_2 a^3}{3}\right)\psi = 0 \qquad (19)$$

The end conditions are

$$\left.\begin{array}{ll} x = 0, & \psi = 0; \\ x = L, & \dfrac{\mathrm{d}\psi}{\mathrm{d}x} = 0 \end{array}\right\} \qquad (20)$$

Equation (19) is the 'frequency equation' and when solved with the end conditions (equation (20)) gives an expression from which Ω^2 may be found.

32. The system is defined as stable for combinations of $\bar{\tau}$, C, β_2, a and d which give real positive values for Ω^2. For all other combinations it is unstable.

33. Equation (19) has a singularity similar to that of equation (13). Hence the combination of parameters selected must not make $C - I_0\sigma_x = 0$ if torsional buckling of the type discussed in connection with equation (13) is to be avoided.

34. Equation (19) can be solved by finite differences. A standard latent root problem then results. The accuracy of the solution is checked by using successively smaller

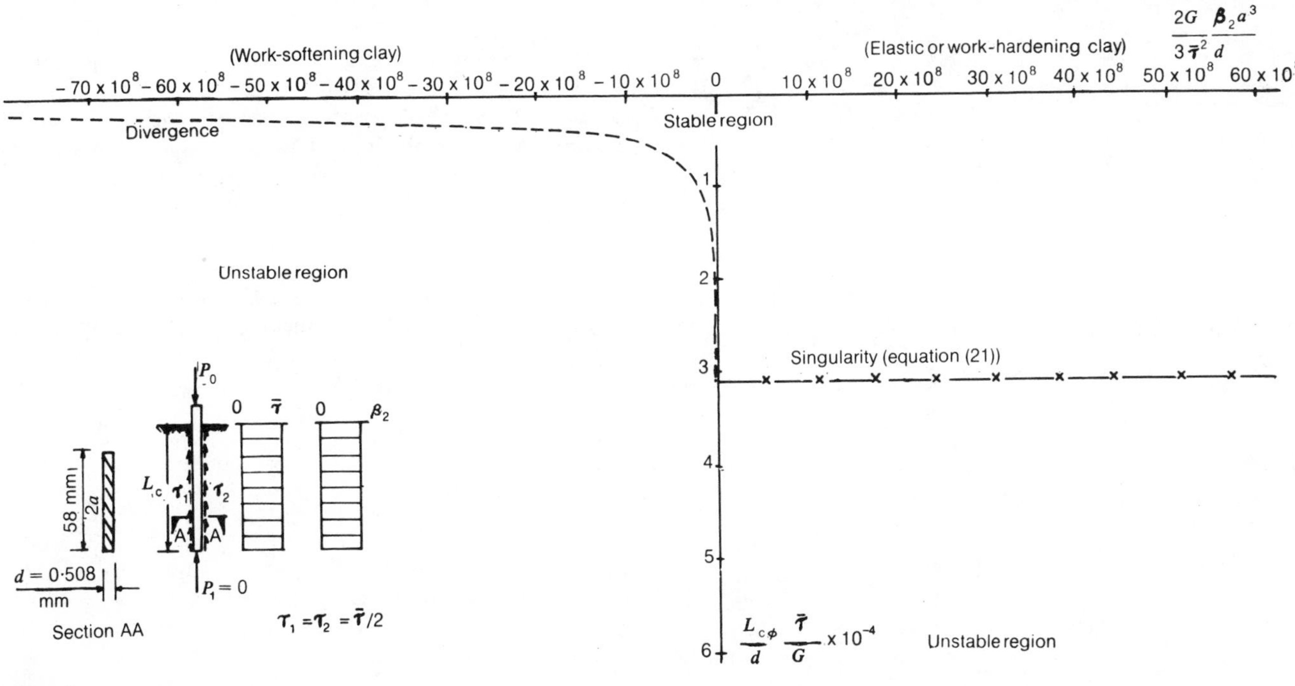

Fig. 4

difference meshes. In general it has been found that 20 points is sufficiently accurate. An independent check on the results may be made by using Galerkin's method. Functions of the form $1-\cos(k\pi x/L)$ which satisfy the boundary conditions may be used. It is practical to take a maximum of five functions (i.e., $k = 1, 2, 3, 4, 5$). The accuracy is slightly inferior to that which can be obtained by using finite differences.

Fig. 5

35. Equation (19) has been solved for the case of $\bar{\tau}$ and β_2 constant with penetration. A cross-section 58 mm × 0.508 mm has been taken since this corresponds to one of the laboratory test piles. The combinations of physical parameters which produce instability are shown in Fig. 4. This graph shows that for elastic or work-hardening soils torsional buckling at the top of the pile is the limiting factor. The critical length in this case is given by

$$L_{c\phi} = \frac{G}{\bar{\tau}} \frac{4d^3}{4a^2 + d^2} \qquad (21)$$

Soils which soften with deformation have critical lengths significantly less than this.

LABORATORY TEST ARRANGEMENTS

36. It has been shown that piles may bend and twist during driving. To confirm this it is necessary to observe the shape of a pile as it advances into the soil. A convenient method of doing this in the laboratory is provided by X rays. The deflexion of the image in the plane being radiographed will indicate bending and the width of the image will indicate the presence of twist.

37. Due to the limited thickness of soil which X rays can penetrate it is necessary to use small piles and ensure that they deflect in one plane only. The requirement for bending in one plane only can be arranged by using piles of thin rectangular cross-section. The double symmetry of a rectangle will result in twist about the centroidal axes. This has advantages when obtaining twists from image widths.

38. Laboratory experiments to investigate stability phenomena are notoriously difficult to perform. The testing of struts is a well known case. Similar problems arise with piles where factors such as initial twist or curvature and eccentricities in the driving mechanism make it difficult to determine precisely when stability is lost. There are in addition problems associated with the preparation of soil samples and the determination of the stress–strain characteristics of the soil.

39. To minimize these problems and at the same time meet the requirements for the X ray technique the equipment shown in Fig. 5 was developed. It consists of a glass-sided box (A) 450 mm high and 220 mm wide which

Table 1

Test	Pile	Section, mm	Surface	$\tau_{1,2}$, kN/m²	$L_{c\phi}/L_c$	L_c (theory), mm	L_c (experiment), mm
T39	M2	25×0·508	Rough	29·0	4·1	145	120
T30	M2	25×0·508	Rough	13·6	6·8	186	190
T32A	M2	25×0·508	Rough	16·4	6·0	175	145
T24	M1	25×0·635	Rough	19·8	8·3	205	160
T25	M1	25×0·635	Rough	24·8	7·1	191	245
T26	M1	25×0·635	Rough	24·6	7·2	191	220
T28A	M1	25×0·635	Rough	16·6	9·3	218	185
T35	M1A	25×0·635	Rough	22·0	7·6	200	210
T36	M1A	25×0·635	Rough	28·6	6·5	182	165
T37	M1A	25×0·635	Rough	18·6	8·6	210	200
T34	M4S	25×1·016	Smooth	17·4	23·1	342	310

contains the sample. The pile is free to deflect in the plane of the box and by radiographing the soil the position of the pile may be located. Due to the limited ability of the X rays to penetrate soil the box is only 80 mm thick. This restricts the width of pile section which can be tested without interference from the sides of the box.

40. Mounted on the sample box is the driving mechanism B. The function of this is to provide directional constraint at the point where the pile enters the soil and at the same time allow the pile to be advanced at a constant speed into the soil. The constant speed is provided by a commercial compression testing machine acting through a rack and pinion device.

41. The accuracy with which the driving mechanism advances the pile along a given path and aberrations of the images on radiographs have been assessed.[13]

42. The piles were made from strip steel. Cross-sections varying from 25 mm to 58 mm wide and 0.38 mm to 1.02 mm thickness were used. Polished and sand-blasted surface finishes were used.

TEST RESULTS FOR CLAY

43. Each box of clay could be used for a number of tests depending on how much disturbance it had suffered from previous tests. At the conclusion of testing, samples were taken at various depths and at different plan locations and the undrained shear strengths determined. Values showed the usual scatter and gave an average value of 35 kN/m^2. It was concluded from the shear strength tests that c_u was roughly constant with depth and that the horizontal subgrade modulus was also constant with depth. The clay did not appear to work-soften during deformation and so the critical length is given[13] by $L_c = [39.93\ EI/\bar{\tau}]^{1/3}$.

44. The skin friction was found by measuring the force required to extract the pile at the end of the test. Values $\tau_{1,2}$ are given in Table 1. The pull-out tests indicated that the skin friction was fairly constant with depth. The average value for all tests was 9 kN/m^2. The adhesion factor corresponding to this is 0.26. This is on the low side and is thought to be due to the low horizontal stresses in the clay, the small volume displaced by the pile, and minute gaps between the pile and clay caused by vibration during insertion.

45. Figure 6 shows a typical tracing of radiographs obtained for three tests on pile M1A. A major difficulty in the interpretation of these curves is establishing the point where a significant loss of stability was experienced by the pile. The method used was to divide the pile into a number of equal segments and obtain the curvature by means of a difference table. The point at which the curvature started to increase significantly was taken to indicate the critical length. Since there are limitations in this procedure each radiograph was checked by eye for rapid changes in curvature.

46. Difficulties of this type are common in stability work. For example, in strut testing, lateral deflexions become greatly magnified as the critical load is approached and to aid in the interpretation of the data the Southwell plot is generally used.

47. Table 1 gives the results of eleven tests where flexural instability was observed. There is fair agreement between the theoretical and experimental critical lengths.

48. In none of the tests in Table 1 did the piles show much tendency to twist. There were, however, a considerable number of tests where twisting was significant. Three typical cases are shown in Fig. 7. For such cases torsion masks the flexural behaviour. It was found that in the

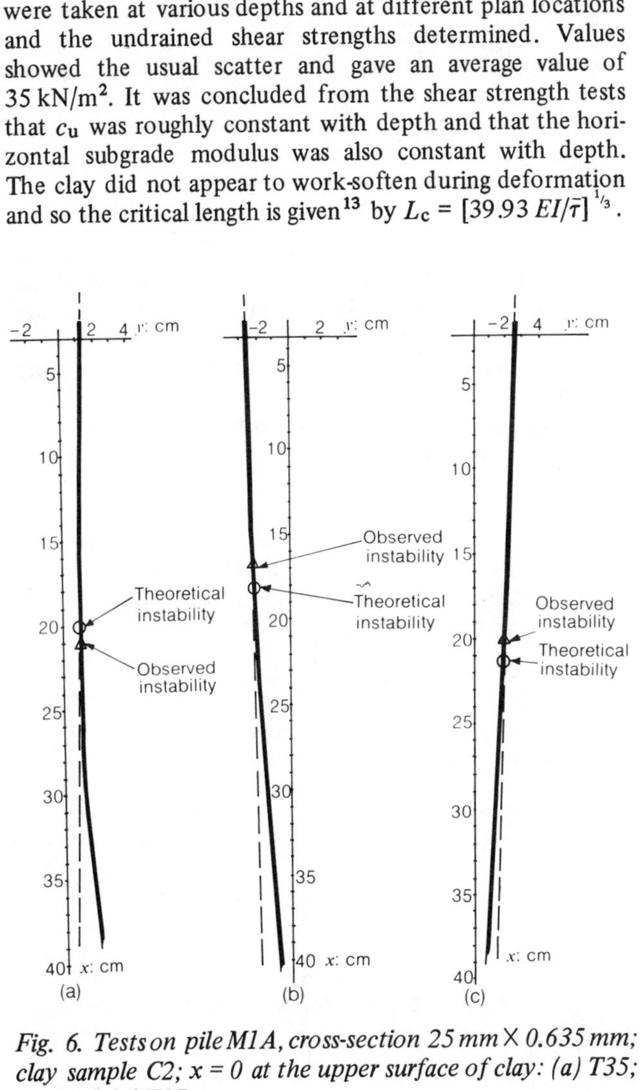

Fig. 6. Tests on pile M1A, cross-section 25 mm × 0.635 mm; clay sample C2; x = 0 at the upper surface of clay: (a) T35; (b) T36; (c) T37

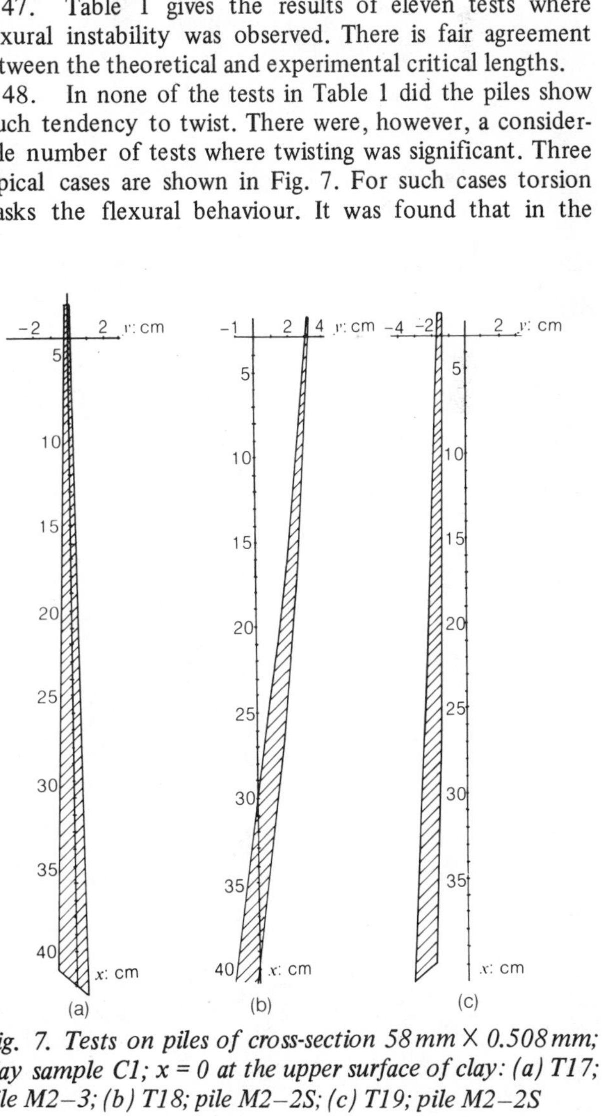

Fig. 7. Tests on piles of cross-section 58 mm × 0.508 mm; clay sample C1; x = 0 at the upper surface of clay: (a) T17; pile M2–3; (b) T18; pile M2–2S; (c) T19; pile M2–2S

majority of tests which were performed this proved to be the case and the detection of flexural instability proved difficult.

49. The results shown in Fig. 7 are due to the growth in twist resulting from the use of an initially twisted pile.

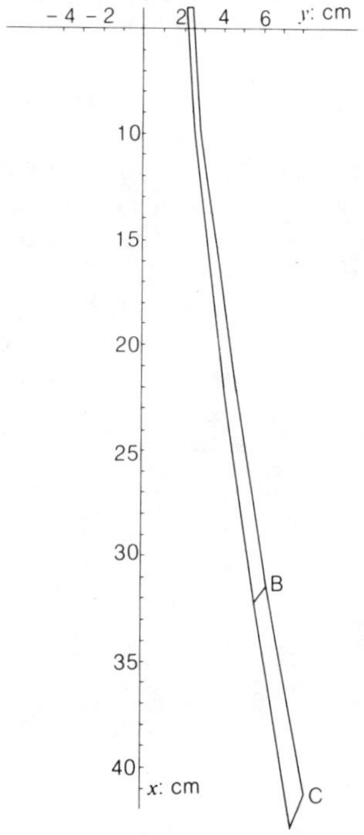

Fig. 8. Test T43; pile M22, cross-section 60 mm × 0.508 mm; clay C3; x = 0 at the upper surface of clay

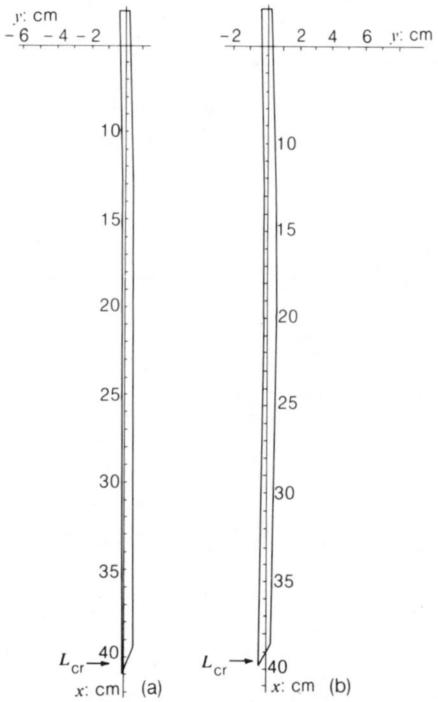

Fig. 9. Pile M2–2S, cross-section 58 mm × 0.508 mm, smooth surface; x = 0 at the upper surface of sand: (a) T6; dense sand; (b) T9; loose sand

They are not due to torsional instability. This is confirmed by the uniform manner in which the width of the image increases with penetration. The presence of torsional instability would cause a rapid change in the image width.

50. Tests were performed in which a pile was radiographed at different stages of penetration. A typical result is shown in Fig. 8. This shows that when the tip has reached B the image width at B is the same as when the tip reaches C. It therefore appears that the pile travels along the cavity formed by the point.

TEST RESULTS FOR SAND

51. The sand used for the tests had a uniformity coefficient of 2.3 and G_s of 2.65. Maximum and minimum densities obtained by Kalbuszewski's method were 1.92 g/cm^3 and 1.51 g/cm^3, corresponding to porosities of 27.3% and 47.8%. The average properties of the sand used are given in Table 2.

52. Two piles 58 mm × 0.508 mm cross-section, one with a smooth and the other with a rough surface, were used for the tests. Pull-out tests on these were used to determine the skin friction and K_2. The forces were very small and consequently the determinations were not very accurate. Average values of K_2 are given in Table 3, together with measured angles of friction between the piles and the sand.

53. Attempts to measure the rate of increase of the horizontal subgrade reaction η_s with depth proved unsuccessful. A lower bound to the critical length was therefore obtained by assuming η_s nearly zero. This gives[11]

$$L_{cr} = [53.94\, EI/\gamma K_2]^{1/4}$$

Values of L_{cr} are entered in Table 3. These are of the same order of magnitude as the greatest length of pile which can be accommodated in the sample box. The piles when fully driven should therefore be stable. Fig. 9 shows results for the smooth pile when driven into dense and loose sand. The piles are reasonably straight.

54. To demonstrate the importance of initial imperfections in determining the path taken by a pile, a new pile manufactured with great care to reduce the initial imperfections was made and tested. The results are shown in Fig. 10. These indicate that for the first few insertions flexural imperfections grew in magnitude and tended to dominate the deflected shape. At the same time torsional

Table 2

	γ, N/cm^3	e	n, %	D_r, %	ϕ_s^0
Loose	0·0163	0·619	38·3	34·6	33·1
Dense	0·0169	0·529	34·6	58·6	40·4

Table 3

Pile	Loose sand		Dense sand		δ^0
	K_2	L_{cr}, cm	K_2	L_{cr}, cm	
M2–2s (smooth)	0·156	39·3	0·139	40·2	16·2
M2–3 (rough)	0·161	39·0	0·184	37·4	19·3

imperfections also grew and so towards the end of testing torsional/flexural behaviour occurred. In the final test torsional behaviour dominated. This pile was subsequently used for tests in clay and torsional behaviour continued to dominate (Fig. 7(a)).

CONCLUSIONS

55. The investigation has shown the importance of initial imperfections and soil characteristics in determining the path taken by a driven pile. In following the path the pile bends and if it is of a thin-walled open cross-section

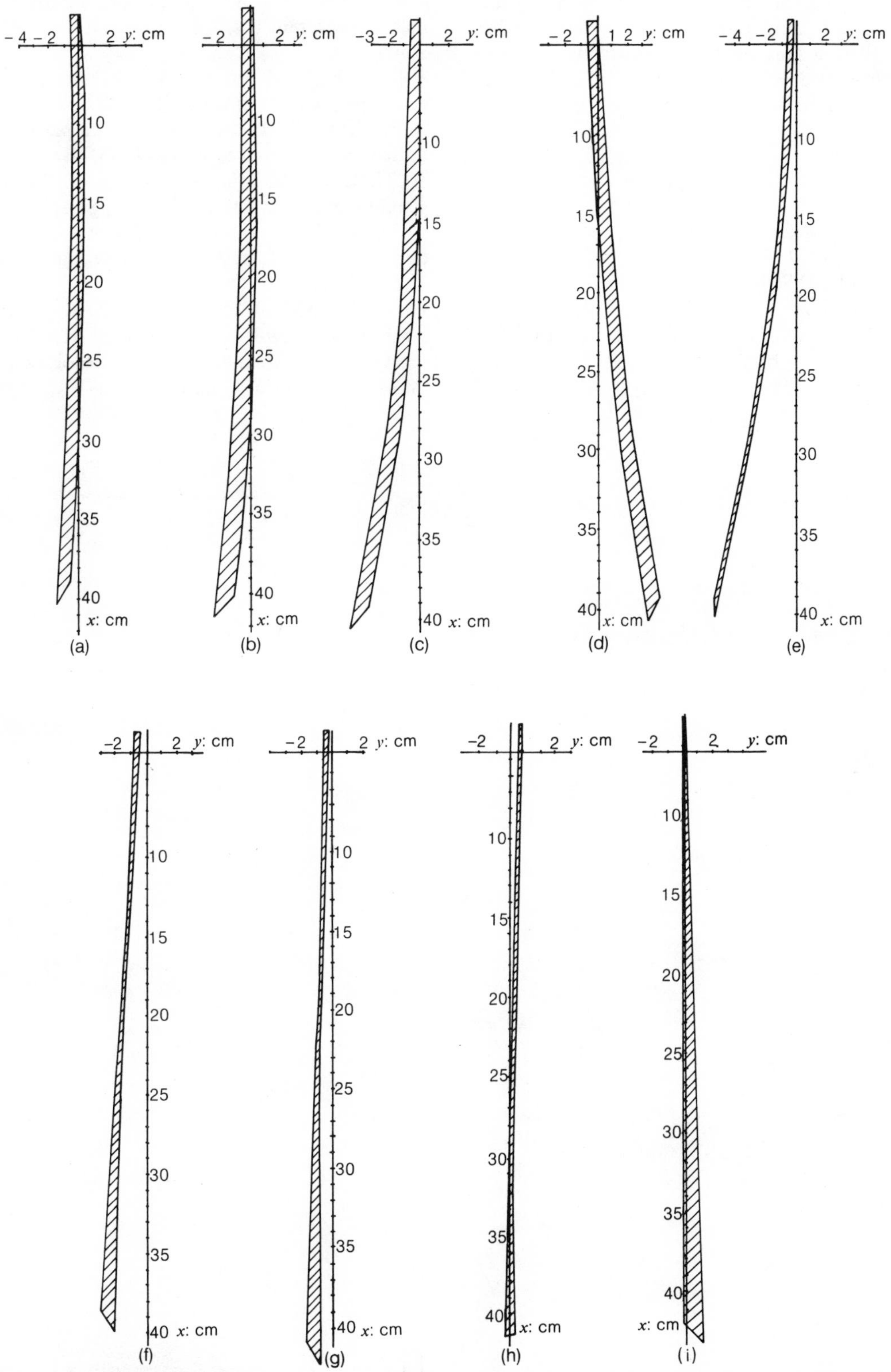

Fig. 10. Pile M2–3, cross-section 58 mm × 0.508 mm, rough surface; x = 0 at the upper surface of sand; (a) T7; dense sand; (b) T8; loose sand; (c) T10; loose sand; (d) T11; loose sand; (e) T12; loose sand; (f) T13; loose sand; (g) T14; loose sand; (h) T15; loose sand; (i) T16; dense sand

(such as an H pile) it will also twist. Since excessive departures of a pile from its assumed course may affect its carrying capacity, it is of some importance to predict when this will occur. A classical stability analysis has been used for this purpose. This gives values for non-dimensional combinations of the physical parameters at which instability occurs. These depend on skin friction and horizontal subgrade modulus, and the distribution of these along the pile. The problem is of most importance with slender piles or for piles installed in soils which work-soften with deformation. For thin-walled open cross-section piles torsional buckling may occur at the top.

REFERENCES

1. BJERRUM L. Norwegian experience with steel piles to rock. Géotechnique, 1957, 7, 73–96.
2. PARSONS J. and WILSON S. Safe loads on dog-leg piles. Trans. Am. Soc. Civ. Engrs, 1956, 121, 695.
3. JOHNSON S. M. Determining the capacity of bent piles. Jnl Soil Mech. Fdn Engng Div. Am. Soc. Civ. Engrs, 1962, 88, No. SM6, 65–79.
4. BROMS B. B. Allowable bearing capacity of initially bent piles. Jnl Soil Mech. Fdn Engng Div. Am. Soc. Civ. Engrs, 1963, 89, No. 5, 73–90.
5. HANNA T. The bending of long H-section piles. Canadian Geotech. J., 1968, 5, No. 3, 150–172.
6. YORK D. Structural behaviour of driven piling. Highw. Res. Rec., 1971, 333, 60–72.
7. FELLENIUS B. Bending of piles determined by inclinometer measurements. Can. Geotech. J., 1972, 9, No. 1, 25–32.
8. BURGESS I. W. A note on the directional stability of driven piles. Géotechnique, 1975, 25, No. 2, 413–416.
9. POSKITT T. J. Discussion: A simple understanding of critical loads. Struct. Engr, 1977, 55, Mar., No. 3, 139–141.
10. TIMOSHENKO S. and GERE J. Theory of elastic stability. McGraw-Hill, New York, 1961.
11. OMAR R. M. Discussion: The stability of slender piles during driving (Burgess I. W.). Géotechnique, 1978, 28, No. 2, 234–239.
12. TIMOSHENKO S. and GERE J. Theory of elastic stability. McGraw-Hill, New York, 1961, 228.
13. OMAR R. M. Instability of piles during driving. PhD thesis, University of London, 1977.

4. Pile drivability predictions by Capwap

G. G. GOBLE, PhD (Chairman, Department of Civil and Architectural Engineering, University of Colorado, USA) and F. RAUSCHE, PhD (President, Goble and Associates, Cleveland, Ohio, USA)

The CAPWAP analysis is performed on data obtained during the installation of a conductor pipe. Dynamic soil properties are derived and are used for analysing the drivability of the jacket piles. A case study is described in which the driving characteristics of jacket piles were predicted and compared with the results obtained during platform installation.

INTRODUCTION

A challenging task in the design and installation of offshore drilling platforms is the design of the piling and the driving system so that the applied loads are adequately supported and the piles can be efficiently installed. A new approach to this problem is discussed in this Paper, with emphasis placed on the latter aspect.

2. The problem of analysing a pile and its driving system for drivability is one that has received increasing attention in the past few years. In the case of offshore piles, the proportions of the system make the problem quite unusual when compared with piles that are used on land. Typically the ram is quite light compared with the weight of the pile, the piles are long, and they commonly have a variable cross-section. Generally they are driven open-ended, and they derive most of their strength from skin friction. The problem is made more difficult by the fact that the soil data available at a given site is usually limited and frequently not particularly quantitative in character. And yet, structures must be designed and installed under these circumstances; these structures costing huge sums of money.

3. The approach to predicting pile drivability presented in this Paper is based on the use of an analysis (CAPWAP) which is performed on a pile that is driven as a conductor pipe; from this analysis dynamic soil properties are obtained which are then used for analysing the drivability of the jacket piles. The conductor pipe data would be obtained during the installation of exploratory well conductor pipes. If the decision was made to develop the area where the exploratory well was drilled, and the permanent platform was not located on exactly the same location, then a soil boring would be made at the site of the platform and also at the location of the exploratory well so that comparative information could be obtained regarding the characteristics of the materials at the two sites. It is assumed that the platform would be installed near enough to the location of the exploratory well for the soils to be similar.

ANALYSIS ACCORDING TO CAPWAP

4. During an extensive research project conducted at Case Western Reserve University, Cleveland, Ohio, over a period of several years, the capability was developed to measure force and acceleration at the pile top during driving, and record this information on analogue magnetic tape. The accomplishment of these measurements had become a routine matter by the end of the project. Light, portable transducers were developed for direct attachment near the top of the pile to measure the strain a short distance under the hammer, and attachment devices were also developed for accelerometers that were located at the same cross-section. Alternatively, in some applications, a large transducer was fixed to the top of the pile for the same measurements. A variety of procedures was developed for verifying the validity and accuracy of the measurements.

5. The acceleration measurements were accomplished using low output impedance, piezoelectric devices and the strain transducers were designed to accommodate the use of resistance strain gauges. Both of these transducers were reusable. After signal conditioning, this data is recorded by a four-channel analogue magnetic tape recorder together with a voice record for noting the unusual events during the operation. The measurement system is shown schematically in Fig. 1.

6. The Case Pile Wave Analysis Program (CAPWAP) was developed to determine the soil resistance forces and their distribution using the measured force and acceleration record. Actually, the velocity record of the top is the information that is used, so in the first operation of data processing, the acceleration is integrated to obtain velocity: the velocity so obtained is referred to in this Paper as the measured velocity. It is useful to perform additional data processing in the field, and to this end a special-purpose analogue computational device, known as the Pile Driving Analyzer, has been developed for routine field use. This device is not discussed in this Paper. The computational procedures embodied in the Pile Driving Analyzer are quite different from the CAPWAP method.

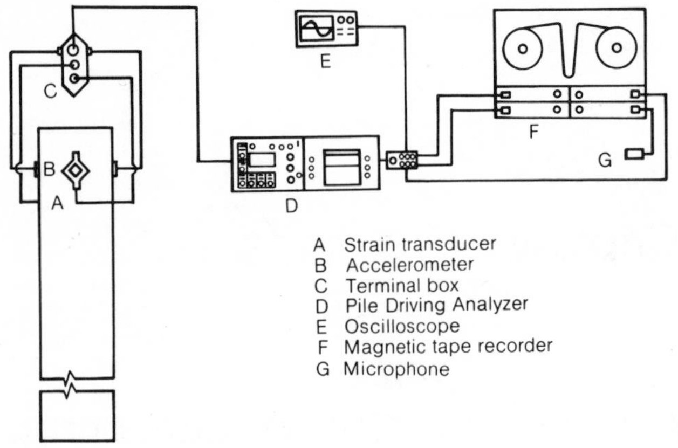

Fig. 1. Pile driving data acquisition system

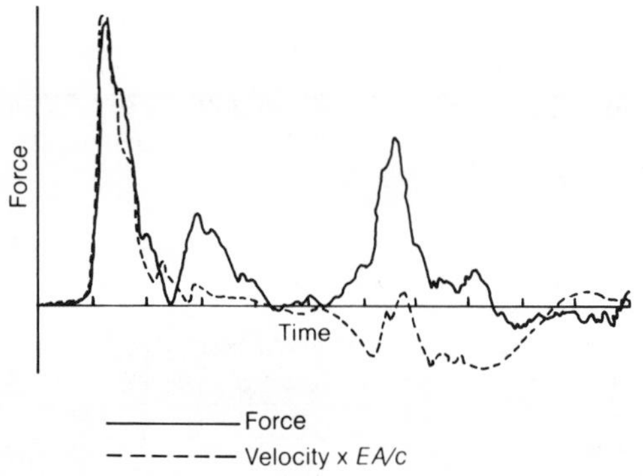

Fig. 2. Example of force and velocity record

7. The measurements stored in analogue form on magnetic tape can be processed automatically by electronic digital computer. The presence of a blow is sensed, and the important part of the signal is converted to digital form using an analogue-to-digital converter controlled by the computer. The resulting digital record can be stored on some sort of peripheral storage device such as digital magnetic tape, or disc, and can then be used in computation or recreated with a digital plotter. An example of a force and velocity record plotted by digital plotter is shown in Fig. 2.

8. In order to perform the CAPWAP analysis the pile, below the point where the transducers are attached, is modelled in the form of a series of lump masses and springs, and the soil resistance is modelled both along the side and at the toe as an elastic–plastic spring and linear dashpot. This model is similar to that proposed by Smith,[1] and later used in the development of a number of wave equation programs for the analysis of pile driving.[2,3] It differs from the wave equation models in that it does not include the driving systems and it excludes all of the pile above the location of the measurements. The model is illustrated in Fig. 3. Consider now the problem which must be solved. In this description computational details are avoided in order that the larger concept of the analysis can be described. The measured velocity at the pile top, treated as an input quantity, is imposed on the top element in the model. The resistance characteristics (i.e., the magnitude of the damping at each element, and the two parameters required to describe the elastic–plastic soil resistance at each element) must now be determined so that when the applied

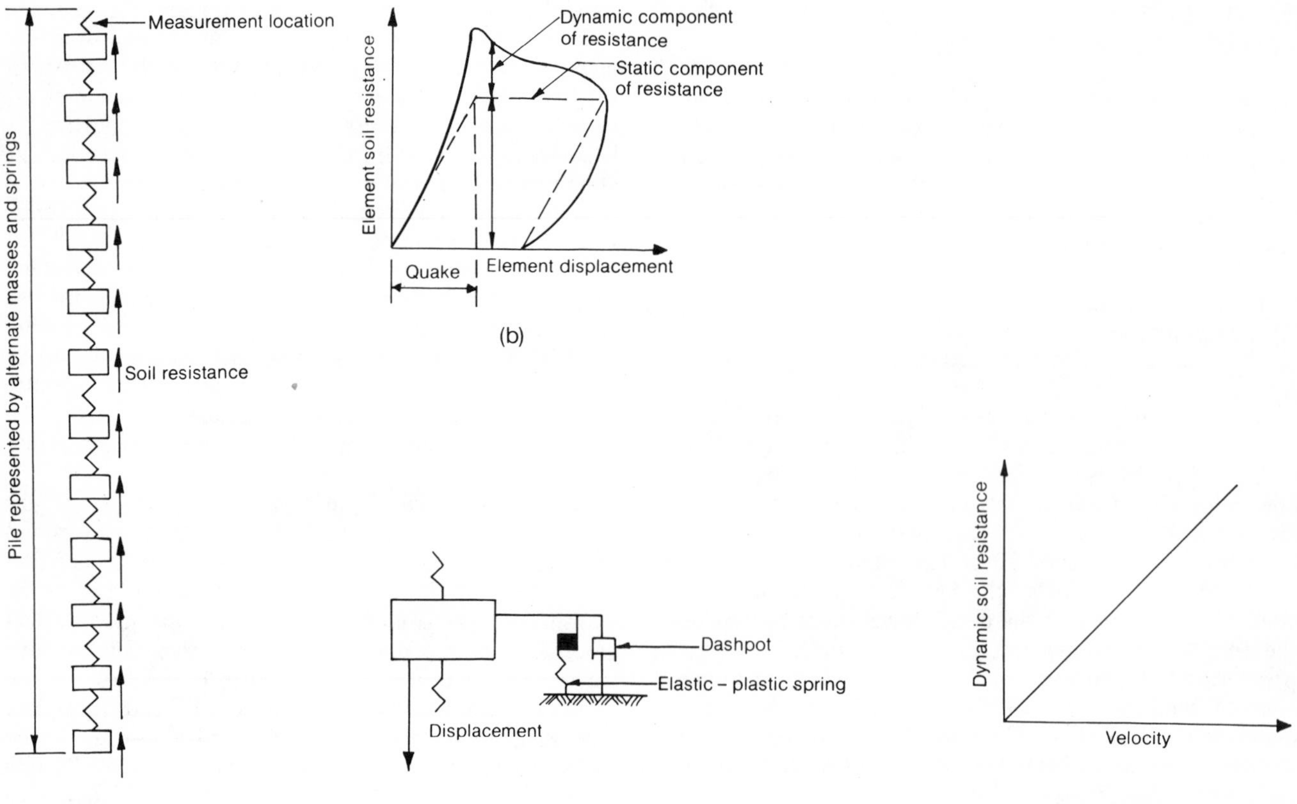

Fig. 3. Mathematical model for CAPWAP: (a) pile model; (b) total soil force–displacement relationship; (c) pile element/soil model; (d) dashpot resistance

velocity is imposed at the top element of the pile the force calculated at the top element will be the same as the measured quantity. In the original version of this method, a precedure was developed for automating the computational process.[4] For a pile of uniform cross-section the force and velocity must be proportional so long as there are no reflections coming from soil resistance. If the pile is of variable cross-section with no soil resistance, the top force can be calculated from the input velocity. Likewise, the force associated with an input velocity, including the reflection from the free end of the pile, can also be directly and readily determined. This quantity is referred to as the free pile solution. When the measured force deviates from the measured velocity (or the free pile solution) it must be concluded that this difference has been caused by the reflection of a soil resistance force. The presence of a deviation at some time interval after impact indicates the location at which the resistance first occurs. This type of information gave a basis for a first estimate of resistance distribution. Subsequent modifications of the resistance were based on the deviation of the calculated and measured force curves. An iterative approach was used to obtain a solution.

9. The automatic computational procedure for resistance distribution was reasonably satisfactory for use with the relatively short piles that are commonly encountered on land (i.e., piles probably about 20 m long and only rarely over 35 m long). When this computational procedure was applied to offshore piles with their very great length it was found that the cost of performing the analysis became excessive. Therefore, the program has been modified to compute the resistance forces and their distribution using an interactive mode. In the interactive mode the measured force and velocity records are input and held available in core storage. The velocity is applied to the pile together with an assumed resistance distribution. For that resistance distribution and magnitude, the force at the pile top as a function of time is calculated and this force record is compared with the measured force. With the calculated and measured force on display and an understanding of one-dimensional wave mechanics it is possible to enter a new resistance distribution. Thus, by successive analysis the resistance distribution can be found that gives the smallest difference between the measured force curve and the calculated force. This, then, is the correct resistance. By matching the calculated and measured force over $4L/c$ of the record (where L is the pile length and c is the stress wave transmission speed) it is possible to separate the static and dynamic resistances.

APPLICATION OF CAPWAP IN A DRIVABILITY ANALYSIS

10. The use of soil constants obtained in a CAPWAP analysis for predicting driving characteristics can best be presented by the description of a test case. In 1977 additional wells were drilled on a platform that had been installed some years earlier in the Gulf of Mexico. During the driving of the conductor pipe for one of those wells dynamic measurements as described above were made. A CAPWAP analysis was performed on those measurements at three depths of penetration to obtain soil parameters. The soil parameters were then scaled up to apply to the much larger jacket piles, and wave equation analyses were made using the driving system that was actually used to drive the jacket piles. Blow counts obtained from the wave equation analysis for the jacket piles were compared with the driving record recorded during the installation of the jacket.

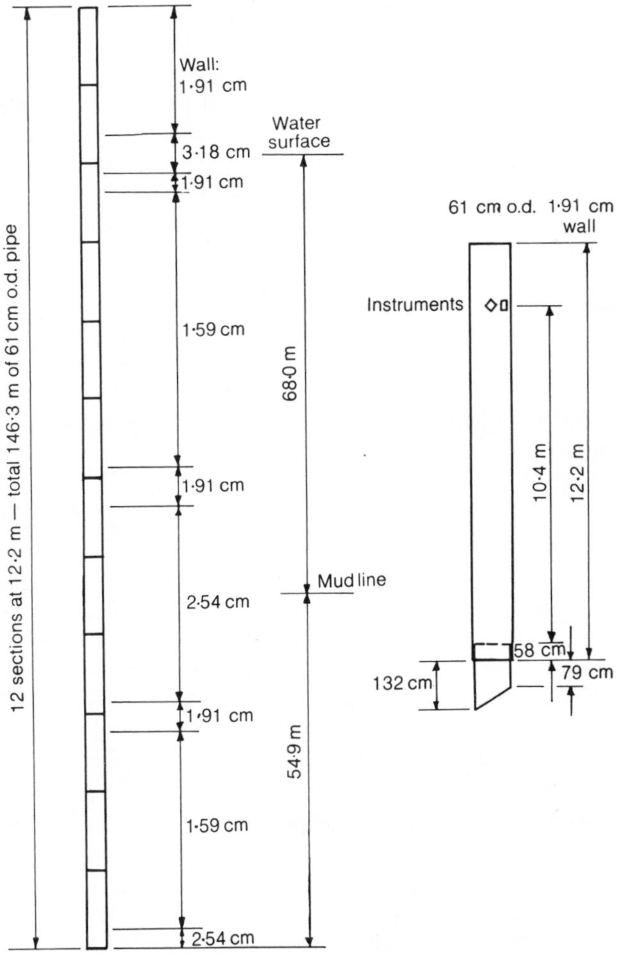

Fig. 4. Conductor pipe details

11. The conductor pipe on which the measurements were made is shown in Fig. 4. It was 61 cm in diameter and 146.3 m long. It penetrated the ocean floor 54.9 m. When measurements were made an instrumented drive nipple was added at the pile top. The driving was accomplished with a DELMAG D–30 hammer.

12. In order to show the performance of CAPWAP some steps are reproduced for one of the blows that was analysed from data obtained at the end of driving. A total of 26 analysis cycles were performed. The resistance parameters used are given in Table 1 and the comparisons between calculated and measured force records are shown in Fig. 5. In the first trial (Fig. 5(a)) the agreement between calculated and measured force is poor. Much too large a resistance has been assumed and it has been applied too far up the pile. There are also deviations between calculated and measured forces in the early part of the record. Since there is no soil resistance in this region and at later times the agreement is nearly perfect these differences must be ascribed to measurement inaccuracies. It is inappropriate to attempt to improve the agreement.

13. In trial 2 (Table 1 and Fig. 5(b)) the resistance was reduced and moved down the pile. The match is substantially changed in the region of the tip reflection.

Now there is too little resistance somewhat above the tip. The result is a dramatic decline in the calculated force to where it is now too low in the region above the tip. In trial 3 (Fig. 5(c)) static resistance was moved up the pile with the total resistance only slightly changed. The agreement is substantially improved up to the $2L/c$ time. In trial 5 (Fig. 5(d)) further adjustments were made in the static resistance to try to improve the match just after the $2L/c$ time.

14. Up to this point a typical damping value has been used and no attempt has been made to improve the agreement after the $2L/c$ time. A poor agreement generally exists in this part of the record. Before discussing damping adjustments it is necessary to discuss the various damping constant definitions. In this Paper the damping force is defined by the relationship

$$R_D(t) = Jv(t) \quad (1)$$

where J is referred to as the viscous damping constant and may carry the subscript s to refer to side damping or t to denote tip damping. This constant carries the units kN s/m. In work reported previously[4] a damping constant j_c was defined by the relationship

$$R_D(t) = j_c(EA/c)\,v(t) \quad (2)$$

where E is the pile material modulus, A is the pile cross-sectional area, c is the velocity of wave propagation in the pile and j_c is the Case damping constant. With this definition the damping constant j_c is dimensionless. Smith[1] suggested a damping constant stated as

$$R_D(t) = j_s R_u v(t) \quad (3)$$

where R_u is the element ultimate resistance and j_s is the Smith damping constant. The damping constant j_s has units s/m. Extensive experience with CAPWAP analyses shows that none of these relations gives fundamental soil properties. Additional research is necessary in this area.

15. In Table 1 the damping constants shown are Case constants. However, the damping force is distributed in the same fashion as the static resistance.

16. Between trials 5 and 12 the agreement between the two curves was not substantially changed by modifying the static resistance distribution. In trial 12 the damping magnitude was increased and the distribution modified, giving an improvement in the match near the tip. Further modifications in damping were made up to trial 14 and now a longer time interval is being considered. However, a troubling positive spike in the calculated force record cannot be eliminated. In trial 15 a change was made in the mass distribution, with the addition of more pile weight primarily at the pile tip. In trial 15 the undesirable force characteristic has been eliminated.

17. Further modifications, primarily in damping magnitude and distribution, were attempted up to trial 27, the final trial.

18. The final results are shown in Table 2. The results from the other two blows analysed are given in Tables 3 and 4. As might be expected, the capacity increased with depth of penetration. The static resistance distribution is illustrated in Fig. 6.

19. The conductor pipe analysed above was driven at an operating platform. Information is available on the driving of three of the leg piles. They were driven using a Vulcan 0–60 hammer with a capblock of alternating layers of 1 in steel cable and ¼ in steel plates. The leg pile characteristics are shown in Fig. 7 and the driving records in Fig. 8. The driving record is shown only to somewhat below the depth penetrated by the conductor. Unfortunately a soil profile and other soils data is not available for this site.

Table 1. CAPWAP interactive data input

Iteration	R_u, kN	j_{cs}	j_{ct}	Q_s, cm	Q_t, cm	Remarks
1	4248	0.5	0.5	0.25	0.25	Element 0–60=0; bottom trapezoidal distribution
2	2825	0.5	0.5	0.25	0.25	Element 0–60=0, 61–65=54 kN, 66–70=200 kN, 71–74=345 kN, 75=165 kN
3	2811	0.5	0.5	0.25	0.25	Element 0–47=0, 48–60=93 kN, 61–65=58 kN, 66–70=98 kN, 71–75=165 kN
5	1890	0.5	0.4	0.25	0.25	Element 0–48=0, 48–53=89 kN, 54–58=107 kN, 59–60=18 kN, 61–66=13.3 kN, 67–69=89 kN, 70–74=71 kN, 75=89 kN
12	1401	0.2	0.1	0.25	0.25	Element 0–48=0, 49–54=27 kN, 55–58=40 kN, 59–70=49 kN, 71–75=98 kN; damping replaced element 49–54=69 kN s/m, 55–58=71 kN s/m, 59–66=15 kN s/m
14	1539	0.2	0.1	0.05	0.05	Element 0–48=0, 49–54=31 kN, 55–58=49 kN, 59–70=58 kN, 71–73=116 kN, 74–75=58 kN; damping replaced element 49–54=69 kN s/m, 55–58=71 kN s/m, 59–66=15 kN s/m
15	1557	0.2	0.1	0.05	0.05	Element 0–48=0, 49–54=36 kN, 55–58=44 kN, 59–70=53 kN, 71–75=107 kN; damping same as iteration 14; weight added at element 5=2.8 kN, 74=4.4 kN
25	1548	0.2	0.1	0.13	0.13	Element 0–48=0, 49–54=9 kN, 55–61=18 kN, 62–65=9 kN, 66–75=133 kN; damping element 49–54=75 kN s/m, 60–61=40 kN s/m, 62–65=6 kN s/m, 66–70=79 kN s/m, 71–74=28 kN s/m, tip=147 kN s/m
Final	1548	0.2	0.1	0.13	0.13	Element 0–48=0, 49–54=9 kN, 55–61=18 kN, 62–65=9 kN, 66–75=133 kN; damping element 49–54=75 kN s/m, 55–59=106 kN s/m, 60–61=40 kN s/m, 62–65=6 kN s/m, 66–70=79 kN s/m, 71–74=28 kN s/m, tip=147 kN s/m; added weight unchanged

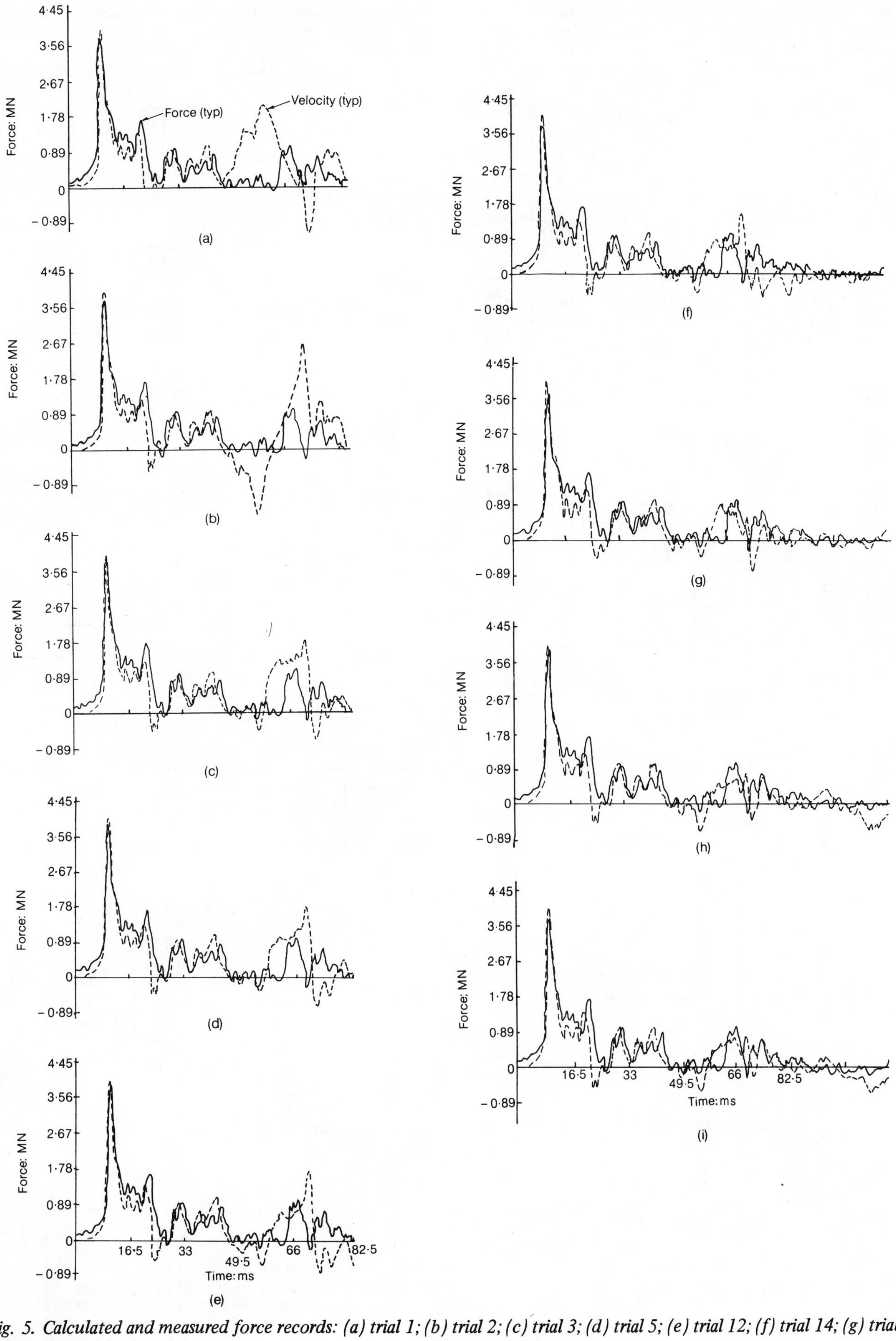

Fig. 5. Calculated and measured force records: (a) trial 1; (b) trial 2; (c) trial 3; (d) trial 5; (e) trial 12; (f) trial 14; (g) trial 15; (h) trial 25; (i) final trial

Table 2. Resistance distribution—54.9 m penetration

Element	Penetration below mud line, m	Quake, cm	Static resistance, kN	*J*, kN s/m
48	0	0.13	0	0
49	2.1	0.13	9	75
50	4.2	0.13	9	75
51	6.3	0.13	9	75
52	8.4	0.13	9	75
53	10.6	0.13	9	75
54	12.7	0.13	9	75
55	14.8	0.13	18	106
56	16.9	0.13	18	106
57	19.0	0.13	18	106
58	21.1	0.13	18	106
59	23.2	0.13	18	106
60	25.3	0.13	18	40
61	27.4	0.13	18	40
62	29.5	0.13	9	6
63	31.7	0.13	9	6
64	33.8	0.13	9	6
65	35.9	0.13	9	6
66	38.0	0.13	133	79
67	40.1	0.13	133	79
68	42.2	0.13	133	79
69	44.3	0.13	133	79
70	46.4	0.13	133	79
71	48.5	0.13	133	28
72	50.6	0.13	133	28
73	52.8	0.13	133	28
74	54.9	0.13	133	28
75	54.9	0.13	133	147
			1548	

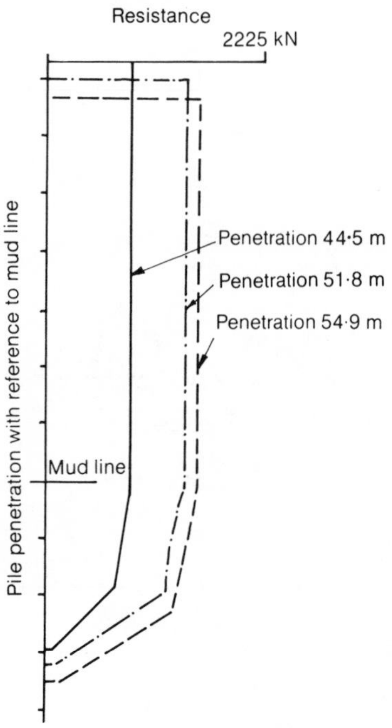

Fig. 6. Force in the pile under ultimate conditions, for the soil resistances calculated from CAPWAP

Table 3. Resistance distribution—50.6 m penetration

Element	Penetration below mud line, m	Quake, cm	Static resistance, kN	*J*, kN s/m
50	0	0.13	0	0
51	2.1	0.13	14	91
52	4.2	0.13	14	91
53	6.3	0.13	14	91
54	8.4	0.13	19	136
55	10.6	0.13	19	136
56	12.7	0.13	19	136
57	14.8	0.13	24	93
58	16.9	0.13	24	93
59	19.0	0.13	24	93
60	21.1	0.13	5	45
61	23.2	0.13	5	45
62	25.3	0.13	5	45
63	27.4	0.13	5	45
64	29.5	0.13	5	45
65	31.7	0.13	5	45
66	33.8	0.13	123	71
67	35.9	0.13	123	71
68	38.0	0.13	123	71
69	40.1	0.13	123	71
70	42.2	0.13	123	71
71	44.3	0.13	123	28
72	46.4	0.13	123	28
73	48.5	0.13	123	28
74	50.6	0.13	123	28
75	50.6	0.13	123	147
			1431	

Table 4. Resistance distribution—46.4 m penetration

Element	Penetration below mud line, m	Quake, cm	Static resistance, kN	*J*, kN s/m
52	0	0.13	0	0
53	2.1	0.13	5	8
54	4.2	0.13	5	8
55	6.3	0.13	5	8
56	8.4	0.13	5	8
57	10.6	0.13	14	23
58	12.7	0.13	14	23
59	14.8	0.13	14	23
60	16.9	0.13	14	23
61	19.0	0.13	9	15
62	21.1	0.13	9	15
63	23.2	0.13	9	15
64	25.3	0.13	9	15
65	27.4	0.13	9	15
66	29.5	0.13	9	15
67	31.7	0.13	80	132
68	33.8	0.13	80	132
69	35.9	0.13	80	132
70	38.0	0.13	80	132
71	40.1	0.13	80	132
72	42.2	0.13	80	132
73	44.3	0.13	80	132
74	46.4	0.13	80	132
75	46.4	0.13	80	147
			850	

20. The assumption was made that the soil resistance forces at a particular depth were a fixed value and related to the pile surface area. Therefore, the static resistance values were all multiplied by the ratio of the pile diameters (2.5). The same values of quake that were obtained from CAPWAP were used on the jacket piles. The viscous damping J was also multiplied by the same ratio. Since there was no scaling on pile impedance this implies that perhaps the Smith concept is relevant. It seems reasonable to assume that the damping resistance generated by the soil is independent of pile impedance. However, this study in no way supports the Smith concept since both quantities (static and dynamic resistance) were scaled by the same amount.

21. The wave equation analysis was made using the WEAP system.[3] The results for the three penetrations are shown in Table 5. At the two deeper locations the agreement is excellent while at the shallow depth the prediction is somewhat high. The predicted driving characteristics are shown in Fig. 8.

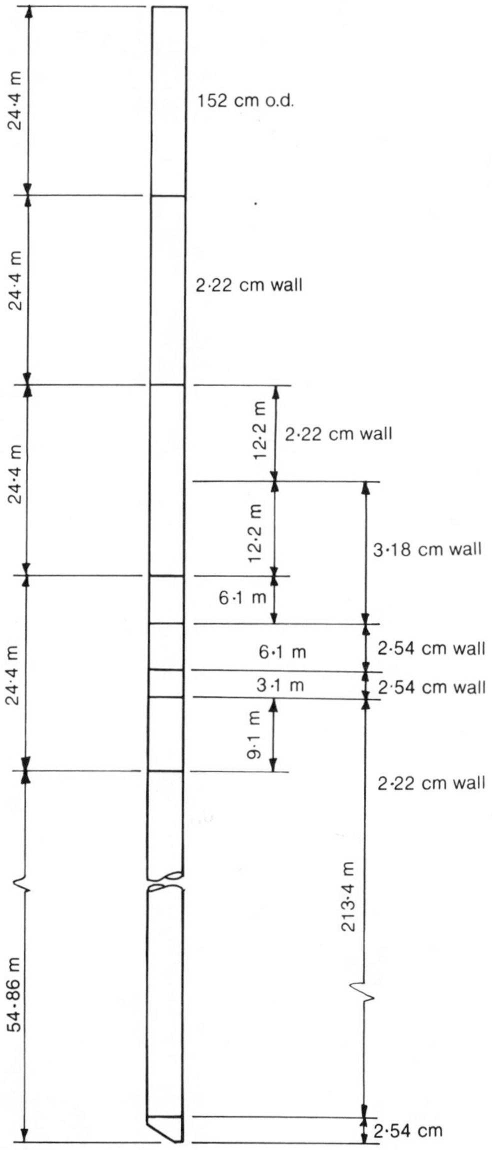

Fig. 7. Jacket pile details

Table 5. Results of WEAP analysis of leg pile

Penetration, m	Maximum stress, MPa	ENTHRU, kJ	Blow count, blows/ft
44.5	139	143.2	17
51.8	139	142.6	29
54.9	139	142.8	30

CONCLUSIONS AND COMMENTS

22. The use of an exploratory well conductor pipe as a 'penetrometer' for predicting pile drivability has been demonstrated. The performance in this case is very good. However, it is easier to correlate at these relatively low blow counts than it is with higher blow counts.

23. The Smith damping values obtained from the CAPWAP analysis are, in some cases, very large; they are commonly over 2.0 where expected values for a soil with high damping would not normally exceed 0.2. Changes in damping can substantially affect the blow count and therefore the problem is a serious one.

24. No mention has been made of set-up effects. If driving is interrupted in these soils a substantial strength increase results. An example of this phenomenon is seen in Fig. 8 where blow counts increase substantially and then decline with additional driving. This occurred when driving was interrupted for splicing. It is possible to measure these set-up effects using the system described here if controlled interruptions are used in driving the penetrometer pile. If this type of data were available it would be possible to

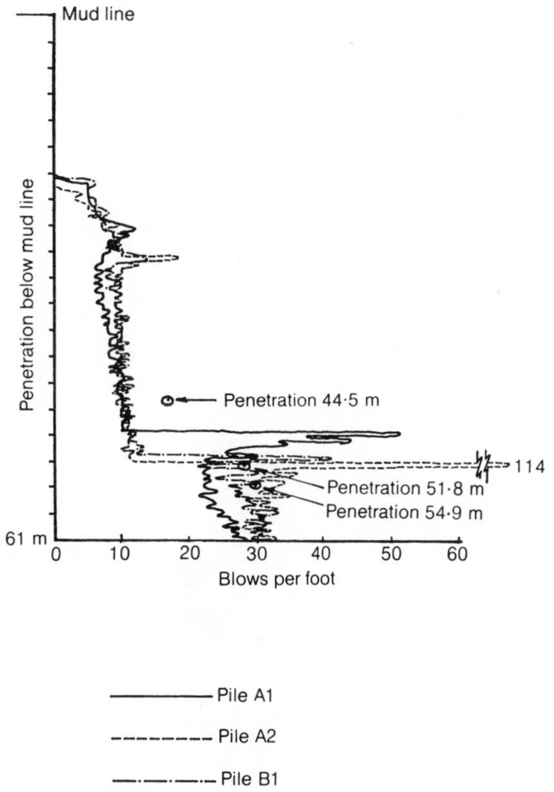

Fig. 8. Driving records for jacket piles

engineer the driving operation and possibly increase the accuracy of the pile capacity determination.

25. It seems, based on this study, that the approach presented can provide a much more reliable means of predicting drivability, and further study is justified.

REFERENCES

1. SMITH E. A. L. Pile driving analysis by the wave equation. *J. Soil Mech. Fdns Div. Am. Soc. Civ. Engrs,* 1960, **86**, Aug.

2. HIRSCH T. J. et al. *Pile driving analysis – TTI program.* Texas Transportation Institute, 1976, Dept of Transportation report FHWA–IP–76–13. 1.

3. GOBLE G. G. and RAUSCHE F. *Wave equation analysis of pile driving – WEAP program.* Goble and Assoc., Inc., Cleveland, 1976, Dept of Transportation report FHWA–IP–14.2.

4. RAUSCHE F. *et al.* Soil resistance predictions from pile dynamics. *J. Soil Mech. Fdns Div. Am. Soc. Civ. Engrs,* 1972, **98**, Sept., No. SM9.

5. An advanced wave equation computer program which simulates dynamic pile plugging through a coupled mass-spring system

E. P. HEEREMA and A. de JONG (Heerema Engineering Service, The Hague, Netherlands)

The standard wave equation computer program cannot distinguish between inside and outside friction of open-ended offshore piles. As the dynamic behaviour of outside friction and inside friction are by nature very different, it was considered desirable to develop a program that could simulate the inside soil column as a mass–spring system inside the mass–spring system of the pile. In particular, the aspect of plug formation in the pile during driving can be investigated in such a true-to-life program. This Paper describes the program (Dynpac) that was developed for this purpose. Experimental runs illustrate the conditions under which dynamic plugging will occur; it is shown that dynamic and static plugging are quite different phenomena. The suitability of the standard wave equation program is investigated through the Dynpac program. The validity of methods for determining permanent set of the pile in the computer run are discussed.

INTRODUCTION

The wave equation[1] computer program has been proved to be a most valuable tool in assessing pile drivability. It very adequately simulates the stress conditions in the pile during passage of the wave caused by hammer impact. As regards soil resistance simulation, however, of course the validity of the program is wholly limited by the validity of the model that is to simulate dynamic soil behaviour and the resistance parameters, which the user puts in.

2. Mainly by systematic post-analysis of many pile driving data, it is possible to learn enough about soil resistance values in order to be able to make reliable drivability predictions.

3. Such post-analysis procedures immediately face the investigator with one vital question: is an open-ended steel pile to be considered as 'plugged' during driving or not? The problem here is that the pile is actually almost always in an in-between condition. This has been shown during many plug level measurements. The soil column inside the pile sets, but usually less than the pile itself.

4. The normal wave equation program can only distinguish in a friction profile along 'the' pile wall, and point resistance on 'the' pile tip. Fig. 1(a) illustrates this simulation. If the pile is considered as plugged, only the outside friction is put in as wall friction, and on the tip the point resistance of the gross point area is put in. Conversely if the pile is considered as coring, the inside and outside friction profiles are added, and only the point resistance acting on the pile annulus is counted at the pile tip.

5. It is obvious that either of the two alternatives has its shortcomings. The 'plugged' pile is in the real situation being hampered by the friction of the inside soil, but the first model does not account for that; and the 'unplugged' or 'coring' pile in the real situation does not quite have as much hindrance from the inside friction as in the second model, because the inside soil column gives way to some extent.

6. It therefore occurred to the Authors that an important contribution could be made to the understanding of the driving behaviour of open-ended piles if a wave equation program could be developed that incorporated the inside soil column as an independent mass–spring system, coupled to the pile through the friction it exerts on the pile wall.

7. This Paper describes the result of this effort. A program (named Dynpac) has been written that can cope with the combined effects of outside friction on the pile wall, inside friction on the pile wall exerted by the internal soil column, point resistance on the pile annulus, and 'point resistance' below the internal soil column. Fig. 1(b) shows schematically the 'pile inside the pile'.

8. It is the intent in this Paper to highlight mainly the results of analyses with this computer program, in which parameters are varied in such a manner that the specific qualities of the program are best illustrated.

SMITH'S WAVE EQUATION MODEL

9. The Dynpac program is based on the principles of Smith's wave equation model[1] (i.e., a discrete place as well as discrete time model). Place discretization is through division of the hammer–pile system in masses and springs, as shown in Fig. 1.

10. Time discretization is done by computing the state

values (displacement and velocity) of all elements in the system at discrete times, at the end of very small time steps, say 0.1 ms. Integration of state values is linear: only first order time effects are taken into account, because at such small time intervals the gain in accuracy by using higher order integration is very small and in no relation to many other effects, especially concerning accuracy of soil parameter input.

11. The wave equation model is simple and versatile. Stresses in the pile itself are reproduced accurately. Intermediate results can easily be obtained, at any level in the pile or during any time step. Non-linear conditions can be built in well; for example, any type of cushion block, including gas cushions of hydraulic hammers; gravity connections, unable to transfer tension; quake (elastic deformation of the soil); and any complex soil resistance model, for example with a non-linear velocity dependence.[2]

PILE PLUGGING

12. To describe the dynamics of pile plugging, paragraphs 13–15 are partly quoted from a previous paper.[3]

13. Pile plugging during driving has little to do with pile plugging in the bearing situation. Generally a pile will sooner be plugged in bearing than during driving.

14. The explanation of this phenomenon is as follows. The hammer impact induces a stress and displacement wave in the pile. As it travels down, it effectuates a dynamic friction on the inside soil column, so that a stress and displacement wave is also induced in the soil column. The displacement wave in the steel pile, however, travels much more rapidly than the wave in the soil column, due to the great difference between the elasticity moduli of steel and saturated soil. The consequence is that the pile wall shoots past the inside soil column; the inside soil column lags behind. Only after the steel pile tip has reached more-or-less

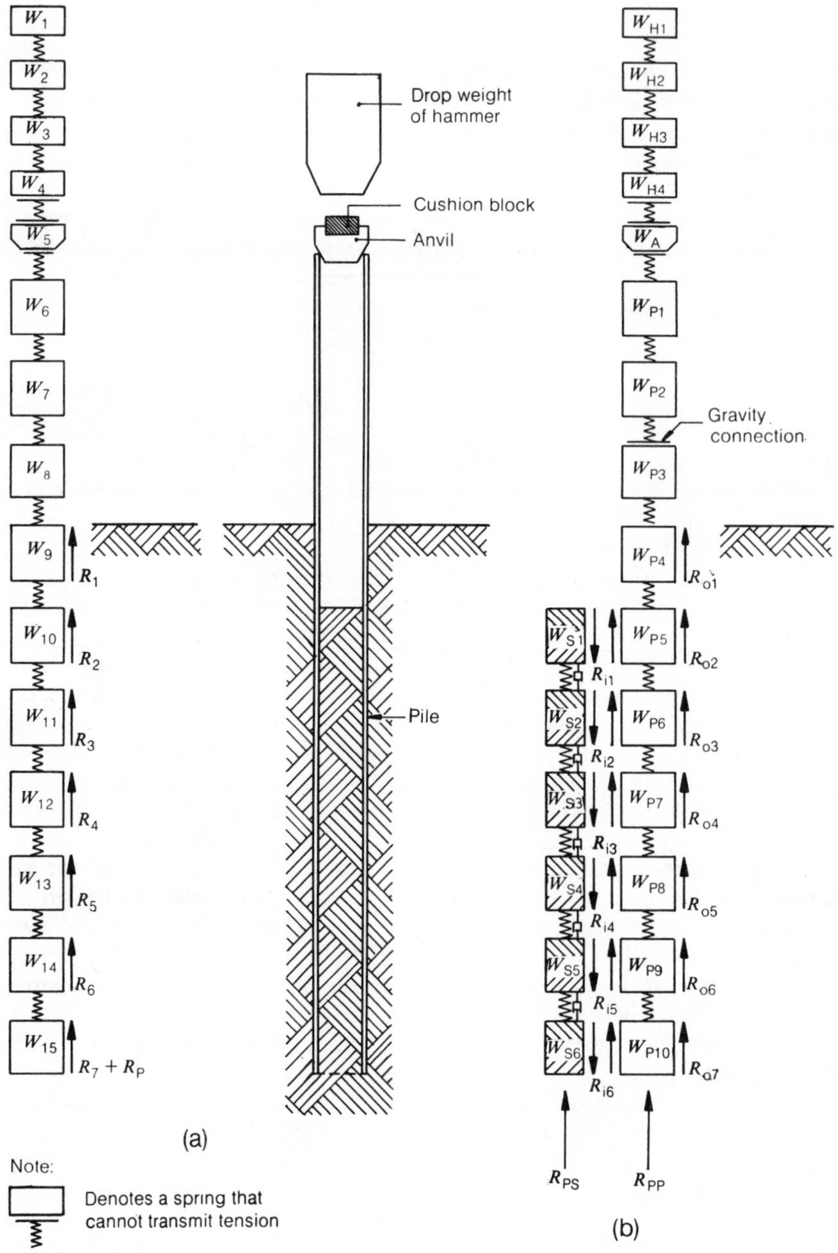

Fig. 1. (a) Conventional hammer–pile model according to Smith; (b) hammer–pile–soil column model as used in the Dynpac program

its final set, the soil column determines its own final set dependent on the resistance it encounters at the tip of the pile. In practice, this is almost always only a smaller part of the set of the pile tip itself; so one can only speak of 'partial' plugging.

15. The static plugging equilibrium is quite a different condition. It is simply the question whichever is smaller: the accumulated inside friction, or the resistance of the soil below the cross-sectional area of the soil column. Thus a pile will sooner plug in the static bearing condition than during driving.

16. It is necessary in the standard wave equation computer program to choose for either a plugged or a non-plugged pile. The above reasoning illustrates that the inside friction is always felt by the stress wave whatever the resistance below the tip is, which justifies the choice of a non-plugged pile (i.e., making the wave equation analysis with outside friction, inside friction, and point resistance on the pile annulus).

17. Plug measurements done after driving in the North Sea normally show that the level of the internal soil column varies from slightly above the mud line to roughly one seventh of the pile's penetration depth below the mud line.

18. One of the purposes of this Paper is to illustrate the validity of the Authors' choice to generally consider a pile as unplugged during driving, when using the standard wave equation program. It may be noted here that it is not common practice among users of the wave equation program to consider a statically plugged pile as unplugged during driving.

EXPERIMENTS WITH THE DYNPAC PROGRAM

Description of the basic case

19. To illustrate the behaviour of the internal soil column during driving, a basic case is compared with variants which each differ only in one respect.

20. For the basic case, the following combination has been chosen: an open-ended steel pile, 48 in. (1219 mm) OD × 2 in. (51 mm) wall thickness; inclination 8:1; length 120 m, penetration 40 m; steam hammer Menck 4600 (rated energy 69 m t). Soil resistance was chosen somewhat arbitrarily as follows: ouside friction uniformly 75 kN/m^2; inside friction uniformly 150 kN/m^2; tip resistances (pile annulus and soil column) 35 MN/m^2. These resistances are all put into the program without the usual amplification factor due to displacement velocity ('damping'[1] factors). Therefore, static and dynamic values are equal. This is only correct for wall friction in sand,[2] but as the intention is here mainly to make comparisons, this inaccuracy does not matter. The reason for not using the damping factors here is that it is then easy to quantify 'the' dynamic resistances involved in the plugging action. For normal analyses, damping factors should be used.

21. In the basic case given above, the total inside friction is 21 MN whereas the resistance below the soil column is 34 MN. The pile is therefore, in the static sense, unplugged.

22. The Dynpac wave equation run provides as a result a blow count of 71 bl/ft; the permanent set of the pile tip is 4.3 mm.

23. The internal soil column does not set. As the pile was unplugged even statically, this result was to be expected.

Investigating plugging: reduction of tip resistances

24. By reduction of the resistance below the pile tip, the pile's tendency to plug during driving can be increased. Table 1 shows the results of such analyses. In case 2 the only difference from the basic case 1 is in the reduction of the point resistance below the soil column. For a clear comparison it is necessary to keep the resistance below the annulus the same as in the basic case. The resistance below the soil column is then equal to the accumulated inside friction. The pile is therefore in the static sense on the brink of plugging. The result is almost identical to the result of case 1.

25. Case 3 has an overall reduced point resistance; the blow count drops to 64 bl/ft.

26. Case 4 is interesting. The point resistance below the soil column has been reduced to such an extent that the pile is in the static sense convincingly plugged. However, during driving the pile appears to be wholly unplugged! And the blow count is hardly affected: it has dropped from 71 bl/ft to 68 bl/ft.

27. The static ultimate capacity of the pile in the basic case is 6.5 MN (point resistance on annulus) + 11.5 MN (outside friction) + 21.1 MN (inside friction) = 39.1 MN; in case 4 it is 6.5 MN (point resistance on annulus) + 11.5 MN (outside friction) + 10.8 MN (point resistance below soil column) = 28.8 MN. Although the pile of case 4 has 74% of the calculated capacity of the pile in the basic case, the blow count is hardly lower. This is a clear illustration of the fact that 'feeling the inside friction' is much more important, in the driving action, than having a low resistance below the soil column. This result supports the reasoning of paragraphs 12–18.

28. Case 5 has the low resistance of case 4 below the soil column, now under the pile annulus as well. The blow count is only 52 bl/ft. It is noteworthy that although inside friction and resistance below the soil column are both the same as in case 4, the pile plugs to a fair extent, in contrast to case 4. This might be explained as follows: if the pile dissipates much energy in its high tip resistance, as in case 4, then it cannot transfer as much energy to the lower part of the soil column.

Increase of tip resistances

29. Case 6 is a case with an obvious result; it could be expected also from a static point of view. The resistance below the soil column is doubled as compared with the basic case, so that the pile is 'still more' unplugged. The blow count hardly reacts.

30. Case 7 has an overall increased point resistance. The effect of the point resistance increase on the annulus is considerable. For the first time, a negative set of the soil column occurs. That is, the soil column is going upwards relative to the mud line. This may seem strange, but it is observed in practice occasionally, particularly in piles with internal driving shoes. The driving shoes reduce the internal friction during driving, and the ratio of the resistance below the soil column to the inside friction becomes larger, as in case 7.

Checking the standard wave equation program through Dynpac

31. In paragraph 26 it was shown that, for driving, the resistance below the soil column was relatively of lesser importance (comparison of cases 4 and 1). Thus a result has

been found which suggests that the standard wave equation program (without internal soil column feature) could work satisfactorily as long as it is used in the non-plugging mode.

32. Case 8, run in the standard wave equation mode, shows that this suggestion is correct. The blow count is almost identical with that of the basic case, run with the internal soil column feature.

33. It is noteworthy that two such different computations lead to almost identical results. But this result becomes logical when one notes that the soil column does not set: the standard wave equation mode, after all, also has a 'non-setting' wall friction.

34. Case 9, therefore, is a real test case for the standard wave equation program. Here is a case of a pile that is plugged in the static sense, and should therefore according to the soil engineer's common practice be run as a plugged pile.

35. It is, however, run as an unplugged pile, in correspondence with the reasoning of paragraphs 12–18. The blow count result is 60 bl/ft. Comparing the result with case 5 (52 bl/ft), this result is considered very acceptable. Considering the soil column sets by 13%, so that part of the energy which the soil column has obtained from the pile is given back to the pile, the fact that the blow count is greater than 52 bl/ft is understandable for case 9, computed with the standard wave equation program, in which such return of energy cannot be simulated.

36. For comparison, case 10 is the same problem as case 9 but run in the plugged mode in the standard wave equation computer program. The blow count is 27 bl/ft, which is not in acceptable agreement with case 5, solved with Dynpac using the internal soil column option.

37. It must thus be concluded that even in cases with full static and partial dynamic plugging, such as case 5, the standard wave equation program leads to reasonable results if it is used in the non-plugging mode.

38. It is the virtue of the Dynpac program that, besides being more accurate, it can indicate the degree of plugging during driving.

DETERMINATION OF PERMANENT SET OF THE PILE

The quake subtraction trick

39. As the impact wave travels back and forth through the pile, continuously deteriorating, the most realistic way to determine the permanent set of the pile tip in the standard wave equation program is to allow the program to run for such a length of time that all displacements of any significance have damped out.

40. The duration of this process, however, is long in terms of computing time. To save computational costs, it has been attempted to determine the permanent set of the pile tip from its maximum dynamic displacement. This maximum dynamic displacement is reached in an early stage, and it is generally possible, then, to cut off the computation after the impact wave has travelled through the pile twice, or less.

41. Smith[1] has presented the rule to determine the permanent set of the pile by subtracting from the maximum dynamic tip displacement the point quake value.

Table 1. Experiments with the Dynpac program

Case	Unit point resistance on annulus, MN/m²	Point resistance below column		Accumulated inside friction, MN	Statically plugged/ unplugged	Blow count, bl/ft	Premanent set of pile tip, mm	Permanent set of soil column, mm	Dynamically plugged/ unplugged (plugging rate)
		Unit, MN/m²	Total, MN						
1. Basic case	35	35	34.3	21.1	Unplugged	71	4.3	0	Unplugged
2. Reduced point resistance only below column	35	21.5	21.1	21.1	Neutral	*71*	4.3	0	Unplugged
3. Evenly reduced point resistance	21.5	21.5	21.1	21.1	Neutral	64	4.8	0	Unplugged
4. Further reduced point resistance only below column	35	11	10.8	21.1	*Plugged*	*68*	4.5	0	*Unplugged*
5. Further reduced point resistance	11	11	10.8	21.1	Plugged	52	5.8	0.8	13% plugged
6. Increased point resistance only below column	35	70	68.7	21.1	Unplugged	*72*	4.3	0	Unplugged
7. Evenly increased point resistance	70	70	68.7	21.1	Unplugged	114	2.7	−0.01	Unplugged
Standard program without soil column feature: 8. Basic case run in unplugged mode	35	–	–	21.1	Unplugged	*73*	4.2	(0)	Assumed unplugged
Standard program without soil column feature: 9. Statically *plugged* case, run in unplugged mode	11	–	–	21.1	Plugged	60	5.1	(0)	Assumed unplugged
Standard program without soil column feature: 10. Statically plugged case, run in plugged mode	11	11	10.8	–	Plugged	27	11.5	(11.5)	Assumed plugged
11. Case 5 with quake variations ($q_{ins.}$ = 1 mm, $q_{pt\ col.}$ = 12 mm)	11	11	10.8	21.1	Plugged	49	6.3	0.05	Unplugged

(Quake is the term for the elastic desplacement of the soil, before slippage between pile wall and soil starts.) This measure, which seems a logical one, has been checked by the Authors' research group.

42. It then appeared at first that the best approximation was found by subtracting only half the quake value applied over the lower third of the pile. This rule would allow for some variation of quake values.

43. The model was then improved by registering the residual stresses in the soil after damping out, and using these values as an input for a second hammer blow. This is an important improvement; for the first blow, only the static equilibrium between pile weight and soil resistance is computed before the blow, but it appears that actually after the blow the residual stresses in the soil are very different from that initial situation.

44. These residual stresses appear to influence the final damped-out set to a fair extent.

45. By running several blows in a row, every time using the residual ground stresses as input to the next, and allowing each blow to damp out fully, a dependable final set value is determined. It then appears that if the full side quake value over the lower third of the pile is subtracted from the maximum dynamic displacement of the first blow, a very acceptable approximation of the correct value is found.

46. This shows that when applying a constant side and point quake, the common practice of determining the permanent set is correct: maximum dynamic displacement of the pile tip during the first blow, minus the point quake value. It is likely that this practice is the result of earlier investigations similar to those described above.

47. Another investigation concerned the effect of the magnitude of the quake on the permanent set. It appeared that as the quake was taken larger, the maximum dynamic displacement of the pile tip increased (weaker soil springs), but the return displacement to the damped-out value was also larger. As a result the permanent set is only weakly dependent on the choice of the quake value.

48. Of course, the practice of subtracting the point quake value from the maximum dynamic displacement of the first blow becomes incorrect when the side quake and the point quake are assumed to be different. For example, if the point quake is increased to twice the side quake, the maximum displacement of the pile tip only slightly increases (unless the point resistance is the greater part of the total resistance) due to the weaker point soil spring, and subtracting precisely the pcint quake then leads to too small a value for the final displacement.

49. This illustrates that it is unnatural to vary only the point quake in order to find a scatter in expected driving results, as is sometimes seen in drivability analyses. If the side and point quake values are varied equally, the scatter found is much less spectacular.

Determining permanent set in Dynpac

50. The permanent set calculation in a complex program such as Dynpac is difficult. The quake subtraction trick does not work, particularly for the internal soil column. But it is the set of the soil column which is important for determination of the 'plugging rate'. The soil column generally sets much less than the pile and in most cases the maximum dynamic displacement of the soil column will be less than the quake value at the point, which would allow no permanent set at all if the quake subtraction rule were followed.

51. There is obviously a great influence of the static pile–soil column interaction through inside friction on the final set of soil column and pile, which is not taken into account in the quake subtraction trick.

52. As mentioned above, the logical way of determining the permanent set is by running the program for a sufficient length of time to allow the blow to damp out fully. This, of course, is expensive in terms of computation time. An accurate, time-saving solution was found in the consideration that from the moment that plastic deformations no longer occur on the pile/soil interfaces, the static equilibrium of the mass–spring systems can be computed immediately. This rapid computation method has the advantage over the quake subtraction method that differences in quake values along the pile have little effect on the found blow count result. Another advantage is that all residual stresses in the soil–pile system are known for an input in the next blow.

53. These residual stresses appear to have a significant influence on the blow count result. A number of consecutive hammer blows is necessary before differences between results become negligible.

54. It appears, generally speaking, that per blow the wave must be allowed to travel four to five times through the pile before plastic deformations cease to occur. The necessary number of consecutive hammer blows appears to be three to four, before differences between results become negligible.

55. The only disadvantage of the process described is the computer time required. The Authors' research group has a large in-house computer available; but for the user who does not have this facility, the necessary executing time could lead to unacceptably high computation costs. Therefore, the search for a computer time-saving method for permanent set calculation is being continued.

VARIATION OF QUAKE VALUES IN DYNPAC

56. In Table 1, case 11 is case 5 run with different quake values. In case 5 all quake values are 4 mm. In case 11, more realistic guesses of quake values have been chosen: outside quake = 4 mm; inside quake (determined by the shear modulus of the soil, not the compressive reaction among soil column parts) = 1 mm; pile annulus point quake = 4 mm; soil column point quake = 12 mm.

57. The resulting blow count is 49 bl/ft, which is very near to the 52 bl/ft of case 5. However, the soil column hardly sets, as opposed to case 5.

FORCES IN THE PILE AND IN THE SOIL COLUMN

58. Figure 2 shows the (mainly compressive) force waves travelling through the pile and through the soil column at three levels in the pile, which has a penetration of 40 m. With reference to the time scale, the time it takes the impact wave to travel once through the full length of the pile is about 27 ms.

59. The build-up of compressive force with depth in the soil column is noteworthy. This build-up is caused solely by the dynamic friction which the pile exerts on the inside soil column, and the reaction of the soil underneath it. The constant value of the soil column's compressive

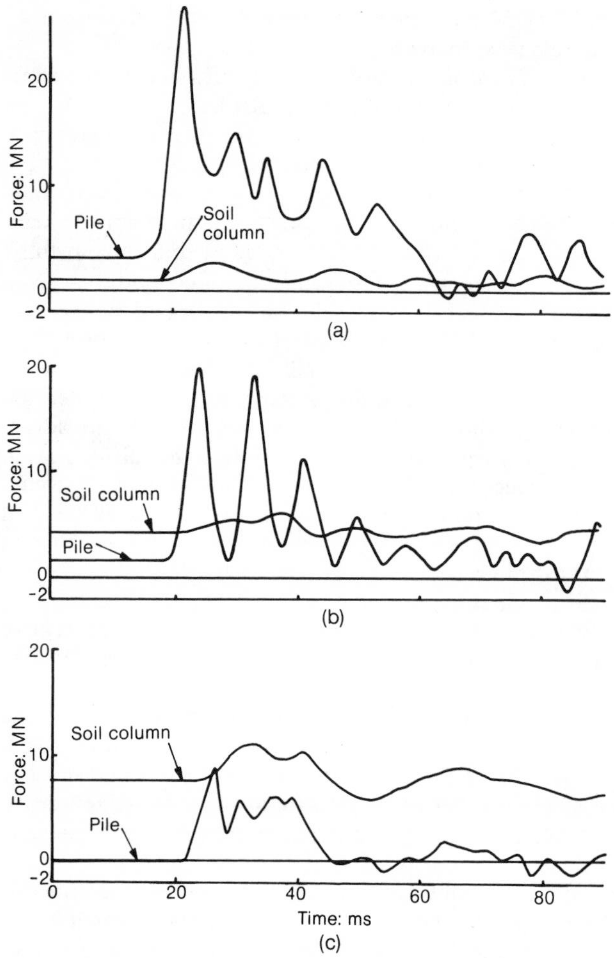

Fig. 2. Forces in the pile and in the soil column as computed by Dynpac (pile penetration 40 m): (a) level 10 m below mud line; (b) level 25 m below mud line; (c) level 37.5 m below mud line

force before the stress wave has reached it is due to precompression by the previously run blows.

AN ADDITIONAL FEATURE OF THE DYNPAC PROGRAM IN DEVELOPMENT

60. In Dynpac, the soil column mass–spring system is coupled to the pile mass–spring system through the soil column's friction against the pile wall. This friction has so far been considered as constant, which is a simplification: the soil column actually arches to some extent. If the soil column is dynamically compressed through the inside wall friction, a horizontal stress build-up results.

61. The Dynpac program is therefore being provided with an additional feature. The internal soil column parts can exert a horizontal stress on the pile wall equal to k times the occurring vertical stress in the soil column parts. This horizontal stress leads to an internal friction through multiplication by a friction coefficient. The factor k (= σ_h/σ_v) and the friction coefficient can be chosen freely.

62. At this stage it is too early to present results. However, the results obtained so far are promising.

CONCLUSIONS

63. Recognition of the differences in dynamic behaviour between inside and outside friction of open-ended piles, and the uncertainty about pile plugging during driving, have led to the development of the Dynpac program. These aspects cannot be studied in the standard wave equation computer program.

64. The Dynpac program simulates the inside soil column as a mass–spring system inside the pile's mass–spring system; the soil column is coupled to the steel pile through the inside skin friction; the pile and the internal soil column each determine their own permanent sets.

65. Pile plugging during driving and in the static bearing condition are two quite different phenomena; it is shown that piles will much sooner plug in the bearing condition than during driving.

66. It is shown that the standard wave equation program is best used in the 'unplugged' mode even where the pile is statically plugged.

67. The custom in the standard wave equation program of determining the permanent set of the pile by subtracting the point quake value from the maximum dynamic pile tip displacement is a satisfactory approximation under certain conditions.

68. In the Dynpac program, this practice does not give correct results. An accurate alternative has been found in computing static equilibrium of the mass–spring system at the moment that plastic pile/soil deformations no longer occur. The residual stresses from this computation are used as an input to the next blow, of which several are necessary to stabilize the results. This method requires much computer time, but is fully true to life.

69. The newly developed Dynpac program has yet unexplored possibilities, which will lead to further understanding of the driving behaviour of open-ended piles.

REFERENCES

1. SMITH E. A. L. Pile driving analysis by the wave equation. Trans. Am. Soc. Civ. Engrs, 1962, 127, Part I, Paper 3306.

2. HEEREMA E. P. Relationships between wall friction, displacement velocity, and horizontal stress in clay and in sand, for pile driveability analysis. Ground Engng, 1979, 55–65.

3. HEEREMA E. P. Predicting pile driveability: Heather as an illustration of the 'friction fatigue' theory. European Offshore Petroleum Conference, London, 1978, Paper 50.

6. Post-analysis of full-scale pile driving tests

P. VAN LUIPEN and G. JONKER (Menck Division of Koehring GmbH, Western Germany)

Two important pieces of pile driving equipment have been developed in the past few years by Menck—a hydraulic hammer and a special anvil. The hydraulic hammer has a net striking energy of 1050 kN m and can be operated above and under water. The anvil does not need hardwood or other solid capboard material for transfer of the kinetic ram energy into the pile. The developments were accompanied by model tests and full-scale pile driving tests. The final test programme was performed on a 72 in o.d. pile. A drive with the steam hammer MRBS 8000 was included for comparison purposes. Many measurements were performed: firstly to check on design considerations, secondly for judging the driving capabilities of the various pile driving systems. Analysis of these measurements, done partly with the aid of the wave equation computer program, resulted in an increased knowledge of the losses which decrease the hammer energy to actual available driving energy in the pile head. Special attention is given in this Paper to the properties of the Bongossi capblock as a function of the total accumulated number of blows. A method is presented to post-analyse the two parameters used for the Bongossi spring for wave equation analysis; i.e., the modulus of elasticity and the coefficient of restitution. The method has been applied to measured test data, resulting in evaluated values for the two parameters as a function of accumulated blows.

INTRODUCTION

Single-acting steam hammers of their own design have been built by Menck since 1883. Several new design concepts have been introduced and the construction of existing models has been constantly improved. However, most important technical steps forward, both in size and basic design, have been made during the last twelve years of offshore development in the North Sea. Further developments and improvements of the single-acting steam hammer are always in progress. It is expected that this hammer type will remain a most important and economical tool for pile installation work for many years. The principle of offshore operation of this hammer is shown in Fig. 1.

2. Nevertheless, to meet the demands of the offshore industry, the design capacity at Menck has also been concentrated on two new developments. These are a hydraulic hammer and a special anvil. Extensive test programmes, varying from small-scale model tests to full-scale pile driving, have been performed to support and check the design work done.

HYDRAULIC HAMMER

3. In above-water application, a hydraulic hammer is an alternative to a conventional steam hammer. A hydraulic hammer does not need major modifications for underwater application. This is mainly due to the many control functions already built into a hydraulic hammer and its closed construction. Due to the apparent offshore trend to deeper waters Menck decided in 1974 to develop a hydraulic hammer for use above and under water. The developed prototype has the designation MRBU 6000. The underwater working principle is given in Fig. 2. The hammer has one central double-acting hydraulic cylinder. The weight of the ram is 600 kN (60 t). The net impact energy is 1050 kN m (105 t m) at a minimum blow rate of 50 bl/min. The design features of this hammer and several test programmes are described by Kühn.[1]

SPECIAL ANVIL

4. In 1974 a concept was developed for a new type of anvil, transferring the impact energy without the need for hardwood or other solid capblock material. The new design should fulfil the following requirements:

(a) protection of both hammer and pile against excessive peak stresses;
(b) efficient transfer of the impact energy into the pile head;
(c) the force–time diagram for the pile should have a shape for effective driving;
(d) maintaining properties (a)–(c) over a long period of driving time;
(e) acceptable investment and maintenance costs.

5. The design which evolved from the 1974 concept and which was expected to satisfy the above requirements is schematically shown in Fig. 3. For comparison a standard anvil with Bongossi capblock is pictured in Fig. 4.

6. The special anvil comprises a hollow anvil housing with a cylindrical bore in which is guided a piston which has limited axial movement and which seals a cavity containing a body of liquid and a pressurized gas cushion. The piston protrudes a fixed small distance beyond the impact surface of the anvil housing. The impact energy is transferred into the pile mainly via the steel body of the anvil housing. At impact the ram first hits the piston, forcing it down into the anvil housing. When the piston has been

depressed to a position level with the impact surface, the ram will hit with its outer area directly on the anvil housing. This steel-to-steel impact causes a sharply rising peak in the force–time diagram. Due to the small protruding distance of the piston, the ram even at minimum striking energy hits the anvil housing for direct impact transfer to the pile, so that there is obtained in any case a peaked force–time diagram similar to that of a standard Menck anvil with Bongossi capblock, which is considered near to the ideal for the driving conditions generally encountered offshore.

FULL-SCALE DRIVING TEST

7. The last stage in the development programme of the hydraulic hammer and the special anvil was the driving of a full-scale test pile. The prototype of the hydraulic hammer MRBU 6000 was used in combination with a standard Bongossi anvil as well as with a prototype of the special anvil.

8. There were four objectives in the test programme:

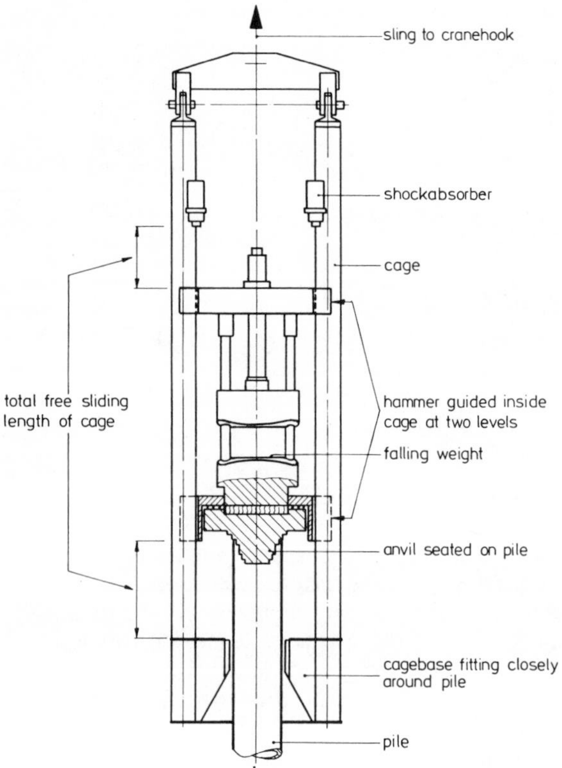

Fig. 1. Principle of operation of Menck steam hammers

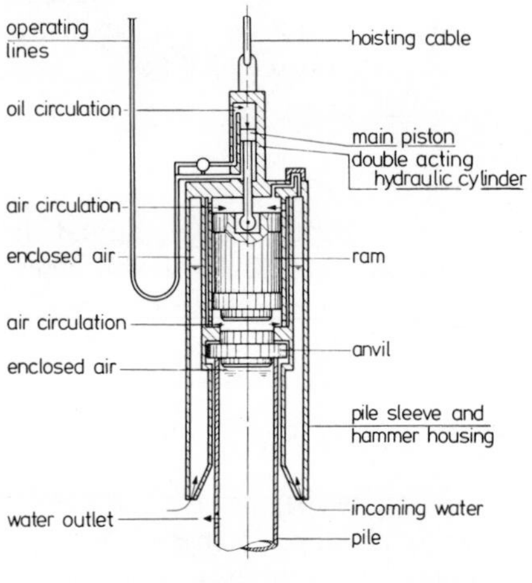

Fig. 2. Underwater operating principle of hydraulic Menck hammer MRBU; situation for hammer moving downwards

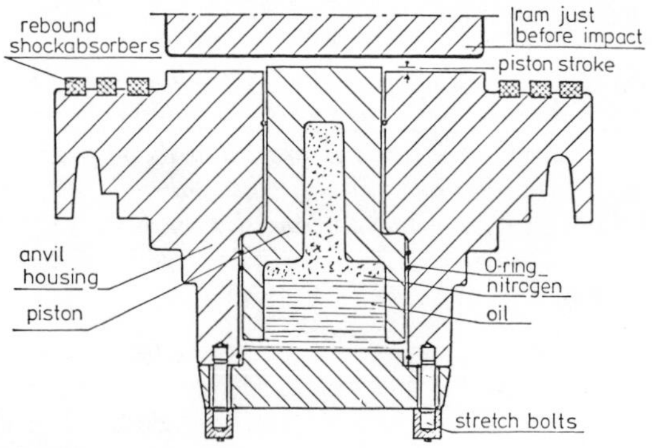

Fig. 3. Special anvil

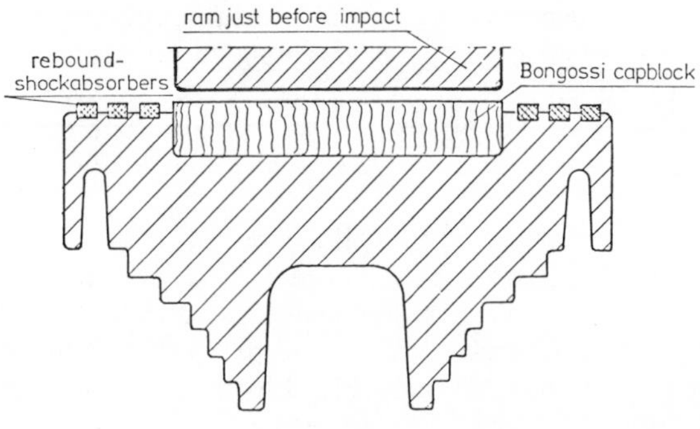

Fig. 4. Standard anvil (with Bongossi capblock)

Table 1. Data on the steam hammers used in the full-scale pile driving test

	MRBS 1500/4	MRBS 8000
Ram weight, kN	150	800
Maximum stroke, m	1.25	1.50
Maximum rated striking energy, kN m	187.5	1200
Anvil weight, kN	100	390
Total weight, including anvil and cage, kN	539	2770
Capblock diameter, m	1.05	1.98
Capblock height, m	0.20	0.30

(a) testing the Menck hydraulic hammer MRBU 6000 under all possible operating conditions;
(b) testing the special anvil under all possible operating conditions;
(c) comparing the impact transfer efficiencies of a standard Bongossi anvil and the special anvil of the MRBU 6000;
(d) proving that the driving capabilities of the hydraulic hammer and the steam hammer MRBS 8000 were similar.

9. The test site was in a harbour basin. The water depth was approximately 10 m. The test pile was an open-ended steel pipe pile with an outer diameter of 72 in (1.83 m) and a wall thickness of 1.875 in (48 mm). One pile length of 30 m was set vertically and predriven with an MRBS 1500/4 to a penetration of 15.3 m. The hammer (for data see Table 1) was operated with compressed air. During predriving the pile penetration and number of blows were recorded. In the course of the test programme, three pile sections with a length of approximately 20 m each were welded on to the test pile and driven to a final penetration of 55.5 m. Final pile length was approximately 87 m. The MRBU 6000 was operated making alternating use of the standard and special anvils. Many different drives had to be done to test the hammer extensively.

10. For comparison purposes (cf. (d) above) the offshore steam hammer MRBS 8000 (for data see Table 1) made one drive from 51.0 m to 53.0 m.

Soil conditions

11. A boring was made at the actual test pile location to investigate the soil conditions at the test site. At 0–10 m penetration mainly sand was found. At 10–26 m 'Geschiebemergel' was encountered, and from 26 m to the end of the boring at 35 m 'Glimmerton' was found. These two layers have undergone glacial preconsolidation. Testing the Geschiebemergel in the laboratory resulted in a particle distribution of 10% clay, 20% silt and 70% fine and medium sand. The Atterberg limits were determined as 13% plastic limit and 27% liquid limit. Water content was 10%. The best estimate for the undrained shear strength of this layer is mainly based on unconfined compression testing and is approximately 450 kN/m^2. For the Glimmerton the particle distribution was 30% clay, 60% silt, and 10% fine sand. Plastic limit was 27% and liquid limit 58%. Water content was 20%. The best estimate of the undrained shear strength was approximately 250 kN/m^2. The soil profile below the penetration of the boring was extrapolated as being the same Glimmerton. This was based on results of other borings in the area and on geological evidence.

Testing MRBU 6000

12. All equipment needed for operating and testing the hydraulic hammer was installed on a pontoon of 13.5 m × 32.0 m. The largest units were two diesel generators, the powerpack, the control cabin, and a living container in which also the measuring equipment was installed. A hydraulic crane with maximum capacity 170 kN did most of the lifting work. A foundation frame for the hammer was made in the centre of the pontoon, so that the hammer could be stored vertically in case of repairs or anvil changes. Handling of the hammer and anvils was done with harbour cranes with 1 MN or 2 MN lifting capacity. As the hammer was free-riding on the pile, the harbour crane could leave the test location after finishing the lifting operation. In the case of modifications to the hammer, the hydraulic crane aboard was often sufficient.

13. Besides the normal data such as pile penetration, blow count, date and times of starts and stops of the drives, the following specific items were recorded during driving: anvil type, accumulated number of blows, blow rate, force–time diagram for the pile head, and several measurements on the hammer itself. The force–time measurements were made 4.0 m below the top of the pile with strain gauges welded inside the pile. Use was made of a carrying frequency (5 kHz) amplifier. Calibration was done electrically. Many more measurements were made on the hydraulic hammer and special anvil to check design computations on points which were thought to be critical.

14. The MRBU 6000 has been operated during testing at net striking energies from the minimum value of 525 kN m to values well above the design maximum net striking energy of 1050 kN m, thus overloading the hammer. Most of the drives were done at a net striking energy of approximately 1050 kN m and a blow rate of approximately 50 bl/min. Efficiency determinations and comparisons of resulting blow counts could thus be done at this design maximum net striking energy, in which one is most interested. The complete blow count/penetration graph is given in Fig. 5. The type of anvil which was used in the hydraulic hammer is indicated in the figure. A general gradual increase in blow count was met. A significant difference between the two anvil types could not be found from the point of view of a blow count comparison. The somewhat higher blow count from 42 m to approximately 46 m is thought to be due to the relatively long stop of 48 days at 42 m. At deeper penetrations very high blow counts were recorded, due to the high soil resistance.

15. In Fig. 6 two examples of measured force–time diagrams using the Bongossi anvil are given. The diagram for 570 accumulated blows, of which 500 were at a reduced energy, shows a higher peak force than that for 2400 accumulated blows. The former diagram also has a higher energy content. These differences in the diagrams are due to the changes in characteristics of the Bongossi capblock with the accumulated number of blows (cf. paragraphs (43–45). The measured force–time diagram using the special anvil is shown in Fig. 7. With the special anvil there are no significant changes in the force–time diagram depending on accumulated blows.

16. The diagrams for the special and Bongossi anvils of the MRBU 6000 both have the quick increase to the peak height and then a gradual decrease in force. The peak heights are somewhat different. Energy determinations showed that the anvils had similar efficiencies for impact energy transfer (85–90%). No significant differences in blow count were found during driving. Wave equation studies with different peak heights confirmed these results for the range of soil resistances encountered. With these results the object of achieving equal impact transfer capabilities for the two anvils could be satisfactorily met (cf. (c) of paragraph 8).

Comparison drive MRBS 8000

17. For fulfilling object (d) of paragraph 8, the test pile was driven over a distance of 2 m with the steam-operated offshore pile hammer MRBS 8000 (for data see Table 1). This hammer was selected because it is thoroughly

proven and because the net striking energy is comparable with that of the MRBU 6000.

18. The MRBS 8000 was set over the pile by a floating derrick. An old whale-hunter vessel, with large water-space boilers aboard, delivered the necessary steam. The steam pressure at the boiler was approximately 1.2 N/mm^2. The steam pressure at the hammer inlet just before the moment of steam intake was approximately 0.8 N/mm^2. The steam supply over two lines of 60 m length with an inner diameter of 6 in (150 mm) was sufficient to achieve full stroke at a blow rate of 36 bl/min. A steam-regulating valve was installed aboard the pontoon for control of the hammer.

19. Besides the normal data as quoted in the preceding paragraph, the following items were recorded during driving: accumulated number of blows, blow rate, displacement–time diagram of the ram, visual estimate of the shortfall, force–time diagram for the pile head, and steam pressure measurements on the hammer. The force–time diagram was measured as described in paragraph 13. The displacement–time diagram of the ram was recorded with an electrical camera which followed a white/black sign on a moving object and transferred the displacement into an electrical signal. The impact velocity of the ram could be determined in this way.

20. Before starting the comparison drive the hammer was thoroughly warmed up at a stroke from height 0.9 m to 1.2 m. Much more time was taken for this procedure than is needed during offshore operations to bring the hammer into optimum condition for the test comparison drive. The drive was interrupted halfway for almost 2 h to transport the electrical camera and its foundation from land on to the pontoon. In the first part of the drive a shortfall of about 0.03 m was estimated. The MRBS 8000 was working in a very regular rhythm for a steam hammer. This was also shown by the measurements of blow rate and peak height of the force–time diagram. During 1500 blows the blow rate was within a 2% margin around the average. The peak height was decreasing due to changes of the characteristics

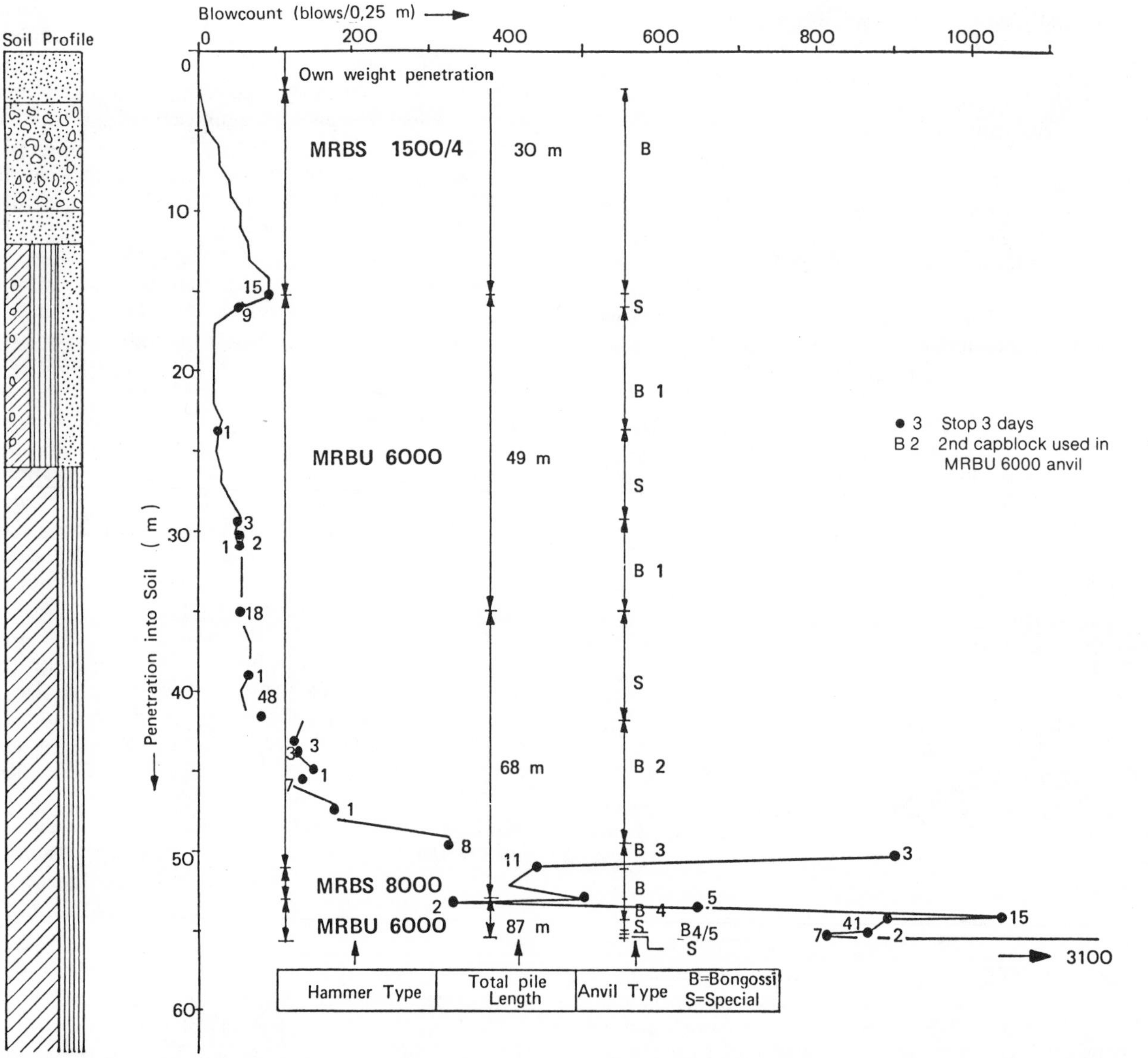

Fig. 5. Full-scale pile driving test: blow count versus penetration: pile o.d. 72 in, wall thickness 1.875 in, water depth approx. 10 m; test results at other than maximum net striking energy of MRBU 6000 are omitted from blow count/penetration curve

of the Bongossi capblock. Short term variations in the peak height were less than 1% around the trend line. The kinetic energy of the ram at impact was measured as 1125 kN m, which represents 94% of the maximum rated striking energy of the ram. The shortfall of 0.03 m means a maximum potential energy of the ram of 1176 kN m. The mechanical hammer efficiency (defined as the ratio of kinetic energy at impact and maximum actual potential energy) was thus 96%.

21. During the second part of the drive the electrical camera recorded the displacement–time characteristic of the pile at the level where the strain gauges were situated. There was no direct measurement of the impact velocity of the ram. The impact energy was determined by drawing trend lines through the measured peak forces (cf. Fig. 16). At the points of discontinuity there is a change in shortfall and in net striking energy. Changes in the characteristics of the Bongossi capblock are not abrupt and are therefore not the reason for such a discontinuity. At these points the foregoing efficiency was multiplied by the square of the ratio of peak forces after and before the discontinuity to arrive at the actual efficiency. It was found that during the greater part of the second drive an impact energy of 1040 kN m was brought onto the capblock, corresponding to 87% of the maximum rated striking energy. Two of the measured force–time diagrams are given in Fig. 8. After approximately 10 ms the rebound wave makes a large contribution to the force level in the pile head. The force–time diagram of the special anvil of the hydraulic hammer has a similar shape, with approximately the same peak height (Fig. 7).

22. The energy content of the MRBS 8000 diagram up to 10 ms is around 900 kN m, which is approximately equal to the energy content of the diagram with the special anvil. The similar shapes, peak heights and energy contents should result in similar blow counts. The results of wave equation runs showing soil resistance during driving versus blow count are given in Fig. 9 for the MRBU 6000 and MRBS 8000. A higher modulus of elasticity should have been taken for the MRBU 6000 capblock (cf. paragraphs 41 and 42). However, this higher value would not much change this wave equation curve at a blow count of around 300 blows per 0.25 m. The blow counts resulting from the drives with the steam hammer are given in Fig. 5. They follow the trend of the blow count curve for the MRBU 6000. The driving capabilities of the hydraulic hammer MRBU 6000

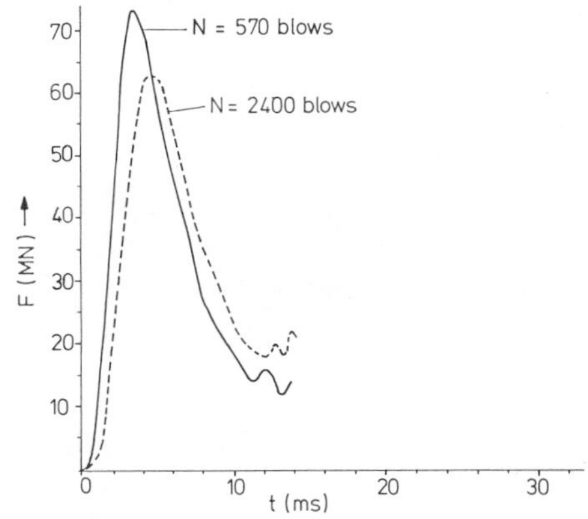

Fig. 6. Measured force–time diagram: MRBU 6000 with Bongossi anvil

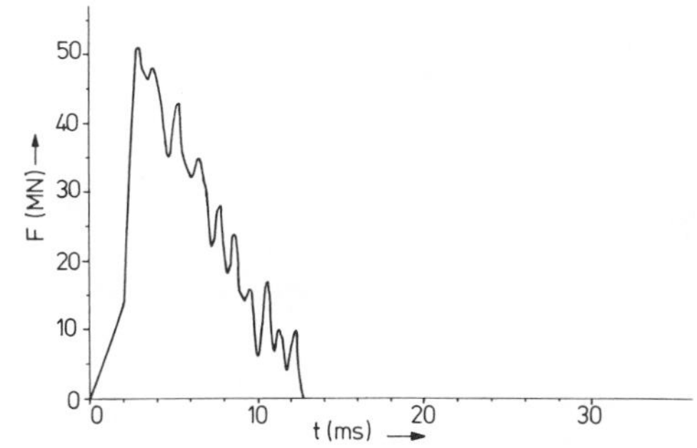

Fig. 7. Measured force–time diagram: MRBU 6000 with special anvil

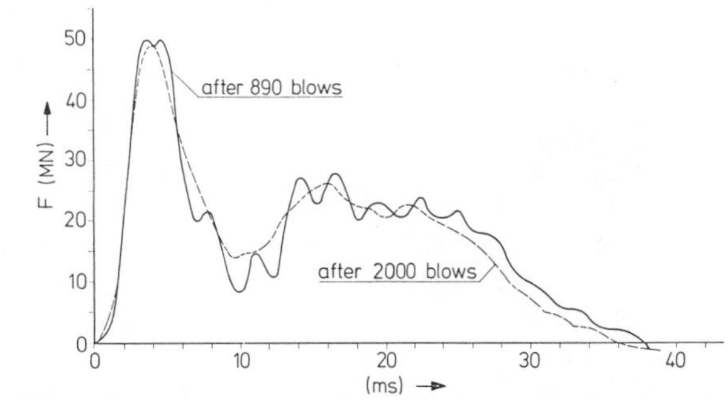

Fig. 8. Measured force–time curves: MRBS 8000 (mechanical efficiency ≈ 96%; wave equation efficiency 94%; capblock Bongossi; pile 72 in × $1^7/_8$ in; pile rake vertical; pile length 67.8 m; pile penetration 51.2 m and 51.9 m; blow count 350 and 380 blows per 0.25 m)

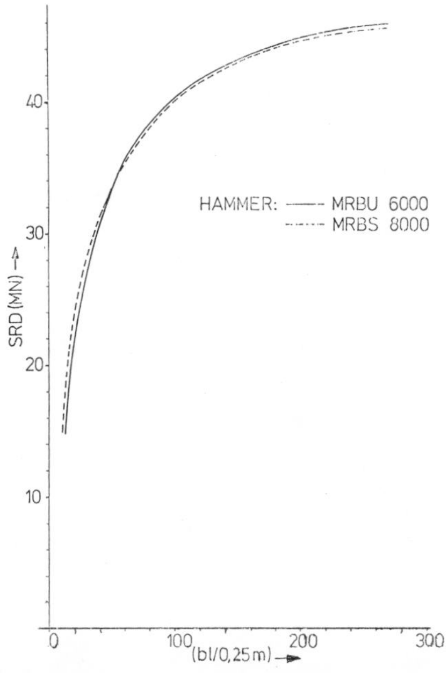

Fig. 9. Soil resistance at time of driving versus blow count: pile 72 in × 1.875 in; for input see Table 2; E = 2 kN/mm², e = 0.75

with either the special or the Bongossi anvil are thus the same as those of the steam hammer MRBS 8000 with its standard Bongossi anvil. This fulfils object (d) of paragraph 8.

POST-ANALYSIS OF BONGOSSI CHARACTERISTICS

23. Reliable input parameters are of utmost importance for good drivability analysis with the wave equation program. Most analyses are made calculating the effect of the ram's impact. The computational model includes the ram and its impact velocity, capblock, anvil, pile and soil. The recommended and mostly used capblock material in the Menck offshore pile hammers is Bongossi hardwood. During driving the properties of the Bongossi change. The measurements made during the driving test as described in the preceding paragraphs make it possible to establish characteristic values of the Bongossi capblock depending on the accumulated number of blows. For the wave equation analysis the two most important values for the capblock are the modulus of elasticity and the coefficient of restitution. By combining measured and wave-equation-predicted values using a certain procedure, the change of these two parameters as a function of accumulated blows is found.

Method of analysis

24. The modulus of elasticity and the coefficient of restitution of the capblock are semi-empirical values for wave equation analysis. In the present study they refer to the hammer–anvil–pile modelling as given in Fig. 10. The ram is taken as one mass segment. The mass of the capblock itself should be added to the mass directly below the capblock spring (i.e., the anvil). The modulus of elasticity is used to calculate the spring constant of the capblock between the impact area of the ram and the anvil. The characteristic of the capblock spring is given in Fig. 11. The coefficient of restitution determines the unloading spring constant (i.e., the unloading modulus of elasticity in the capblock). This model has among others been described by Smith[2] and Lowery et al.[3] In reality the force–deformation relationship is much more complex[3] than the model consisting of only two characteristic spring constants (i.e., a loading and an unloading one). The correct applicable force–deformation relationship for one certain blow seems difficult to obtain, mainly due to the fact that this relationship will strongly depend on the loading history and the actual loading rate and stress level. However, such an accuracy is not needed. For drivability studies the two parameters are quite adequate for the prediction of a force–time diagram for the pile head that is similar to the real occurring diagram.

25. Strictly speaking, the changes in the spring constant are analysed. However, for easier reference, especially for comparison with hammers of different size, the modulus of elasticity of the Bongossi is used, without making corrections for the reduction in height of the capblock.

26. The procedure used for the present post-analysis is based on comparison of specific characteristics of the measured force–time diagrams for the pile head with the same characteristics of wave-equation-predicted diagrams. The measured force–time diagrams are different for different blow numbers. The wave equation diagrams are different for different combinations of modulus of elasticity and coefficient of restitution. The characteristics selected for the correlation were:

(a) the maximum increase in force per unit of time $(dF/dt)_{max}$ in the first part of the force–time diagrams;

(b) the energy content of the force–time diagrams for the pile head up to the first minimum after the maximum peak force (mentioned as the energy in first peak: ENE_1).

27. The energy content of the first peak of the stress wave was computed with the formula

$$ENE_1 = \frac{c}{E_p A_p} \int_0^{t_1} F_2 \, dt$$

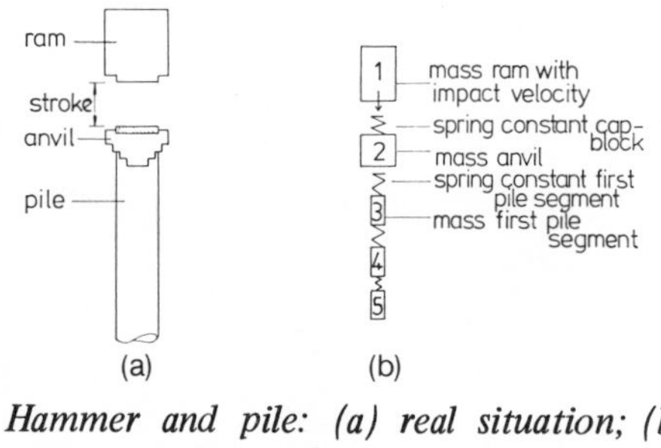

Fig. 10. Hammer and pile: (a) real situation; (b) simulation for wave equation analysis

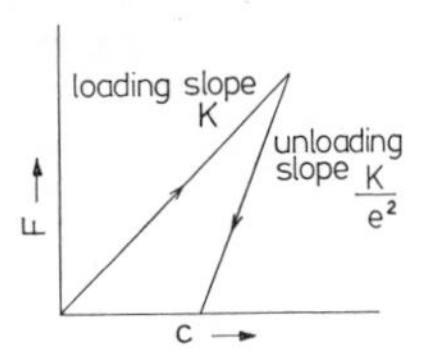

Fig. 11. Wave equation force–deformation lines for Bongossi capblock: K = EA/l = spring constant during loading, e = coefficient of restitution.

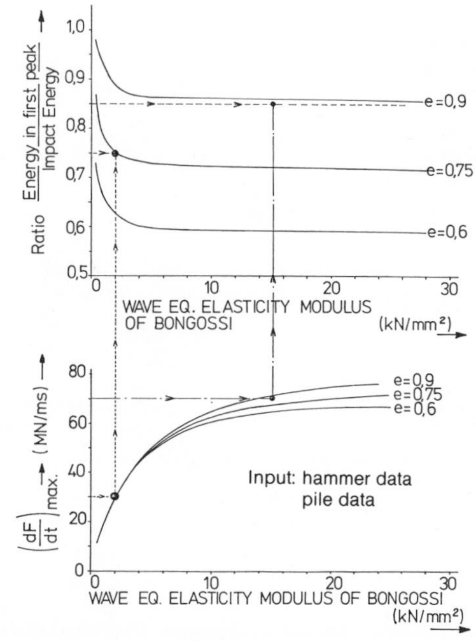

Fig. 12. Method of correlation

where

c = velocity of longitudinal stress wave in pile material
E_p = modulus of elasticity of pile material
A_p = cross-sectional area of pile
F = force at certain level in pile, as a function of time t
t_1 = time of first minimum after maximum peak.

This formula gives the energy of the input wave only as long as no rebound wave has arrived up to time t_1.

28. Typical wave equation correlations for $(\mathrm{d}F/\mathrm{d}t)_{max}$ and ENE_1 as a function of E and e are given in Fig. 12. (The given figures are valid only for this example and may be different for actual cases.) There are two interesting points: at E values lower than 4 kN/mm^2 the relation between $(\mathrm{d}F/\mathrm{d}t)_{max}$ and E is independent of e; and at E values higher than 8 kN/mm^2 the relation between energy in first peak and e is independent of E.

29. The combinations of E and e belonging to the measured force–time diagrams can be determined as follows. For low values of $(\mathrm{d}F/\mathrm{d}t)_{max}$ the value of E can be found directly from the lower diagram of Fig. 12; the upper diagram is then entered with both the derived E value and the calculated energy ratio, to arrive at the value of e. For high values of $(\mathrm{d}F/\mathrm{d}t)_{max}$ the correlation should be started in the upper diagram of Fig. 12 for finding e; the lower diagram is then entered with this e value and the $(\mathrm{d}F/\mathrm{d}t)_{max}$ value, to determine the value of E. For intermediate values, an iteration must be done.

30. This correlation procedure has the following advantages. First, the energy content of the first peak is a most important factor for the calculation of the permanent set. In this correlation method the predicted energy content is set equal to the measured one. For relatively short piles (i.e., piles standing a short distance out of the soil) the further part of the measured force–time diagram may be influenced by the rebound stress wave. For piles standing higher above the top of the soil, whereby the complete force wave is not influenced by rebound, it is recommended to compare the total energy in the measured diagram with that of the diagram predicted by wave equation using the evaluated e. In the present study the pile length was in general too short.

31. The second advantage of the correlation procedure is that the total rise time from zero force to peak force is closely related to the maximum rate of increase of force, and so the correlation will result in a good prediction of the peak force. The peak force is another important factor for the calculation of the permanent set.

32. The third advantage is that the total shape of the predicted force–time diagram will be similar to that of the real force–time diagram. This is also a very important item for calculation of the permanent set.

33. The peak height depends on both E and e over the full range of E and is therefore a more difficult correlation characteristic.

34. The ratio of total energy transferred into the pile head and impact energy of the ram is mainly dependent on e and in a minor degree on the E value of Bongossi. For example, for the MRBU 6000 at a coefficient of restitution of e = 0.75, the wave equation program predicted impact energy transfer efficiencies of 87% for E = 0.5 kN/mm^2 and 88% for E = 24 kN/mm^2.

35. To avoid misunderstandings it should be realized that the energy loss in the capblock due to a coefficient of restitution of e is not equal to $1 - e^2$. If it had been true, a value of e = 0.75 would have meant a loss of 44% of impact energy in the capblock, transferred into heat. The figures already quoted for impact energy transfer efficiencies show much smaller losses. The explanation is that the formula $1 - e^2$ for energy loss is valid only if the Bongossi would have accumulated the full impact energy first, before returning part of this energy. However, in the case of driving, the impact energy is not completely stored in the capblock, but at practically the same moment that impact force (and energy) is entering at the top of the capblock, a force (and energy) is delivered downwards. For such a case this simple energy loss formula does not apply.

Wave-equation-predicted relationships

36. Wave equation computer runs were made to establish the relationships needed for the method of correlation as described in the preceding paragraph. The input data used is given in Table 2. In addition to the hammers MRBU 6000 and MRBS 8000, data is given on the hammer MRBS 750. Some results on a test done with this hammer on a 36 in o.d. pile are presented for comparison with the results of the bigger hammers. No description of the test itself is given in this Paper. The relationships between maximum rate of increase of force $(\mathrm{d}F/\mathrm{d}t)_{max}$, wave equation modulus of

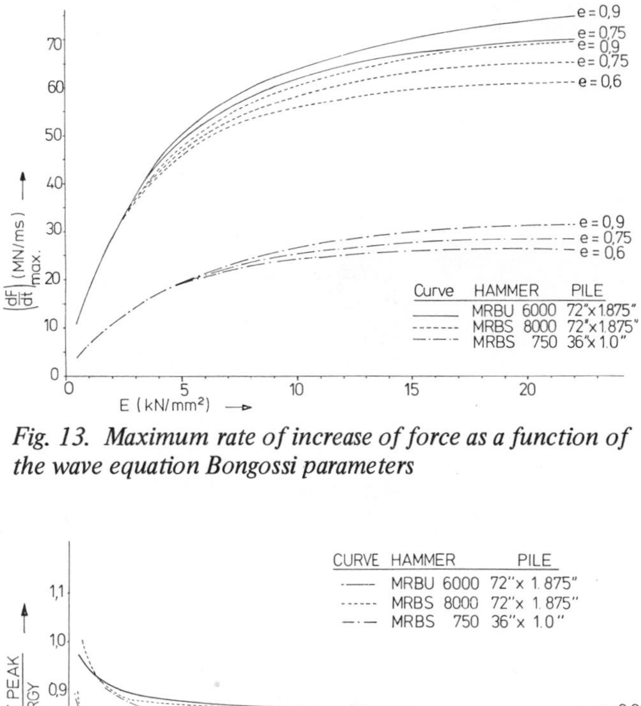

Fig. 13. Maximum rate of increase of force as a function of the wave equation Bongossi parameters

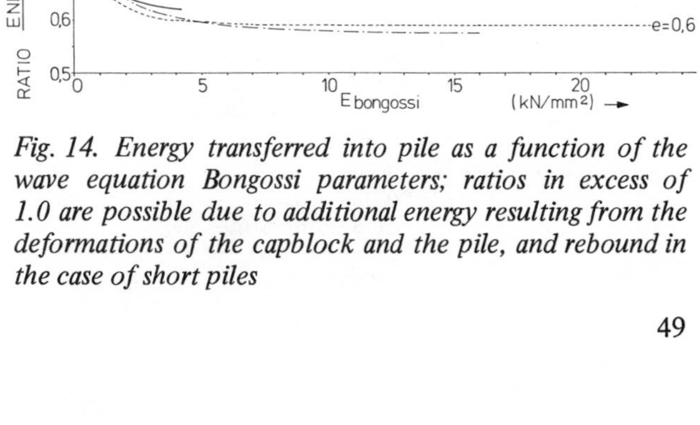

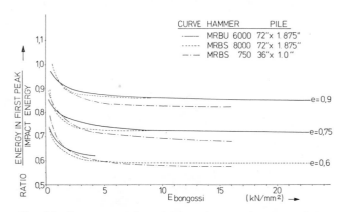

Fig. 14. Energy transferred into pile as a function of the wave equation Bongossi parameters; ratios in excess of 1.0 are possible due to additional energy resulting from the deformations of the capblock and the pile, and rebound in the case of short piles

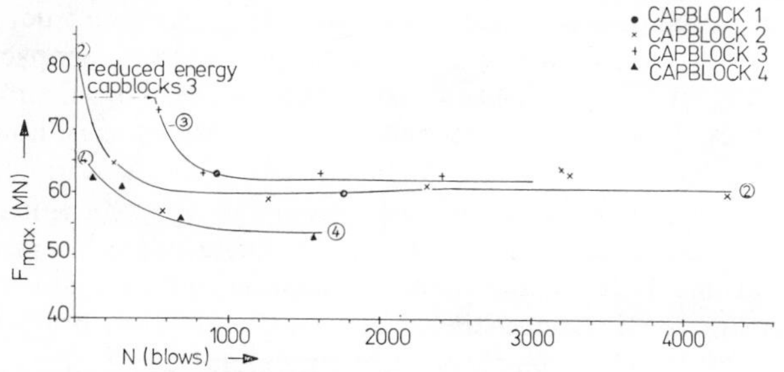

Fig. 15. Measured maximum peak force versus accumulated blows for MRBU 6000 (pile 72 in × 1.875 in)

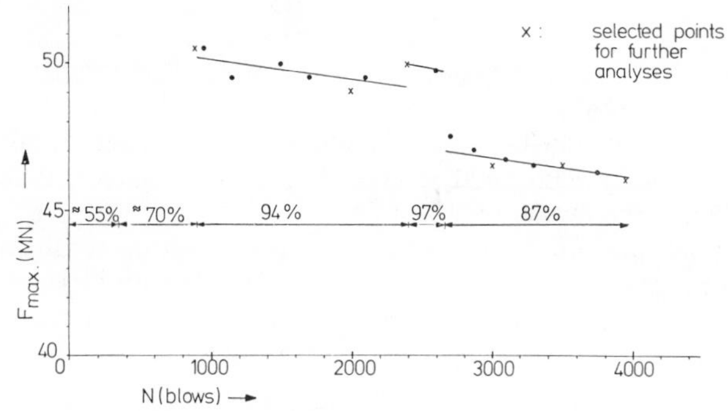

Fig. 16. Measured maximum peak force versus accumulated blows for MRBS 8000 (pile 72 in × 1.875 in)

elasticity E and coefficient of restitution e are given in Fig. 13 for the MRBU 6000, the MRBS 8000 and the MRBS 750. The ratio of energy in the first peak of the force–time diagram and the impact energy, depending on E and e, is given for all three hammers in Fig. 14. For the MRBS 8000 some rebound forces are included in this ratio due to the short free length of the pile above the soil. This short pile length for the wave equation runs was taken intentionally to copy the real test conditions. Assuming that the same mistake (i.e., including rebound forces in the energy content of the first peak) is made in both the measured and predicted diagrams, the correlation method can be applied. The selected soil input parameters were considered to be applicable. To check the possible error, the energy content of two wave-equation-predicted diagrams were compared, one using the relatively short pile and one using a pile length long enough to avoid rebound influence. For E = 0.8 kN/mm^2 and e = 0.72, the relative difference was less than 2%. The possible influence on the correlation results is, of course, smaller.

Measured data

37. Figs 6–8 show some of the measured force–time diagrams for the MRBU 6000 and the MRBS 8000. The peak forces in the pile as a function of accumulated blows for both hammers are presented in Figs 15 and 16. During the many test drives with the MRBU, five Bongossi capblocks were used, but for the fifth capblock only a few blows were given and no relevant data was recorded. Only

Table 2. Input parameters: unless otherwise stated, these values have been used in the wave equation analysis for the correlation study on Bongossi

	MRBS 750	MRBS 8000	MRBU 6000
Hammer:			
Ram weight (one segment), kN	75	800	600
Wave equation efficiency, %	90	94	100
Impact energy, kN m	84.4	1128	1050
Capblock:			
Diameter, m	0.72	1.98	1.58
Height (initial), m	0.18	0.30	0.28
Height of confinement, m	0.13	0.25	0.28
Anvil:			
Weight (including capblock), kN	31.5	390	300
Pile:			
Outer diameter, in	36	72	72
Wall thickness, in	1.0	1.875	1.875
Total length, m	24	66	123
Segment length, m	1.5	3.0	3.0
Soil:			
Pile penetration, m	3.0	51	54
Total resistance, MN	9	45	45
Ratio point/total resistance, %	20	5	5
Unit side friction distribution	*	*	*
Quake side, mm	2.54	2.54	2.54
Quake point, mm	2.54	2.54	2.54
Damping side, s/m	0.164	0.656	0.656
Damping point, s/m	0.492	0.033	0.033

*Triangular.

those points relating to a net striking energy of around 1050 kN m are plotted in Fig. 15.

38. For the MRBS 8000 the impact energy of the ram as a function of the maximum rated striking energy is noted in Fig. 16. The hammer worked in the beginning at reduced energies of about 55% and 70% of the maximum rated energy. The actual test drives were done at 94%, 97% and 87% of the maximum rated energy. Only one capblock was used.

39. Calculated values of $(\mathrm{d}F/\mathrm{d}t)_{\max}$ from measured force–time diagrams are given in Figs 17 and 18.

Results

40. The evaluated values of E as a function of accumulated number of blows are plotted for the MRBU 6000, the MRBS 8000 and the MRBS 750 in Fig. 19. The values for the MRBS 8000 were derived taking the varying wave equation efficiencies into account.

41. There is a significant difference between the E values for the two MRBS types of hammer and the MRBU. Therefore, two separate trend lines were drawn. Both curves seem to start at a value of E which is around 20 kN/mm^2 for the first blow. The value of E decreases very quickly for the first 1000 blows and then begins to stabilize. For the MRBU capblocks it stabilizes at an E value of 7 kN/mm^2 and for the MRBS capblocks at an E value of 1–2 kN/mm^2.

42. The evaluated values of e as a function of accumulated blows are plotted in Fig. 20. Results for capblock 4 of the MRBU are considered to be less reliable. They fall outside the range of values not only for the evaluated E and e values, but already for the measured peak forces as given in Fig. 15. There is some scatter in the evaluated e values. No separate trend lines for MRBU and MRBS seem justified. Therefore, only one trend line has been drawn.

43. There are many factors influencing the values of E and e of a Bongossi capblock depending on the accumulated number of blows. These factors include the physical properties, the method of storage, the fitting accuracy into the anvil recess, the degree of confinement, the loading history, and interruptions in driving. Obviously, the loading history of a capblock depends on the hammer type and the site conditions. All these factors may have had an influence on the scatter of the results of the evaluated capblocks of the MRBU 6000. The tests with the hydraulic hammer were done with four different capblocks, at somewhat varying net striking energies, with intermittent drives at much smaller energies, at varying blow rates, on different pile lengths, with varying duration of stops, and at different penetrations.

44. The diminishing decrease in the modulus of elasticity of Bongossi can be explained as follows. At the first blow the (semi-empirical) wave equation modulus of elasticity is near to the static value, which is approximately

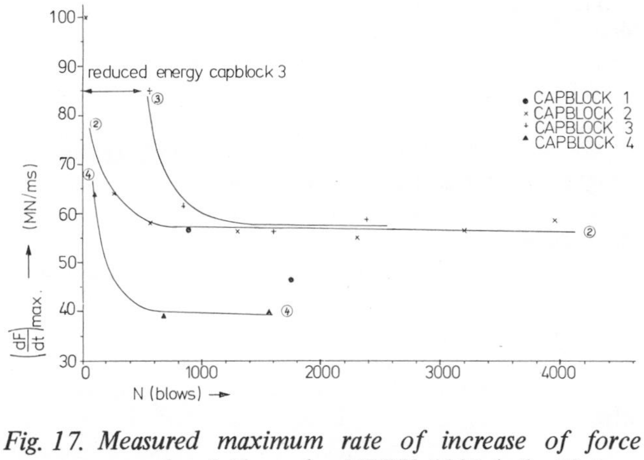

Fig. 17. Measured maximum rate of increase of force versus accumulated blows for MRBU 6000 (pile 72 in x 1.875 in)

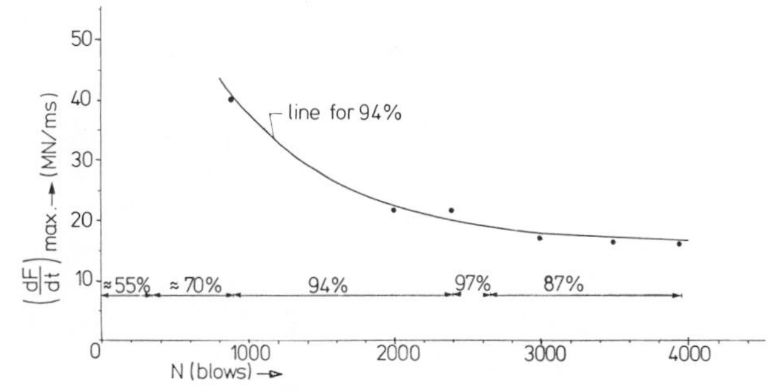

Fig. 18. Measured maximum rate of increase of force versus accumulated blows for MRBS 8000 (pile 72 in x 1.875 in)

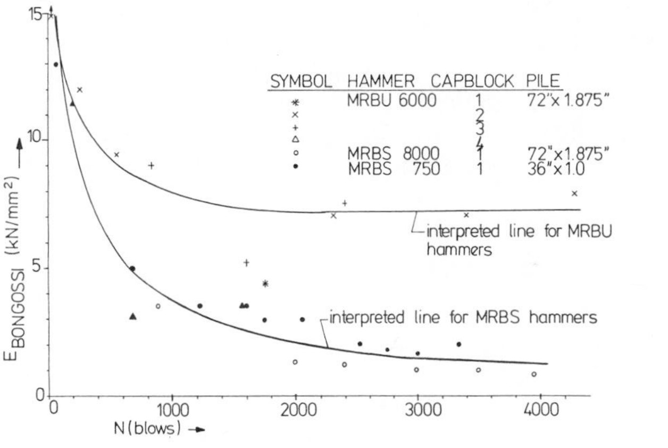

Fig. 19. Post-analysis wave equation modulus of elasticity for Bongossi as a function of accumulated blows

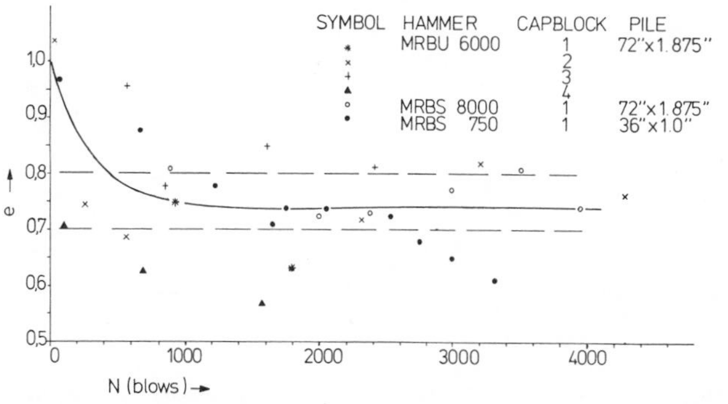

Fig. 20. Post-analysis restitution coefficients as a function of accumulated blows

20 kN/mm². This means a relatively high spring constant for the capblock, resulting at the first impact in high stresses in the Bongossi. These stresses are above the compressive strength. Therefore, the wood structure will partly fail, probably due to buckling of the (vertical) grains. Thereby the modulus of elasticity decreases as well as the compressive strength. Due to the fact that the spring constant has decreased, a second impact will cause less high stresses, which are still above the 'new' compressive strength. The wood structure will again fail partly. This process continues.

45. However, the modulus of elasticity decreases more quickly than the compressive strength. This again results in a gradually stabilizing modulus of elasticity. A similar explanation is possible for the diminishing decrease of the coefficient of restitution. After a high number of accumulated blows the relatively low modulus of elasticity causes relatively low levels of stress in the Bongossi. These occurring stresses are in such a relation to the compressive strength that an almost constant coefficient of restitution is obtained.

46. The explanation given above considers the two parameters to be valid for the full height of the Bongossi capblock as a whole. There are, however, different types of area present in the capblock due to the driving influences. In the present study the capblock has been considered as one unit. This approach is correct as long as the evaluated parameters are able to predict reliable force–time diagrams and exactly this was the approach selected.

47. It is most important to know that relatively large variations in these parameters have only a small effect on the drivability predictions. This is, for one example of hammer, pile and soil, demonstrated in Fig. 21, where three wave equation curves are given showing the relationship between soil resistance during driving and blow count. These three curves were computed using different combinations of E and e as shown in the figure. At 200 blows per

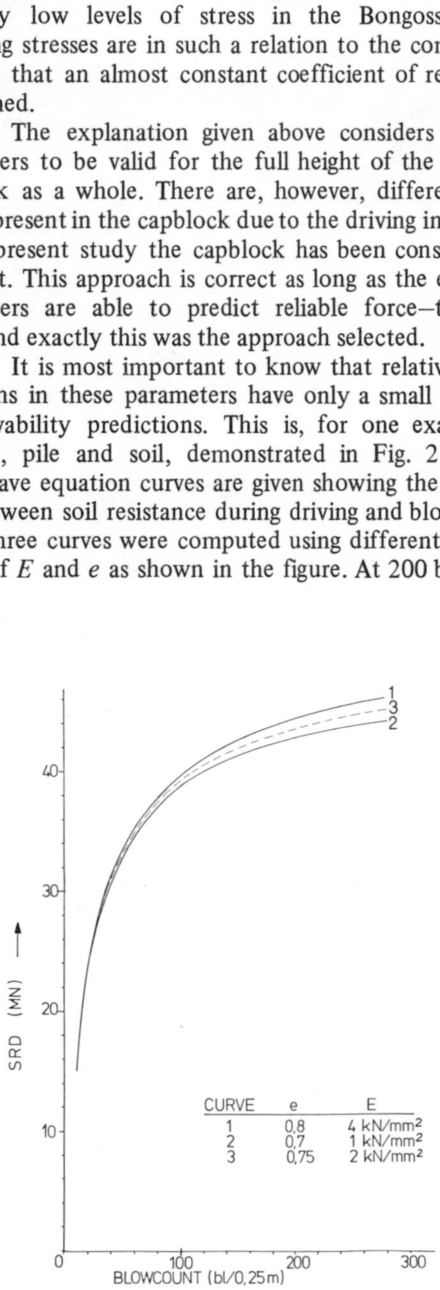

Fig. 21. Soil resistance at time of driving versus blow count for MRBS 8000 (pile 72 in × 1.875 in; input data as in Table 2)

0.25 m, curves 1 and 2 vary from curve 3 by only 2%. This difference is certainly smaller than the influences of other factors on the driving behaviour of the piles. Of course, this is only one example. Nevertheless, it seems justified to recommend values of E = 2.0 kN/mm² and e = 0.75 for most equation analyses for MRBS hammers, and values of E = 7.0 kN/mm² and e = 0.75 for the MRBU 6000. Only in special cases (e.g., post-analysis studies) could deviating values be applied, making use of the published data and taking into consideration the factors influencing the rate of decrease in E and e.

48. It is hoped that more measurements will become available for obtaining more evaluated points for the modulus of elasticity and the coefficient of restitution and thus increasing the knowledge regarding the factors influencing these parameters.

CONCLUSIONS

49. The hydraulic hammer MRBU 6000 and the steam hammer MRBS 8000 showed equal driving capabilities with either the standard Bongossi anvil or the special anvil.

50. A mechanical hammer efficiency (defined as the ratio of kinetic energy of the ram at impact and maximum actual potential energy) of 96% was obtained for the MRBS 8000. This efficiency is expected not to be lower than 90% under most offshore pile driving conditions. This mechanical hammer efficiency is a part of the efficiency factor as used for wave equation studies, applied to the maximum rated striking energy of the hammer.[4]

51. The impact transfer efficiency of the standard Menck anvil with Bongossi capblock was around 90%, depending on the condition of the Bongossi. This efficiency has not to be applied for wave equation studies, because it is incorporated in other parameters.

52. The selected method of determining for the Bongossi capblock the two parameters needed for wave equation studies was found to be adequate in the prediction of force–time diagrams similar to those measured.

53. The modulus of elasticity and the coefficient of restitution for the Bongossi capblock have values of approximately 20 kN/mm² (equal to the static modulus of elasticity) and 1.0 for the first blow. These values decrease quickly during the following 1000 blows and are then relatively constant.

54. The following values may be recommended for most wave equation analyses: a coefficient of restitution of 0.75 for the MRBU and MRBS range of hammers and moduli of elasticity of 7 kN/mm² for the MRBU hammer and 2 kN/mm² for the MRBS hammers.

REFERENCES

1. KUEHN H. How Menck is attacking the hydraulic hammer market. Offshore Engineer, 1978, Mar., 29–30.
2. SMITH E.A.L. Pile driving analysis by the wave equation. J. Soil Mech. Fdns Div. Am. Soc. Civ. Engrs, 1960, **86**, Aug.
3. LOWERY L.L. et al. Pile driving analysis – simulation of hammers, cushions, piles, and soil. Texas A&M University, 1967, research report 33–9.
4. VAN LUIPEN P. Experience with heavy pile driving hammers in the North Sea. Proc. Offshore Brazil 78 Conference, Rio de Janeiro, 1978. CELP, Rio de Janeiro, OB–78.13.

7. Soil response during pile driving

V. N. VIJAYVERGIYA, PhD, PE (President, Geotec Engineering, Inc., Houston, Texas)

The first part of this Paper deals with the effect of pile driving on the soil at the pile tip and around the pile. The stress changes in the soil are examined during pile driving, after the pile is driven, and during a static load test. Based on these stress changes it is demonstrated that a direct correlation between static capacity and the resistance during driving is very difficult. Subsequent sections deal with the development of a rational method to predict the soil resistance during driving and its applications in predicting drivability using wave equation analysis.

INTRODUCTION

The tremendous increase in the use of piles in foundations of both land-based and offshore structures and the development of new pile driving methods have created great engineering interest in finding more reliable methods for analysis and design of piles. Pile driving formulas have been used in the past; yet, most texts in this subject caution the engineer to be wary of the use of dynamic pile driving formulas inasmuch as they cannot be relied on to predict a pile's ultimate capacity (static) with a reasonable degree of accuracy. Similarly, the one-dimensional wave equation analysis has been used to solve complex pile driving problems since its publication by Smith[1] in 1960. The results of the wave equation analysis are typically presented in the form of a relationship between ultimate driving resistance, R_{ud}, versus the rate of pile penetration, N, as shown in Fig. 1. Several investigators have attempted to predict the ultimate static capacity of a pile, Q, from the results of wave equation analysis, but with varying degrees of success. It is believed that future correlations can be improved with better understanding of soil response during pile driving.

2. This Paper examines the effect of pile driving and the stress changes in the soil surrounding the pile. A rational method is developed to predict the soil resistance during driving, R_{ud}, from the basic soil properties. It is suggested that this R_{ud} be utilized in analysing the results of wave equation analysis to solve pile driving problems.

EFFECT OF PILE DRIVING

3. There are two types of driven pile commonly used for supporting superstructures. These are the displacement type and the non-displacement type.

4. The displacement type piles are those in which a significant volume of the soil is displaced from the bottom of the pile during pile driving. Examples of such piles are timber piles, prestressed precast concrete piles and closed-end steel pipe piles.

5. The non-displacement type piles are those in which the soil mass is not displaced significantly during pile driving. Examples of such piles include open-ended pipe piles, steel H piles, sheet piles and cylindrical prestressed precast concrete piles.

6. During pile driving, the soil underneath the tip of the pile is subjected to shear stresses in excess of the shear strength of the soil. Excessive stresses cause deformation of the soil mass and its ultimate failure. The failure of the soil mass during driving is in the form of plastic flow (Fig. 2(a)), which is different from that during static loading in which the soil fails along multiple planes. During flow, the deformed soil mass is pushed downward and laterally away from the pile surface. In the case of open-ended pipe piles, the soil also enters the pile. During flow and deformation, the soil is disturbed and remoulded and excess pore water pressures are generated. As a result of remoulding and generation of excess pore water pressure, the shear strength of the soil around the pile during driving is significantly lower than the shear strength of the undisturbed soil. In general, it may be said that the state of stress in the remoulded zone (Fig. 2(b)) has been altered and a new state of stress has been created.

STRESS CHANGES DUE TO PILE DRIVING

7. In a saturated soil, the effective horizontal and effective vertical stresses can be expressed as

$$\bar{\sigma}_v = \sigma_v - U \qquad (1)$$

$$\bar{\sigma}_h = \sigma_h - U \qquad (2)$$

$$K_0 = \bar{\sigma}_h / \bar{\sigma}_v \qquad (3)$$

where

σ_v and $\bar{\sigma}_v$ are total and effective vertical stresses, respectively

σ_h and $\bar{\sigma}_h$ are total and effective horizontal stresses, respectively
U is the pore water pressure
K_0 is the coefficient of earth pressure at rest.

8. Figure 3(a) shows the effective stresses on an undisturbed soil element at a depth x. Immediately after a pile is driven to its final penetration, the state of stress in a remoulded soil element around the pile surface at that depth can be expressed as follows:

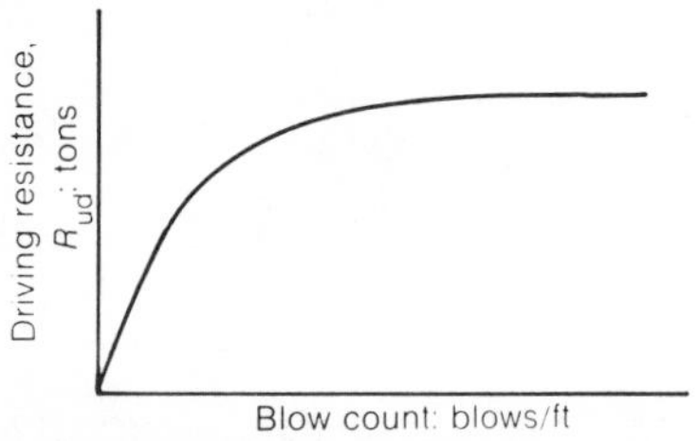

Fig. 1. Typical results of wave equation analysis

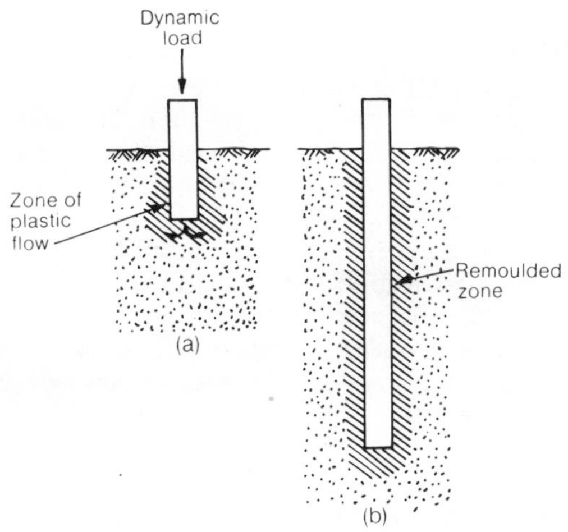

Fig. 2. Effect of pile driving: (a) initial penetration; (b) final penetration

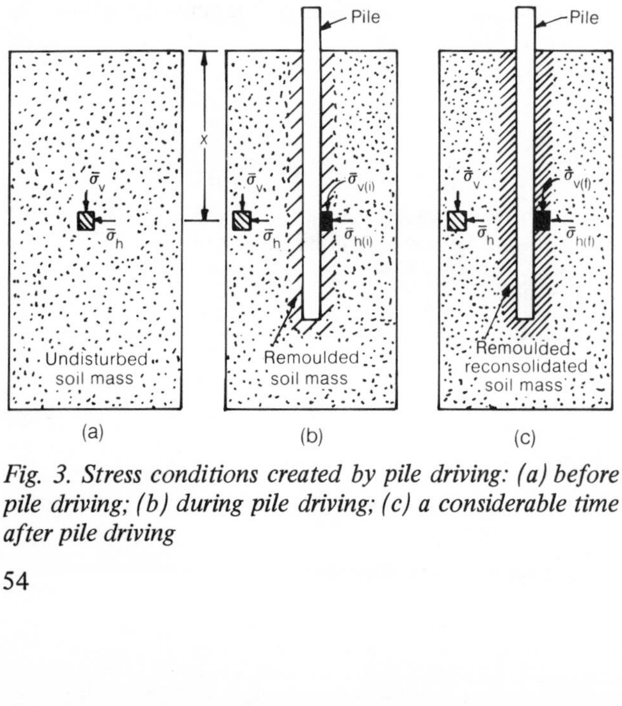

Fig. 3. Stress conditions created by pile driving: (a) before pile driving; (b) during pile driving; (c) a considerable time after pile driving

$$\bar{\sigma}_{v(i)} = \sigma_{v(i)} - (U + \Delta U) \quad (4)$$

$$\bar{\sigma}_{h(i)} = \sigma_{h(i)} - (U + \Delta U) \quad (5)$$

where the subscript i indicates an intermediate stress condition at the end of driving. The shearing of the soil mass and excessive deformations of the soil mass result in excess pore water pressures, designated by ΔU. This stress condition for the remoulded element of soil is illustrated in Fig. 3(b). It can be noted that a soil element at some distance away from the remoulded zone is not affected by pile driving and still has the original stress condition. Field measurements of ΔU immediately after pile driving[2-4] show that the excess pore water pressure can be several times $\bar{\sigma}_v$ near the pile surface.

9. After the pile is driven to final penetration, with passage of time the excess pore water pressure, ΔU, in the remoulded soil mass dissipates. In other words, the remoulded soil reconsolidates and a new stress equilibrium is reached as given by

$$\bar{\sigma}_{v(f)} = \sigma_{v(f)} - U \quad (6)$$

$$\bar{\sigma}_{h(f)} = \sigma_{h(f)} - U \quad (7)$$

where the subscript f indicates final equilibrium condition. This stress condition is shown in Fig. 3(c). This phase of reconsolidation has been confirmed by Clark and Meyerhoff,[5] Cummings et al.[6] and Flaate[7]. According to a hypothesis proposed by Vijayvergiya,[8] the consolidation stress, σ_c, under final equilibrium condition can be expressed in terms of effective horizontal stress, $\bar{\sigma}_h$, in the undisturbed condition as

$$\sigma_c = \bar{\sigma}_h \quad (8)$$

STRESSES DURING STATIC PILE LOAD TEST

10. The stress conditions on a remoulded, reconsolidated soil element near the pile surface during a pile load test are illustrated in Fig. 4. Just prior to the load test, the excess pore water pressure, $\Delta U'$, is zero and, consequently, the effective normal stress, $\bar{\sigma}$, on the vertical failure plane is equal to the consolidation stress, σ_c. Since a pile load test is generally conducted at a deformation rate that does not allow drainage during test, a consolidated, undrained condition is simulated in the field. As the pile load, Q, is increased, the total and shearing stresses increase, causing excess pore water pressure to develop along the failure plane, and the magnitude of the effective normal stress changes. For steel piles, any change in total stress within the failure zone is assumed to be taken by pore water pressure. Field measurements of $\Delta U'$ during static tests in overconsolidated clays by Airhart et al.[9] show that excess pore pressure rises with increased load until the pile plunges.

11. With the increase in pile load, pile movement increases and the shearing stress, τ, reaches a maximum value, τ_f, called peak shear strength. Assuming that hydrostatic stress remains constant during load test, the shear strength, τ_f, can be expressed in terms of effective stress parameters c' and ϕ' as

$$\tau_f = c' + \bar{\sigma} \tan\phi' \quad (9)$$

where

$\bar{\sigma} = \sigma_c - \Delta U' = K_0\, \bar{\sigma}_v - \Delta U'$ (from equations (3) and (8))
$\Delta U'$ is excess pore water pressure at failure.

BASIC DIFFERENCE BETWEEN STATIC AND DYNAMIC CAPACITIES

12. The magnitude of excess pore water pressure generated during a static pile load test, $\Delta U'$, (equation (9)) is relatively small and may be a fraction of $\bar{\sigma}_v$. But the magnitude of excess pore water pressure generated during driving, ΔU, (equation (4) or (5)) is several times $\bar{\sigma}_v$. Consequently, the shearing strength of the soil around the pile is substantially lower during pile driving than it is in the static condition. To develop any meaningful correlations between stress conditions during pile driving and stress conditions during load tests, it is important to obtain information on excess pore water pressures developed in the soil mass both during pile driving and during pile load tests. According to Hagerty and Garlanger,[2] theoretical predictability of excess pore water pressures generated during pile driving and during static tests is questionable. In spite of the basic differences in stress conditions during pile driving and during static tests, many investigators have attempted the prediction of static capacity from R_{ud}. Consequently, such predictions should be viewed with scepticism, as the data may be applicable to the subject site only.

SOIL RESISTANCE DURING DRIVING

13. The wave equation analysis predicts the relationship between R_{ud} and N, the number of blows required to advance the pile per unit distance for a given soil–pile–hammer system. This relationship (Fig. 1) can be used to predict the ultimate soil resistance during driving at a known blow-count value of N or to predict the estimated value of blow-count (N) for a known R_{ud}. Thus by estimating R_{ud} from the soil properties at the site, one can predict the drivability of the pile in the field. This section deals with the formulation of numerical expressions for computing R_{ud}.

14. For displacement piles, where the bottom of the pile is closed, the ultimate soil resistance during driving, R_{ud}, is derived from the shearing resistance along the surface of the pile and the base resistance at the pile tip (Fig. 5). This can be expressed numerically as

$$R_{ud} = T_o + R_b \quad (10)$$

where

T_o is the shearing resistance along the outer surface of the pile during driving
R_b is the base resistance of the pile tip during driving.

15. For open-ended pipe piles, the base resistance may be derived from the cross-sectional area of the wall only. Additionally the shearing resistance on the inner surface of the pile will also be developed. Fig. 6 shows the soil resistance during driving, R_{ud}, for a pipe pile with no soil plugging. Thus

$$R_{ud} = T_o + T_i + R_s \quad (11)$$

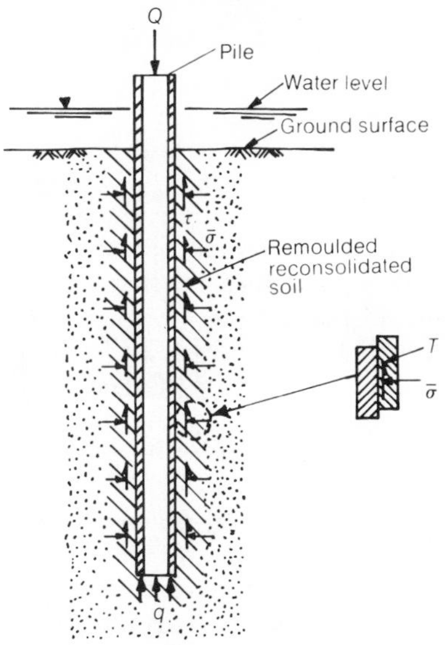

Fig. 4. Stresses during pile load test: inset–details of shearing element

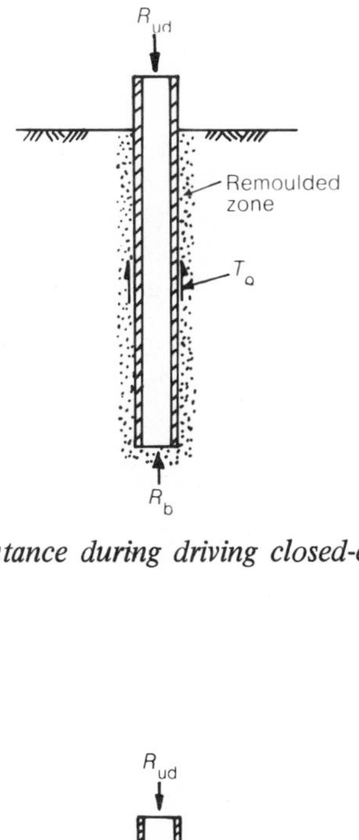

Fig. 5. Soil resistance during driving closed-end pile: $R_{ud} = T_o + R_b$

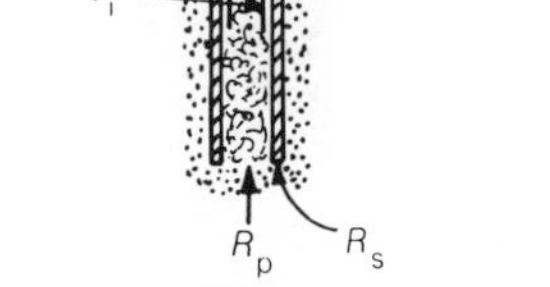

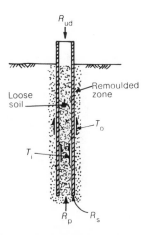

Fig. 6. Soil resistance during driving pipe pile, no plugging: $R_{ud} = T_o + T_i + R_s$, $T_i < R_p$, $R_b = R_s$

where

T_i is the shearing resistance on the inner surface of the pile during driving
R_s is the base resistance at the cross-sectional area of the wall during driving.

As long as the base resistance on the soil core inside the pile during driving, R_p, is greater that the shearing resistance on the inside of the pile, T_i, the soil core will keep on moving upward as the pile is driven, as shown in Fig. 6. However, when T_i becomes equal to or greater than R_p, the soil will stop moving upward and the pile is said to have been plugged. This plugged condition is shown in Fig. 7.

Shearing resistance during driving

16. The magnitudes of shearing resistances during driving, T_o and T_i, depend on several factors. Some of the important factors include sensitivity of the soil, the nature of remoulding of the soil, the surface characteristics of the pile, the magnitude of the confining stresses on the remoulded soil, the effect of reorientation of soil particles (on the upper portion of the pile), the type of pile shoe, densities, grain size distribution, permeability etc. Numerically, shearing resistances T_o and T_i can be expressed as follows:

$$T_o = \alpha_o c_r A_o \tag{12}$$

where

α_o is a reduction coefficient under dynamic conditions for the outer surface
c_r is the average remoulded shear strength of the soil
A_o is the outside surface area of the pile

and

$$T_i = \alpha_i c_r A_i \tag{13}$$

where

α_i is a reduction factor under dynamic conditions for the inner surface of the pile
A_i is the inside surface area of the pile.

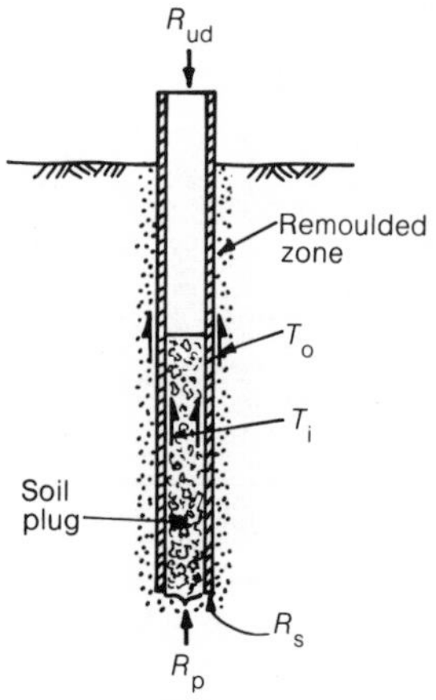

Fig. 7. Soil resistance during driving pipe pile, pile plugged: $R_{ud} = T_o + R_b$, $T_i \geqslant R_p$, $R_b = R_s + R_p$

17. For cohesive soils, $c_r = c/S$, in which c is the undrained shear strength of the clay and S is the sensitivity of the clay. For granular soils, it is difficult to estimate the remoulded shear strength, c_r. It may range from almost zero for liquefaction condition to static unit skin friction on the pile surface, f, which can be determined from

$$f = K\bar{\sigma}_v \tan\delta \tag{14}$$

where

K is the coefficient of lateral earth pressure,
$\bar{\sigma}_v$ is the effective vertical stress,
δ is the angle of friction between pile surface and soil.

Base resistance during driving

18. The maximum base resistance during driving, R_b, occurs when piles are driven closed-ended or are plugged during driving; it can be expressed as

$$R_b = R_s + R_p \tag{15}$$

where

R_s is the base resistance at the steel tip during driving
R_p is the base resistance on the soil core inside the pile during driving.

The dynamic unit end bearing capacity, q_d, at the base of the pile can be expressed in terms of static unit end bearing value, q, as

$$q_d = K_d q \tag{16}$$

in which K_d is the dynamic bearing capacity factor. Thus, the total base resistance for plugging piles can be expressed as

$$R_b = K_d q (A_s + A_p) \tag{17}$$

where A_s and A_p are the areas of steel tip and inside of pile base, respectively. It is known from the dynamic tests on soil samples that the compressive strength of a soil sample is greater under a short term impulse loading than under static condition. This suggests that the value of K_d is greater than unity. In reality, the value of K_d depends on point damping factor and pile penetration velocity.[10, 11]

19. For cohesive soils, the value of q can be determined from $q = N_c c$, where N_c is bearing capacity factor for clay.

20. For granular soils, the value of q may be determined from $q = N_q \bar{\sigma}_v$, where N_q is the bearing capacity factor for granular soils and depends on ϕ, the angle of internal friction of granular soil.

DETERMINATION OF COEFFICIENTS K_d, α_o AND α_i

21. At present, limited data is available to establish realistic values of α_o, α_i and K_d for either granular or cohesive soils. Based on compression tests on soil samples under dynamic conditions,[10, 11] it has been found that the value of K_d may range from 2 to 2.5 for cohesive soils and from 1.5 to 2.5 for granular soils. The remoulded soil around the pile near the upper portions of the pile may undergo a greater degree of remoulding than suggested by

sensitivity S determined in the laboratory. In granular soils such a phenomenon has been suggested by Vijayvergiya[12] for pipe piles. Such considerations would suggest that the value of a_o is generally less than 1.0. For clays the value of a_o may range from 0.5 to 1.0 and for granular soils it may be taken as 1.0.

22. The value of a_i can be derived by considering the plugged conditions of a pipe pile. When a pile is plugged, as shown in Fig. 7, the shearing resistance on the inside of the pile must be equal to or greater than the base resistance on the inside area of the pile tip; thus, plugging will occur when

$$R_p = T_i \tag{18}$$

By using equations (13) and (17), the above expression can be rewritten as

$$\alpha_i = \frac{K_d q A_p}{c_r A_i} \tag{19}$$

For large diameter piles, the difference between d_1, the outer diameter, and d_2, the inner diameter, can be considered negligible and it can be assumed that $d_1 = d_2 = d$. Thus equation (19) can be expressed as

$$\alpha_i = \frac{1}{4}\frac{K_d q d}{c_r L} \tag{20}$$

in which L is the length of the soil plug inside the pile measured from the pile tip.

23. From the pile driving data collected in the Gulf of Mexico, it has been observed that the piles in clay can be driven to penetrations of about 300–400 ft without plugging if no significant delay occurs during driving.[13] Also, the sensitivity for the Gulf of Mexico clays ranges from about 2.5 to 3.5. For a pile diameter of 3 ft the following approximate relationship can be obtained:

$$\alpha_i = \frac{K_d}{15} \tag{21}$$

Equation (21) indicates that the value of a_i is a very small fraction of the dynamic capacity factor, K_d. For $K_d = 3$, the value of a_i is 0.2.

24. The value of a_i depends on several factors. However, a range of 0.1–0.3 may be reasonable for the normally consolidated to slightly overconsolidated clays.

25. It has been the Author's observation that in granular soils in the Gulf of Mexico when open-ended piles are driven in dense sand formations (silica sand), the piles tend to plug after driving to about generally less than 200 ft penetration below ground surface. For a dense sand with $N_q = 40$ and $c_r = K\,\bar{\sigma}_v \tan\delta$, a typical relationship can be developed between a_i and K_d. For a pile diameter of 3 ft with plugging occurring at about 200 ft penetration, the a_i can be approximately expressed as

$$a_i = K_d \tag{22}$$

Equation (22) suggests that in granular soils a_i is very high and a plugged condition may be used for computing R_{ud}.

APPLICATION OF R_{ud}

26. Using the procedure outlined above, the values of the soil resistance during driving, R_{ud}, can be computed for various pile penetrations. Using these values of R_{ud} and the results of wave equation analysis (Fig. 1), the blow-count data can be predicted for given hammer–cushion–pile combination. Such a predicted relationship between pile penetration and blow count, for a 36 in o.d. pile driven in chalk in the North Sea,[14] is presented in Fig. 8. Since the chalks at the site were cohesive in character, parameters applicable to cohesive soils were utilized in computing R_{ud}. A comparison of predicted blow count and the lower bound of the observed blow count shows good agreement. The upper bound of blow count is related to delays in pile driving caused by several factors. The effect of delay on blow count for this site is described by Vijayvergiya et al.[15]

CONCLUSIONS

27. The effect of pile driving on the soil at the pile tip and around the pile should be evaluated carefully to develop any meaningful correlations between static capacity of pile and the soil resistance during driving. The theoretical considerations indicate that such correlations may not be reliable.

28. A rational method is developed to determine R_{ud}. This method takes into account the effect of dynamic impact and remoulding on the shear strength characteristics of the soil. The computed values of R_{ud} can be utilized in predicting pile drivability in the field by using wave equation analysis. A comparison of the predicted and observed pile driving data shows good agreement. It is suggested that this procedure be used to determine appropriate hammer size for driving piles to design penetrations.

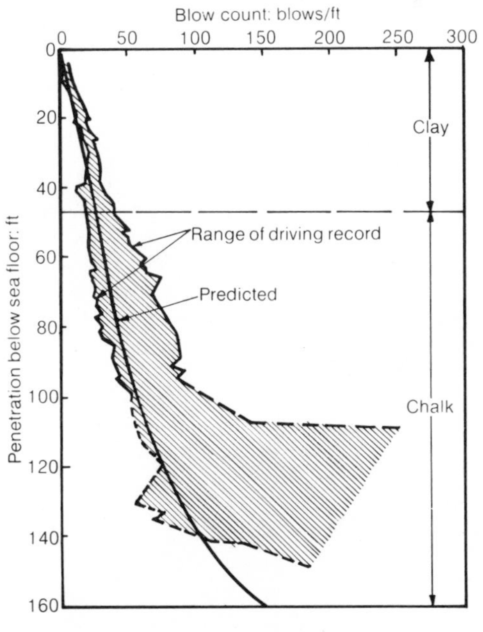

Fig. 8. Observed and predicted driving data in cohesive chalk:[14] 36 in dia. piles; Menck 2500 hammer (broken lines based on an equivalent Menck 2500 hammer)

REFERENCES

1. SMITH E. A. L. Pile driving analysis by the wave equation. J. Soil Mech. Fdns Div. Am. Soc. Civ. Engrs, 1960, 86, Aug., No. SM4, proc. paper 2574.

2. HAGERTY D. J. and GARLANGER J. E. (AMERICAN SOCIETY OF CIVIL ENGINEERS (ed.)). Consolidation effects around driven piles. Proc. specialty conf. on performance of earth and earth supported structures. ASCE, New York, 1972, 1, Part 2, 1207–1222.

3. KOIZUMI Y. and ITO K. Field tests with regard to pile driving and bearing capacity of piled foundations. Soil Fdn, 1967, VII, Aug., No. 3, 30–52.

4. SEED H. B. and REESE L. C. Action of soft clay along friction piles. Trans. Am. Soc. Civ. Engrs, 1957, 122, 731–759.

5. CLARK J. C. and MEYERHOFF G. G. The behavior of piles driven in clay – investigation of the bearing capacity using total and effective strength parameters. Canadian Geotech. J. 1973, 10, 86–102.

6. CUMMINGS A. E. et al. Effect of driving piles into soft clay. Trans. Am. Soc. Civ. Engrs, 1950, 115, 275–285.

7. FLAATE K. Effects of pile driving in clays. Canadian Geotech. J. 1972, 9, No. 1, 81.

8. VIJAYVERGIYA V. N. Friction capacity of driven piles in clay. Proc. 9th Annual Offshore Technology Conf., Houston, Texas, 1977, III, No. 2939, 465–474.

9. AIRHART T. P. et al. Pile–soil system response in clay as a function of excess pore water pressure and other soil properties. Texas Transportation Institute, Texas A & M University, 1967, research report 33–8.

10. GIBSON G. C. and COYLE H. M. Soil damping constants related to common soil properties in sands and clays. Texas Transportation Institute, Texas A & M University, 1968, research report 125–1.

11. REEVES G. N. et al. Investigation of sands subjected to dynamic conditions. Texas Transportation Institute, Texas A & M University, 1967, research report 33–7A.

12. VIJAYVERGIYA V. N. Discussion on tests on instrumented piles, Ogeechee River site by A. S. Vesic. J. Soil Mech. Fdns Div. Am. Soc. Civ. Engrs, 1971, 97, Jan., SM1, 252–256.

13. KINDEL C. E. Mechanism of soil resistance for driven pipe piles. Proc. Ports 77 Conf. American Society of Civil Engineers, New York, 1977.

14. VIJAYVERGIYA V. N. and CHENG A. P. Offshore pile foundations in chalk. Proc. GEOCON–India Conf. Geotechnical Engineering, New Delhi, 1978, 1, 484–491.

15. VIJAYVERGIYA V. N et al. Effect of soil set up on pile driveability in chalk. J. Geotech. Engng Div. Am. Soc. Civ. Engrs, 1977, 103, Oct., No. GT10, 1069–1082, proc. paper 13293.

8. Soil-pile interaction under dynamic loads

M. NOVAK (Professor, Faculty of Engineering Science, The University of Western Ontario, Canada)

Two theories for the calculation of impedance functions of piles are described. Both are based on soil reactions calculated from equations of the viscoelastic medium. The more rigorous one contributes to the understanding of main features of soil–pile interaction but its validity is limited to one homogeneous layer of soil. The more approximate approach is very versatile. It makes it possible to consider soil layering and different types of pile. The piles can be end-bearing or floating, may be of variable cross-section, and may have a pedestal or stick out of the ground. Also, pile separation from the ground can be accounted for approximately. This versatility can improve the agreement between the theory and experiments. A few further factors such as non-linearity, interaction of piles in a group, pile batter and dynamic analysis of flexible pile caps are also discussed.

INTRODUCTION

Pile dynamics has been finding application in such traditional areas as machine foundations, buildings and bridges, and recently new interest in it has been generated by the development of offshore towers. A number of approaches have been formulated for the analysis of dynamic soil–pile interaction. They employ the lumped mass model,[1,2] the continuum model[3–15] and the finite element method.[16–20] Each of these theories has some advantages and some limitations, as discussed in a recent state-of-the-art report by Tajimi.[21]

2. This Paper describes two approaches developed at The University of Western Ontario. They are based on soil reactions derived from the constitutive equations of the linear viscoelastic continuum. Such soil reactions have the advantage of reflecting the nature of wave propagation associated with the vibration of the embedded pile but they make the theories applicable primarily in the region of small strains. For large strains, only approximate corrections can be made. The emphasis is on impedance functions of the embedded part of the pile because they can be introduced in the analysis of any superstructure to represent the soil and pile lying below the ground surface or mud line.[22]

CONTINUUM APPROACH TO SOIL–PILE INTERACTION

3. In all approaches to dynamic response of embedded piles the resistance of soil to the motion of the pile must be established first. In the continuum approach, this resistance can be calculated from the equations of the viscoelastic medium. The inclusion of viscosity is essential because without it the solution can exhibit strong resonant phenomena and give unrealistic descriptions of pile behaviour. The viscosity should be of the frequency-independent hysteretic type as experiments with different soils suggest. This type of material damping can be introduced into the equations of an elastic medium by complementing Lamé's constants, λ, G, with their imaginary (out-of-phase) components, λ', G'. Then, the equations of motion of the linear viscoelastic medium are, in cylindrical co-ordinates,

$$\left.\begin{aligned}
&[(\lambda+2G)+\mathrm{i}(\lambda'+2G')]\frac{\partial\Delta}{\partial r}-\frac{2(G+\mathrm{i}G')}{r}\frac{\partial\omega_z}{\partial\theta}\\
&\quad+2(G+\mathrm{i}G')\frac{\partial\omega_\theta}{\partial z}=\rho\frac{\partial^2 u}{\partial t^2}\\
&[(\lambda+2G)+\mathrm{i}(\lambda'+2G')]\frac{\partial\Delta}{r\partial\theta}-2(G+\mathrm{i}G')\frac{\partial\omega_r}{\partial z}\\
&\quad+2(G+\mathrm{i}G')\frac{\partial\omega_z}{\partial r}=\rho\frac{\partial^2 v}{\partial t^2}\\
&[(\lambda+2G)+\mathrm{i}(\lambda'+2G')]\frac{\partial\Delta}{\partial z}-\frac{2(G+\mathrm{i}G')}{r}\frac{\partial}{\partial r}(r\omega_\theta)\\
&\quad+\frac{2(G+\mathrm{i}G')}{r}\frac{\partial\omega_r}{\partial\theta}=\rho\frac{\partial^2 w}{\partial t^2}
\end{aligned}\right\}\quad(1)$$

in which u, v, w are displacements, Δ is relative volume change, $\omega_{r,z,\theta}$ are components of rotational vector, ρ is mass density of soil, t is time and i is imaginary unit.

4. When a pile vibrates harmonically with frequency ω and amplitude $u(z)=u$ (Fig. 1), it meets soil resistance $p(z)\exp(\mathrm{i}\omega t)$ which must satisfy both the equations of the medium (equations (1)) and the governing equation of the pile motion

$$E_\mathrm{p}I\frac{\partial^4}{\partial z^4}(u\,\mathrm{e}^{\mathrm{i}\omega t})+\mu\frac{\partial^2}{\partial t^2}(u\,\mathrm{e}^{\mathrm{i}\omega t})=-p(z)\mathrm{e}^{\mathrm{i}\omega t}\quad(2)$$

in which E_pI is the bending stiffness of the pile and μ is the mass of the pile per unit length. (Pile damping can also be introduced in equation (2) but it is small compared with the damping of the medium.)

5. Equations (1) and (2) constitute the soil–pile interaction problem for the basic case of a linear homogeneous medium and a vertical pile perfectly bonded to the medium. Even for this rather idealized situation, it is difficult to satisfy all the boundary conditions at the soil surface, the pile surface and the surface of the underlying bedrock. To facilitate the solution, Kobori et al.[3] assumed an infinitely long pile (infinitely deep layer) and derived a solution for horizontal vibration of the pile; their assumption is well justified for long slender piles exposed to horizontal loads but is less adequate for the description of the vertical response.

6. Tajimi[14] suggested that the problem could be simplified by neglecting the vertical component of the motion when solving horizontal response. This idea was exploited by Nogami and the writer[6] who developed an efficient solution to soil–pile interaction for both the horizontal direction[9] and the vertical direction.[5] An example of the results is shown in Fig. 2 in which the horizontal impedance function (complex stiffness) of a point bearing pile is plotted. In Fig. 2, l is the length of the pile, r_0 is the pile radius, $V_s = (G/\rho)^{1/2}$ is the shear wave velocity of soil and D is the material damping of soil defined in this case as

$$D = \frac{G'}{G} = \frac{\lambda'}{\lambda} = \tan\delta \tag{3}$$

where δ is the loss angle.

7. The impedance functions are useful in that they can be introduced into the analysis of the superstructure to represent the pile as well as the soil. Their real part stands for true (in-phase) stiffness while the imaginary part describes the damping of the soil–pile system. A number of observations can be drawn from Fig. 2. If soil material damping is neglected ($D = 0$), the real stiffness exhibits sharp minima at the natural frequencies of the layer and can vanish at the first resonance; the inclusion of soil material damping eliminates these minima and brings the dynamic stiffness close to static stiffness ($a_0 = \omega = 0$) in the lower frequency range. With increasing frequency, soil material damping reduces the stiffness of the pile.

8. The imaginary part of the impedance function grows with frequency in an almost linear manner in the range of higher frequencies. This indicates that in that region the pile damping resembles viscous damping whose major part derives from energy radiation (geometric damping) and a smaller almost constant part stems from soil material damping. In the region of the first layer resonance and below, the role of soil material damping is much more prominent (Fig. 3). Below the first layer resonance no pile damping is generated in the absence of material damping which implies that no progressive wave occurs. Practically all the damping generated below the first resonance is due to material damping of soil and in the case of hysteretic damping is almost constant. If material damping is assumed to be viscous (i.e., proportional to vibration velocity (frequency)), the damping diminishes with decreasing frequency. However, this situation is reversed for the constant of the equivalent viscous damping of the soil–pile system which is often used and defined as $c = \text{Im}\ K/\omega$ (Fig. 3).

9. The continuum approach makes it possible to conduct an extensive yet inexpensive study of all the governing parameters such as stiffness ratio, slenderness ratio, Poisson's ratio, ν etc. It can also be used to evaluate the local soil stiffness which even in a homogeneous medium changes with depth, z, and dimensionless frequency, $a_0 = r_0\omega/V_s$. The local soil stiffness constant relates the local soil resistance per unit of pile length, $p(z)$, with the displacement of the pile. For vertical vibration of the pile, $w(z)$, this local stiffness constant of soil $K(z)$ can be defined as

$$K(z) = \frac{p(z)}{w(z)} = G\,\bar{K}(z) \tag{4}$$

in which $\bar{K}(z)$ is the dimensionless local stiffness of soil. An example of the variation of the local soil stiffness with relative depth, z/H, and dimensionless frequency, a_0, is shown

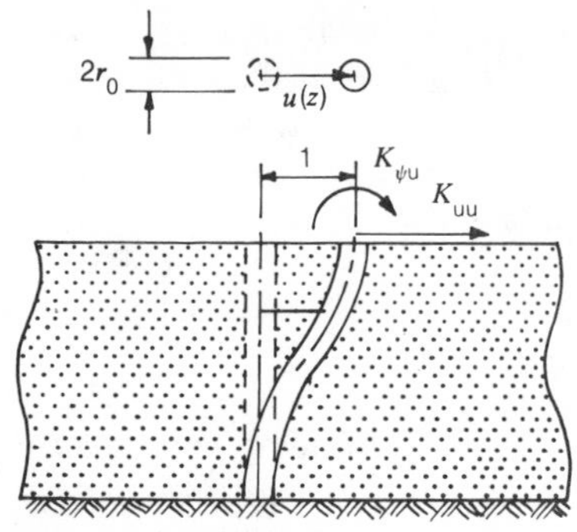

Fig. 1. Soil–pile interaction

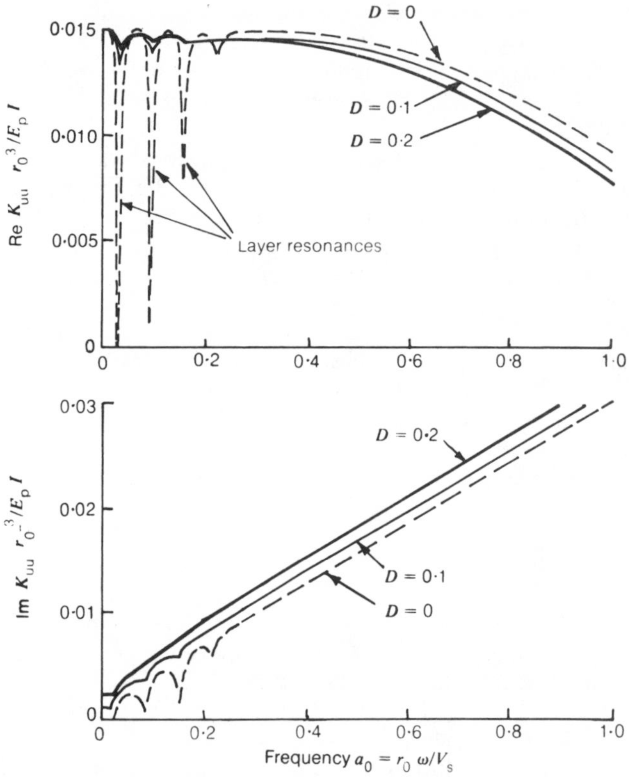

Fig. 2. Real and imaginary parts of impedance function of embedded pile; $l/r_0 = 50$

in Fig. 4. For comparison, soil stiffness evaluated for an infinitely long rigid pile is plotted in dashed lines and denoted as plane strain. The local real stiffness increases towards the surface and approaches the plane strain case as the frequency increases. The imaginary part (damping) does not vary significantly with depth and also approaches the plane strain case with increasing frequency. At very low frequencies, the plane strain case overestimates the damping. In very stiff soils or with very short piles, the variations of local stiffness with depth and frequency may be much more dramatic. (The stiffness ratio may be defined as $\bar{v} = V_s/v_p$ where v_p = longitudinal wave velocity in the pile.)

10. The growth of the local stiffness of the homogenous medium towards the surface, as observed in Fig. 4, can hardly be expected to materialize in real soils. In most soils, the local stiffness actually decreases towards the surface due to the decrease in confining pressure. In addition, soil properties often change very markedly with depth due to soil layering. Further complications can be caused by pile separation from the soil close to the surface (mud line) and imperfect fixity of the pile tip. These factors were found to be very important in experiments even at small strains but cannot be incorporated into any rigorous continuum theory.[10] If they are to be accounted for, simplifying assumptions must be adopted.

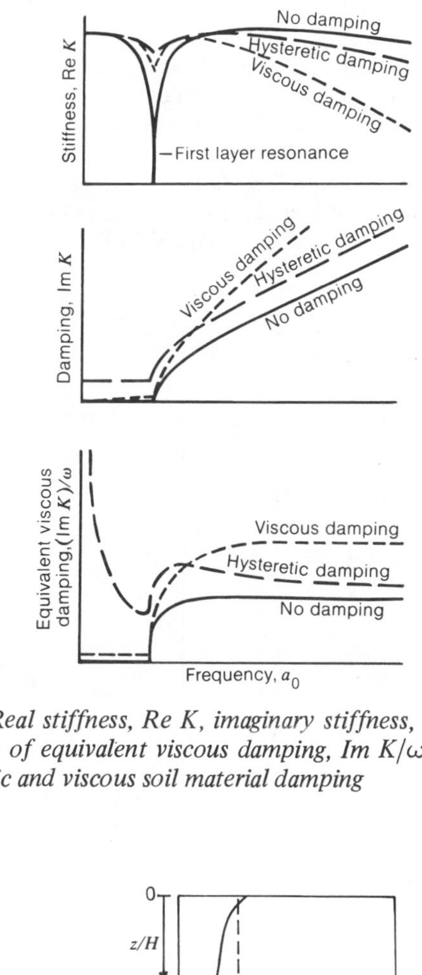

Fig. 3. Real stiffness, Re K, imaginary stiffness, Im K, and constant of equivalent viscous damping, Im K/ω, for both hysteretic and viscous soil material damping

APPROXIMATE CONTINUUM APPROACH

11. An approximate but very versatile theory is based on the following main assumptions. The soil is composed of horizontal layers that are homogeneous, isotropic and linearly viscoelastic with material damping of the frequency-independent hysteretic type. The soil properties can be different in individual layers. However, the soil reactions acting in each layer are taken as equal to those of an infinitely long rigid pile undergoing uniform harmonic vibration in an infinite medium whose properties are equal to those of that layer. Such reactions are derived from equations (1) under the conditions of plane strain, which actually means that only horizontally propagating waves are accounted for. This assumption is not unreasonable because in a layer elastic waves tend to propagate primarily along the interfaces, particularly at higher frequencies. Fig. 4 supports this point. Then, the soil reaction per unit length of the pile is, for horizontal vibration,

$$k_{us} = G\,[S_{u1}(a_0,\nu,D) + \mathrm{i}S_{u2}(a_0,\nu,D)] \qquad (5)$$

Dimensionless stiffness and damping parameters S_{u1}, S_{u2} are given in the literature,[23] together with analogous expressions for all other vibration modes. The soil reaction at depth z, $p(z) = k_{us}u$, is independent of the displacements at other stations like in Winkler's medium. This is, of course, not quite true because cross stiffnesses between individual stations also exist. However, their effect on the numerical results is not great, particularly in deep deposits modelled by a half-space in which layer resonances do not occur. This can be anticipated on the basis of Mindlin's static solution,[24]

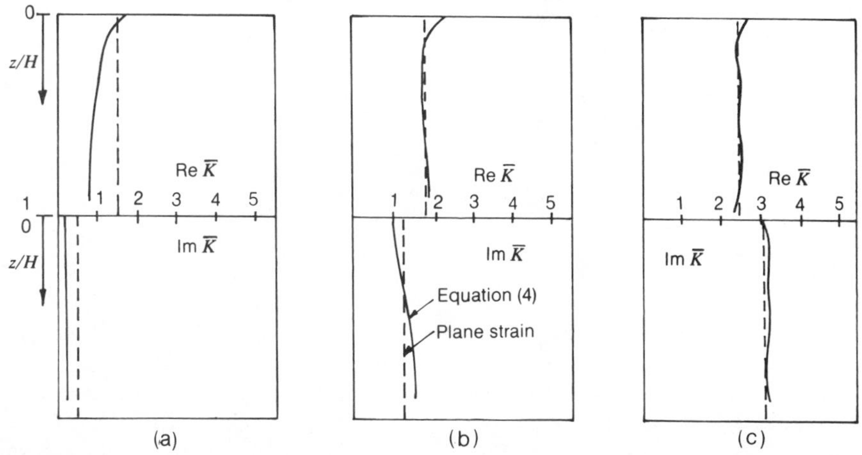

Fig. 4. Variation of local soil stiffness in vertical direction with depth and frequency;[5] $H/r_0 = 100$, $\bar{v} = 0.01$, $\nu = 0.4$, $D = 0.02$: (a) $a_0 = 0.04$, (b) $a_0 = 0.1$, (c) $a_0 = 0.4$

from which soil flexibility and all stiffness constants can be established. An example of static soil displacements calculated using Mindlin's theory for both concentrated and distributed loads is shown in Fig. 5. The displacements decay very quickly with distance from the loading zone, especially for the concentrated load. This indicates that the local (diagonal) stiffness predominates. This assumption has already been adopted by Penzien,[2] and its justification was examined in detail in a similar study involving seismic response of buried pipelines.[25] It was found that the effect of off-diagonal stiffness and damping terms was for all practical purposes negligible. Thus, it may be expected that soil stiffness and damping, defined by equation (5), can adequately describe the soil reaction and can be introduced into the equation of motion of the pile.

12. The pile is assumed to be vertical, linearly visco-elastic with hysteretic material damping and of constant cross-section that may be different in individual layers. The pile is bonded to the soil; if its upper part is separated from the soil or sticks out of the ground the adjacent soil layer is considered as void. The different types of pile and soil profile that can be analysed are shown in Fig. 6. Rigid bodies underlain by layered media can be treated as piles if the column of soil underneath the footing is viewed as a pile of soil (Fig. 6). Such a column is not slender and, therefore, the effects of shear and rotatory inertia have to be accounted for. (This treatment of rigid embedded bodies is similar to that of Tajimi and Shimomura.[19]) For rather rigid foundations and the soil column, the moment soil reactions and hysteretic material damping have to be included also. For slender piles, these factors are not important but the effect of the static load, $N_{st}(>0)$, should be considered. It is advantageous to formulate a unified solution for slender piles and rigid bodies. This can be done by including all the factors mentioned in one differential equation which holds for the elements of the slender pile, rigid body (this means a massive body not necessarily completely rigid) and the soil column beneath the body. With harmonic lateral vibration, $u(z, t) = u(z)\exp(i\omega t)$, the equation of the complex vibration amplitude is

$$\frac{d^4 u}{dz^4} + \beta_1 \frac{d^2 u}{dz^2} - \beta_2 u = 0 \tag{6}$$

in which

$$\beta_1 = \frac{1}{E_p I}\left[r^2 \mu\omega^2 - G r_0^2 S_\psi + \frac{\kappa E_p I}{G_p A}(\mu\omega^2 - G S_u) \right.$$
$$\left. + N_{st}\left(1 + \frac{\kappa N_{st}}{G_p A}\right) \right]$$
$$\left.\begin{array}{c}\\\\\\\\\end{array}\right\} \tag{7}$$
$$\beta_2 = \frac{1}{E_p I}(\mu\omega^2 - G S_u)\left[1 - \frac{\kappa}{G_p A}(r^2 \mu\omega^2 \right.$$
$$\left. - G r_0^2 S_\psi - N_{st})\right]$$

In equations (7), $r = (I/A)^{1/2}$ is the radius of gyration of the cross-section; κ is the constant associated with shear; A, I and μ are the cross-sectional area, moment of inertia and mass per unit length of the body, respectively; and parameters $S_u = S_{u1} + iS_{u2}$, $S_\psi = S_{\psi 1} + iS_{\psi 2}$ are given in the literature.[23] Finally, E_p and G_p are the complex moduli of the pile.

13. For the horizontal displacements, u, and rotations, ψ, of the ends of an element whose length is h (Fig. 6), the solution to equation (6) yields the element stiffness matrix

$$[k_u] = \frac{E_p I}{h} \begin{array}{cccc} u_1 & \psi_1 & u_2 & \psi_2 \\ \left[\begin{array}{cccc} F_6/h^2 & F_4/h & F_5/h^2 & -F_3/h \\ F_4/h & F_2 & F_3/h & F_1 \\ F_5/h^2 & F_3/h & F_6/h^2 & -F_4/h \\ -F_3/h & F_1 & -F_4/h & F_2 \end{array}\right] \end{array} \tag{8}$$

in which F_i are dimensionless frequency functions.[13] These are mathematically accurate and incorporate all the properties of the pile and soil, including mass and damping. Consequently, the dynamic stiffness matrix, equation (8), is complex and frequency-dependent but a separate formulation of mass and damping matrices is not needed.

14. Using the element stiffness matrix, the overall (structure) stiffness matrix, $[K_u]$, is assembled for the whole embedded pile composed of elements separated by the interfaces between the layers. Then, the impedance functions associated with unit displacement amplitudes $u_1 = 1$ and $\psi_1 = 1$ follow from the relationship

$$\begin{array}{cc} 1 & 2 \end{array}$$
$$\begin{bmatrix} K_{uu} & \vdots & K_{u\psi} \\ K_{\psi u} & \vdots & K_{\psi\psi} \\ 0 & \vdots & 0 \\ 0 & \vdots & 0 \\ \cdot & \vdots & \cdot \\ \cdot & \vdots & \cdot \\ \cdot & \vdots & \cdot \\ 0 & \vdots & 0 \end{bmatrix} = K_u \begin{array}{cc} 1 & 2 \end{array}\begin{bmatrix} u_1 = 1 & \vdots & u_1 = 0 \\ \psi_1 = 0 & \vdots & \psi_1 = 1 \\ u_2 & \vdots & u_2 \\ \psi_2 & \vdots & \psi_2 \\ \cdot & \vdots & \cdot \\ \cdot & \vdots & \cdot \\ \cdot & \vdots & \cdot \\ \psi_{n+1} & \vdots & \psi_{n+1} \end{bmatrix} \tag{9}$$

From these equations, all the other displacements of the pile, u_2, ψ_2 etc., are also obtained. An example is shown in Fig. 7. The other impedance functions are calculated analogously.

15. The complex impedance functions can be rewritten as

$$K = k + i\omega c \tag{10}$$

in which k is the real stiffness and c is the coefficient of equivalent viscous damping.

16. To facilitate the calculation of the stiffness and damping constants of the pile embedded in different layered media, an efficient computer program (PILAY) was written and is distributed by Systems Analysis, Control and Design Activity, Faculty of Engineering Science, The University of Western Ontario. The computing costs are negligible. The program calculates the impedance functions for all the cases indicated in Fig. 6 and all vibration modes. An extensive parametric study and more details are given in the literature.[11-13]

17. The theory is mathematically accurate but is

approximate because of the definition of the soil reactions. Therefore, the theory was compared with other solutions and experiments.

18. Comparison with other theories is possible for the basic case of an end-bearing pile embedded in a homogeneous layer. For this case, Kuhlemeyer[17] presented a finite element solution which compares favourably with the approximate analytical solution for one layer[7] as can be seen from Fig. 8 ($f_{1,2}$ stands for dimensionless stiffness and damping in rotation, ψ, translation, u, and the cross term, c.) Comparison with the finite element solution by Blaney et al.,[16] who use a different boundary than Kuhlemeyer, is also very good. Finally, the approach was compared with the more accurate continuum solutions by Nogami and the writer;[5,9] it appears that the two methods approach asymptotically as the frequency increases. For frequency a_0 approaching zero, the stiffness which has been calculated approximately diminishes because the plane strain soil reactions for translations approach zero. This drawback can be rectified, however, by adjusting the soil stiffness at very low frequencies to a constant static value as indicated in Fig. 9. Such an adjustment was built into the computer program PILAY which, therefore, gives realistic values of stiffness even for frequencies approaching zero. This can be seen from Fig. 10, in which the true stiffness computed using the Author's theory is compared with static stiffness calculated by Kotoda and Kazama,[26] whose theory is based on Mindlin's solution. The stiffness shown was calculated for a large reinforced concrete test pile of diameter 2 m.

19. Comparison of the theory with small-scale experiments conducted with small strains was also very favourable.[11, 12] It showed that the dominant factors to be accounted for are the variation of soil properties with depth and particularly the reduction of soil stiffness toward the surface, pile separation at the surface and, for vertical response, imperfect fixity of the pile tip. Further important factors are non-linearity, batter and interaction of piles in a group.

Non-linearity

20. It has been recognized for many years that at large strains, pile behaviour exhibits strong non-linearity. However, its incorporation into any approach poses serious problems.

21. In the lumped mass approach, the soil reactions are modelled by discrete springs and dashpots which can be defined as non-linear and the system can be solved using numerical techniques. This was done by Penzien,[2] Matlock et al.[1] and others. When experimental results are available

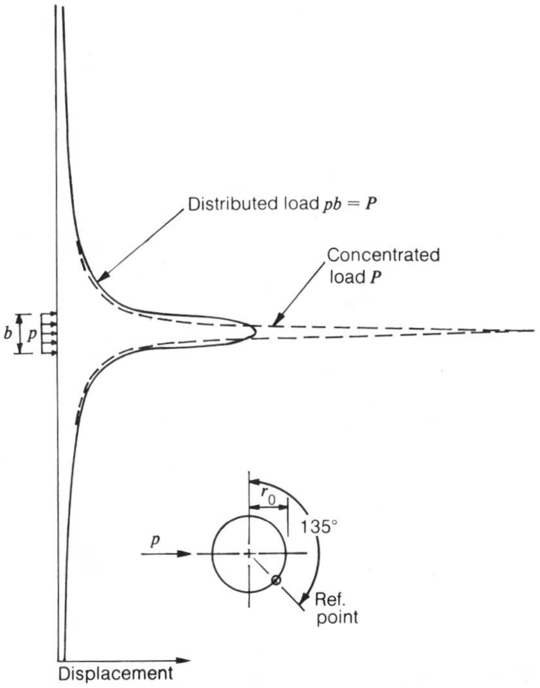

Fig. 5. Static soil deflexions due to both concentrated and distributed loads; $\nu = 0.25$, $b/r_0 = 10$

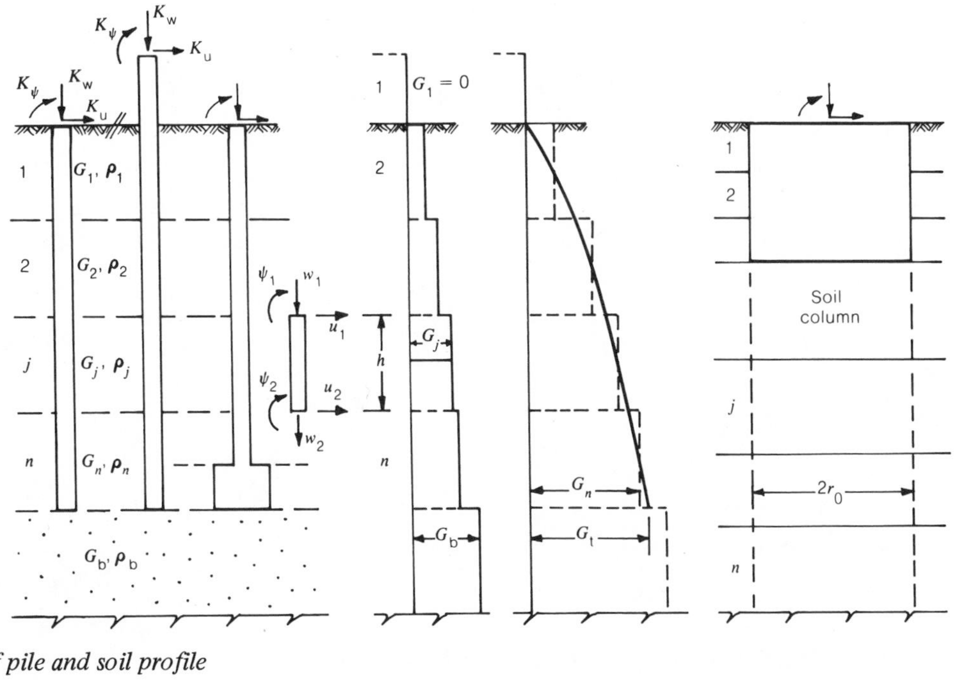

Fig. 6. Types of pile and soil profile

beforehand, it is possible to select non-linear laws for the discrete stiffness and damping elements such that satisfactory agreement between the theory and experiments can be achieved.[1] However, a true prediction of these non-linear laws and the pile response on the basis of given soil properties is very difficult because the stresses in the soil vary with the co-ordinates, r, θ, z. When the soil properties are strain-dependent and hence variable in the co-ordinates, the problem is how to choose the discrete springs and dash-pots to represent the whole stress field in which each station actually has different soil properties. To illustrate this point, Fig. 11 shows the variation of vertical shear stress with distance, r, for the basic case of a rigid, infinitely long pile vibrating vertically with a unit amplitude. This stress follows from the expression[23]

$$\tau_{rz} = -G(1+iD_s)s\frac{K_1(sr)}{K_0(sr_0)} \tag{11}$$

in which

$$s = \frac{\omega}{V_s(1+iD_s)^{1/2}}\, i \tag{12}$$

and K_0, K_1 are the modified Bessel functions. For horizontal response, the variations of the stress are also dependent on the angular co-ordinate θ and are different for shear stress and normal stress.

22. In the continuum approach, non-linearity cannot be incorporated into the dynamic solution of soil–pile interaction with any degree of rigour as yet. The only way

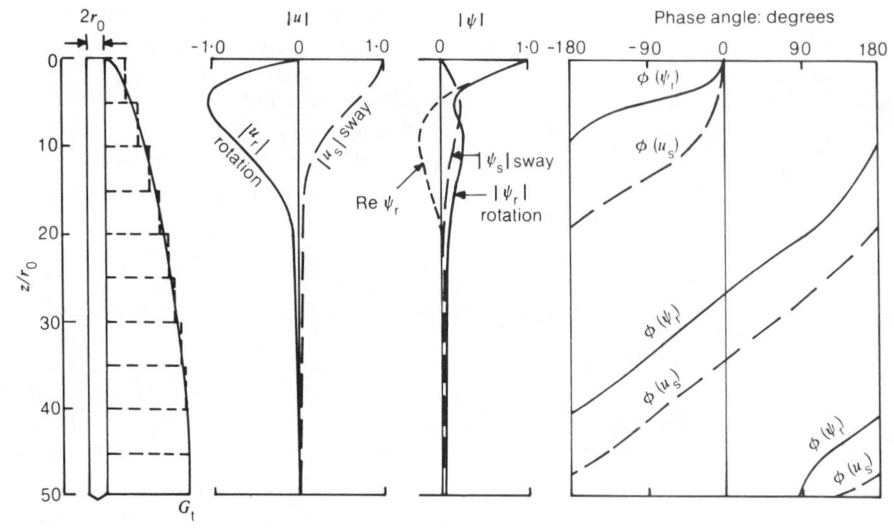

Fig. 7. Displacements, rotations and phase shifts due to unit displacements of pile head; $(G_t/E_p)^{1/2} = 0.03$, $l/r_0 = 50$, $\tan\delta = 0.1$, $\nu = 0.3$

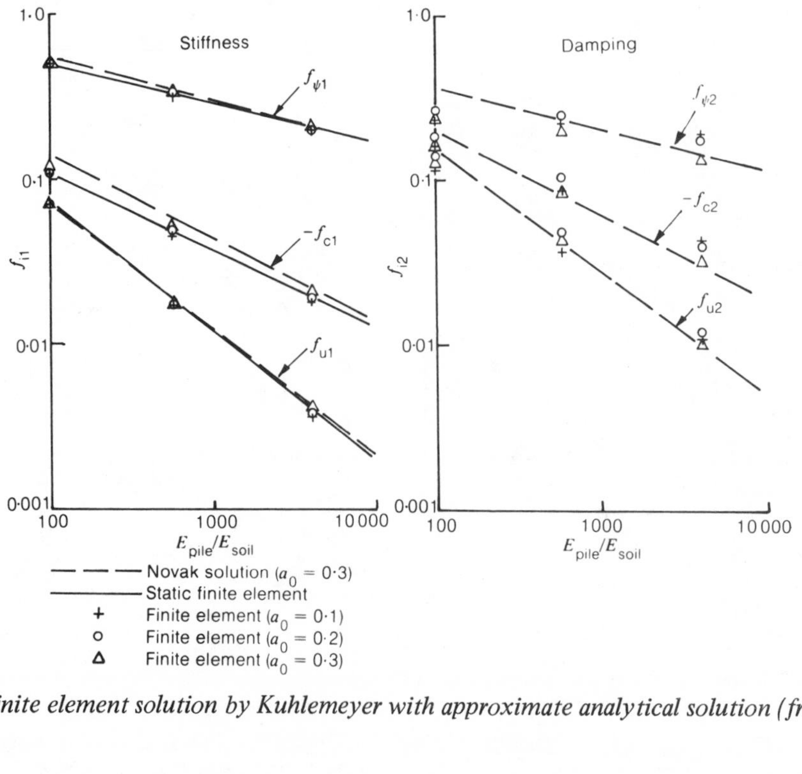

Fig. 8. Comparison of finite element solution by Kuhlemeyer with approximate analytical solution (from Kuhlemeyer[17])

to account for it is to adjust the soil stiffness and damping according to the level of stress in a few iterative steps. However, this can only be done very approximately as already discussed in relation to Fig. 11. Some guidelines for improving this adjustment are being developed.

23. It appears that only the finite element method could treat soil non-linearity according to the local state of stress. However, it is known that such dynamic analysis is costly and inaccurate even under plane strain conditions; therefore, the non-linear solution is usually replaced by an iterative adjustment of the linear soil properties in each element to its level of strain. In the three-dimensional case of piles, the inclusion of non-linearity would eliminate the efficient axisymmetrical elements. Thus, a general three-dimensional dynamic solution would be needed. Such a solution is also required if pile separation is to be accounted for. A general finite element solution of this kind has not been presented thus far.

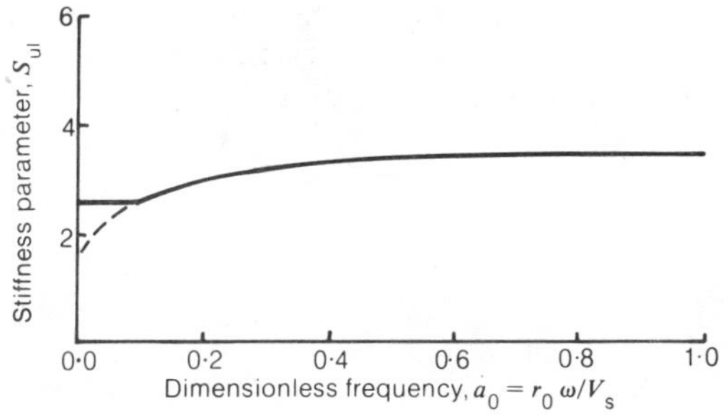

Fig. 9. Adjustment of plane strain soil reaction for frequencies approaching zero; ν = 0.25, D = 0.1

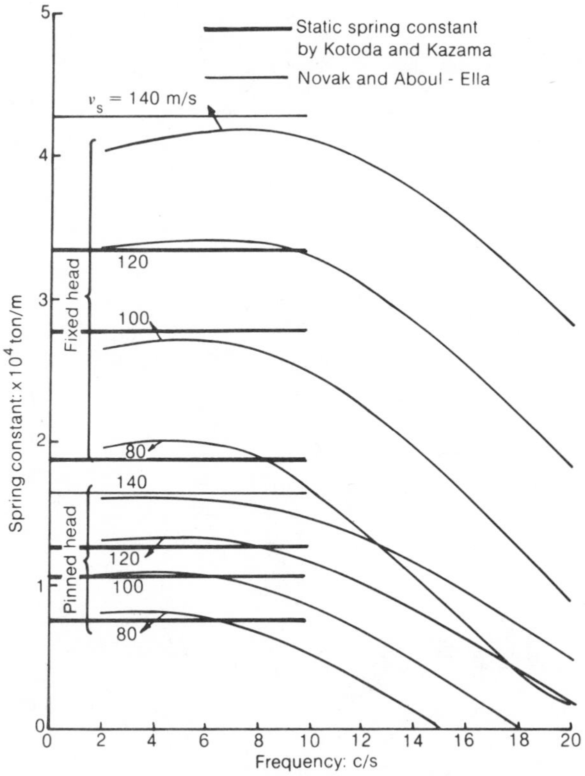

Fig. 10. Comparison of static stiffness by Kotoda and Kazama with Author's theory (from Kazama[26])

Battered piles

24. There is no analytical solution for impedance functions of battered piles but they can be established approximately using an important observation made by Poulos and Madhav[27]. These writers theoretically investigated static displacements of battered piles using Mindlin's equations. They applied axial loads (in line with the longitudinal axis of the battered pile), normal loads (perpendicular to the longitudinal axis) and moments. They found that both the axial and normal displacements are almost unaffected by the batter of the pile. For a batter of 30°, the maximum effect of the batter is about 4%. This finding suggests that the impedance functions of battered piles may be calculated by means of the available dynamic solutions for vertical piles. Two cases may occur.

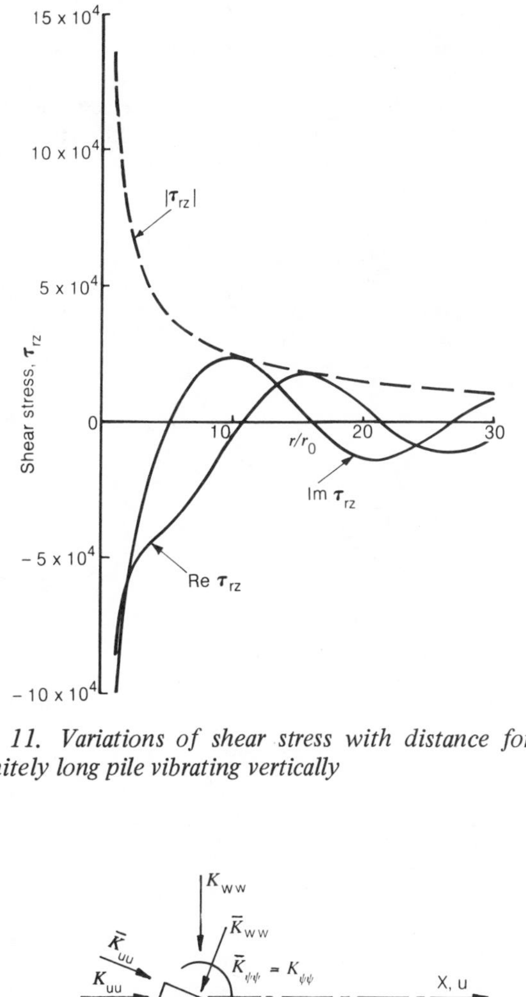

Fig. 11. Variations of shear stress with distance for an infinitely long pile vibrating vertically

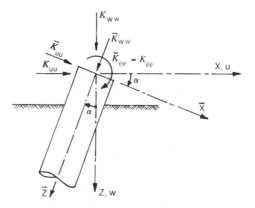

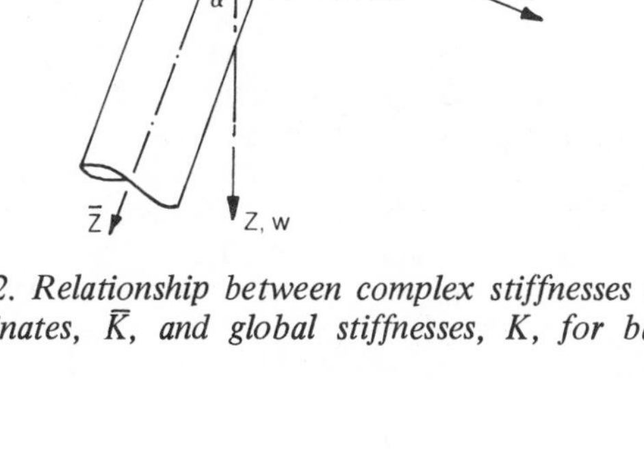

Fig. 12. Relationship between complex stiffnesses in pile co-ordinates, $\bar{K}$, and global stiffnesses, K, for battered pile.

25. When the impedance functions are requested for the directions coincidental with the principal axes of the pile, the complex stiffnesses obtained for the vertical pile are approximately valid for the corresponding stiffnesses of the battered pile; i.e., the vertical stiffness is taken for axial stiffness, the horizontal stiffness for lateral stiffness, etc.

26. When the impedance functions are requested for the global co-ordinate system characterized most often by the vertical and horizontal axes, the impedance functions can be calculated first as for a vertical pile and then transformed into the global co-ordinate system. This transformation is a standard operation by which the element stiffness matrix in element co-ordinates, $[\bar{K}]$, is transformed into the element stiffness matrix in global co-ordinates, $[K]$, by the relationship

$$[K] = [T]^{T}[\bar{K}][T] \tag{13}$$

in which the transformation matrix, $[T]$, depends only on direction cosines.

27. When the horizontal co-ordinate axis lies in the plane of the batter (Fig. 12), the transformation matrix is

$$[T] = \begin{bmatrix} \cos\alpha & \sin\alpha & 0 \\ -\sin\alpha & \cos\alpha & 0 \\ 0 & 0 & 1 \end{bmatrix} \tag{14}$$

and equation (13) becomes

$$\begin{bmatrix} K_{uu} & K_{uw} & K_{u\psi} \\ K_{wu} & K_{ww} & K_{w\psi} \\ K_{\psi u} & K_{\psi w} & K_{\psi\psi} \end{bmatrix} =$$

$$\begin{bmatrix} \cos^2\alpha\,\bar{K}_{uu} + \sin^2\alpha\,\bar{K}_{ww} & \cos\alpha\sin\alpha\,(\bar{K}_{uu} - \bar{K}_{ww}) & \cos\alpha\,\bar{K}_{u\psi} \\ \cos\alpha\sin\alpha\,(\bar{K}_{uu} - \bar{K}_{ww}) & \sin^2\alpha\,\bar{K}_{uu} + \cos^2\alpha\,\bar{K}_{ww} & \sin\alpha\,\bar{K}_{u\psi} \\ \cos\alpha\,\bar{K}_{\psi u} & \sin\alpha\,\bar{K}_{\psi u} & \bar{K}_{\psi\psi} \end{bmatrix} \tag{15}$$

28. The element impedance functions, $\bar{K}$, are calculated assuming that the pile is vertical; thus, $\bar{K}_{uw} = \bar{K}_{wu} = \bar{K}_{\psi w} = \bar{K}_{w\psi} = 0$ and $\bar{K}_{u\psi} = \bar{K}_{\psi u}$.

Pile interaction

29. Piles in a group interact with each other because the displacement of one pile contributes to the displacements of the others. Consequently, the stiffness of the pile group is smaller than a simple sum of stiffnesses of individual piles unless the piles are widely spaced. Under static conditions, these effects were studied by Poulos,[28, 29] who presented charts of interaction coefficients. In the absence of similar readily usable data for dynamic response, such static values may be used to approximately evaluate the group effect on dynamic stiffness and damping. The application of static interaction coefficients to dynamic problems seems possible because the dimensionless frequency, a_0, is usually small for piles and the imaginary part of the impedance function is a complement to its real part.

30. Dynamic analysis of pile groups has been started only recently. Wolf and von Arx[20] presented a finite element solution for a group of piles and studied the effect of pile interaction. Fig. 13 shows their results for horizontal stiffness of a group of four piles, k^g_{uu}, in relation to the sum of stiffnesses of isolated (single) piles, Σk^s_{uu}, as a function of pile separation, $d/2r_0$. It can be seen that horizontal stiffness of the group is reduced considerably by pile interaction for separations of up to about 30 pile diameters. Also plotted in Fig. 13 are the results calculated using the static interaction coefficients taken from Poulos[28]. For the case shown, the two approaches agree quite well.

31. The stiffness of a rigid pile cap can be evaluated in terms of stiffness of a single pile, k^s. The flexibility of the group of piles without a cap referred to pile heads may be written as $F^s[\alpha]$ in which $F^s = 1/k^s$ is the flexibility of a single pile and $[\alpha]$ is the dimensionless flexibility matrix. Applying unit displacements of a rigid cap to the whole group the stiffness of the cap, k^g, is for vertical and horizontal directions

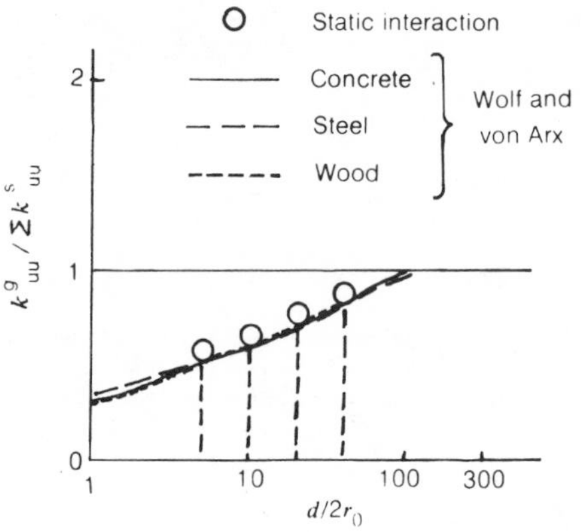

Fig. 13. Effect of pile interaction on horizontal stiffness of group of four piles according to Wolf and von Arx[20] and Poulos[28]

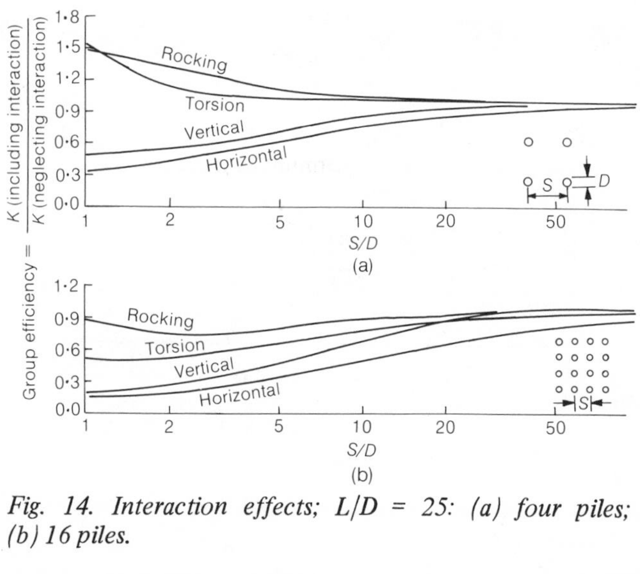

Fig. 14. Interaction effects; L/D = 25: (a) four piles; (b) 16 piles.

$$k^g = k^s \sum_i \sum_j \alpha_{ij}^{-1}$$

in which α_{ij}^{-1} are the elements of the matrix $[\alpha]^{-1}$. Rocking stiffness of the cap derived from the vertical stiffness of single piles, k_{ww}^s, becomes

$$k_{\psi\psi}^g = k_{ww}^s \sum_i \sum_j \alpha_{ij}^{-1} x_j x_i$$

in which x is the horizontal distance of the pile from the axis of rotation. A similar formula can be found for torsion. In the absence of accurate values of the flexibility coefficients, α_{ij}, the interaction coefficients due to Poulos[28,29] or others can be used instead in matrix $[\alpha]$. Fig. 14 shows the interaction effects in various vibration modes for groups four and 16 piles as a function of pile separation S/D. Interaction always reduces the vertical and horizontal stiffnesses but can increase the rocking and torsional stiffnesses of small groups.

32. With flexible caps, the finite element method can be used to model the cap. The impedance functions of the piles are incorporated in the equilibrium equations of the cap quite readily if the finite element mesh is arranged so that the nodal points coincide with the pile heads. With large pile foundations, this approach can, however, lead to a very large number of elements. The number of finite elements can be significantly reduced and made independent of the number of piles if the cap element stiffness matrix is modified so that the piles are allowed to occur anywhere within the element. In such a case, the impedance functions of the piles can be incorporated into the stiffness matrix of the cap element using the same displacement functions as those used for the cap alone. Then, the cap element stiffness matrix, $[K]$, becomes complex and frequency-dependent and can be written in the form

$$[K] = [k] - [M]\,\omega^2 + \sum_i [N]_i^T [A]_i [N]_i \qquad (16)$$

in which $[k]$ is the static stiffness matrix, $[M]$ is the mass matrix, $[N]$ is the matrix of shape functions and $[A]$ is the matrix of pile impedance functions. The summation is taken over all piles lying within the cap element. This approach is described in more detail elsewhere,[30] with examples.

CONCLUSIONS

33. Two theories are described that make it possible to calculate impedance functions of piles. Both theories are based on soil reactions formulated from the constitutive equations of the viscoelastic medium.

34. The more approximate approach makes it possible to consider soil layering, piles with variable cross-sections and pedestals and any degree of fixity of the pile tip; also, the pile can stick out of the ground and pile separation can be accounted for approximately. This versatility considerably improves the agreement between the theory and experiments.

ACKNOWLEDGEMENTS

35. The study was supported by a research grant from the National Research Council of Canada. The continuous assistance of T. Nogami, F. Aboul-Ella and M. Sheta is acknowledged.

REFERENCES

1. MATLOCK H. et al. Simulation of lateral pile behaviour under earthquake motion. Proc Am. Soc. Civ. Engrs Specialty Conference on Earthquake Engineering and Soil Dynamics, Pasadena, California, 1978, II, 704–719.
2. PENZIEN J. (WIEGEL R. L. ed). Soil–pile foundation interaction. In Earthquake engineering. Prentice-Hall, Englewood Cliffs, NJ, 1970, ch. 14, 349–381.
3. KOBORI T. et al. Dynamic behavior of a laterally loaded pile. Proc. 9th Int. Conf. Soil Mech., Tokyo, 1977, specialty session 10, 6.
4. KOBORI T. et al. Torsional vibration characteristics of the interaction between a cylindrical elastic rod and its surrounding viscoelastic stratum. Proc. 24th Japan National Congress for Applied Mechanics. University of Tokyo Press, 1974, 24, 197–205.
5. NOGAMI T. and NOVAK M. Soil–pile interaction in vertical vibration. Int. J. Earthq. Engng Struct. Dynamics, 1976, 4, Jan.–Mar., No. 3, 277–293.
6. NOGAMI T. and NOVAK M. Resistance of soil to a horizontally vibrating pile. Int. J. Earthq. Engng Struct. Dynamics, 1977, 5, July–Sept., No. 3, 249–261.
7. NOVAK M. Dynamic stiffness and damping of piles. Canadian Geotech. J., 1974, 11, No. 4, 574–598.
8. NOVAK M. Vertical vibration of floating piles. J. Engng Mech. Div. Am. Soc. Civ. Engrs, 1977, 103, Feb., No. EM1, 153–168, proc. paper 12747.
9. NOVAK M. and NOGAMI T. Soil pile interaction in horizontal vibration. Int. J. Earthq. Engng Struct. Dynamics, 1977, 5, July–Sept., No. 3, 263–282.
10. NOVAK M. and GRIGG R. F. Dynamic experiments with small pile foundations. Canadian Geotech. J., 1976, XIII, Nov.–Dec., No. 4, 372–385.
11. NOVAK M. and HOWELL J. F. Dynamic response of pile foundations in torsion. J. Geotech. Engng Div. Am. Soc. Civ. Engrs, 1978, 104, May, No. GT5, 535–552, proc. paper 13767.
12. NOVAK M. and ABOUL-ELLA F. Impedance functions of piles in layered media. J. Engng Mech. Div. Am. Soc. Civ. Engrs, 1978, 104, June, EM3, 643–661.
13. NOVAK M. and ABOUL-ELLA F. Stiffness and damping of piles in layered media. Proc. Am. Soc. Civ. Engrs Specialty Conference on Earthquake Engineering and Soil Dynamics, Pasadena, California, 1978, II, 704–719.
14. TAJIMI H. Earthquake response of foundation structures. Fac. Sci. Engng, Nihon University, 1966. (In Japanese).
15. TAJIMI H. Dynamic analysis of a structure embedded in an elastic stratum. Proc. 4th Wld Conf. Earthq. Engng, Chile, 1969.
16. BLANEY G. W. et al. Dynamic stiffness of piles. Proc. 2nd. Int. Conf. Numerical Methods in Geomechanics. American Society of Civil Engineers, New York, 1976, 1001–1012.
17. KUHLEMEYER R. L. Static and dynamic laterally loaded piles. Department of Civil Engineering, University of Calgary, Calgary, Alta, Canada, 1976, res. report CE 76–9, 48.
18. SHIMIZU N. et al. Three-dimensional dynamic analysis of soil–structure systems by thin layer element method (Part 2). Trans. Archit. Inst. Japan, 1977, Apr., No. 254, 39–47.

19. TAJIMI H. and SHIMOMURA Y. Dynamic analysis of soil–structure interaction by the thin layered element method. Trans. Archit. Inst. Japan, 1976, May, No. 243, 41–51. (In Japanese.)

20. WOLF J. P. and von ARX G. A. Impedance functions of a group of vertical piles. Proc. Am. Soc. Civ. Engrs Specialty Conference on Earthquake Engineering and Soil Dynamics, Pasadena, California, 1978, II, 1024–1041.

21. TAJIMI H. Seismic effects on piles. Proc 9th Int. Conf. Soil Mech., Tokyo, 1977. State-of-the art report, specialty session 10, 12.

22. PENZIEN J. Structural dynamics of fixed offshore structures. Proc. BOSS '77 on Behaviour of Offshore Structures. Norwegian Institute of Technology, Oslo, 1977, 12.

23. NOVAK M. et al. Dynamic soil reactions for plane strain case. J. Engng Mech. Div. Am. Soc. Civ. Engrs, 1978, 104, Aug., No. EM4, 953–959.

24. MINDLIN R. D. Displacements and stresses due to nuclei of strain in the elastic half-space. Department of Civil Engineering and Engineering Mechanics, Columbia University, New York, NY, 1964.

25. HINDY A. and NOVAK M. Earthquake response of underground pipelines. Faculty of Engineering Science, The University of Western Ontario, London, Canada, 1978. Research report GEOT-1-1978, 45.

26. KAZAMA S. Vibration behavior of the concrete pile with a large cross section. Research Report, University of Michigan, 1978.

27. POULOS H. G. and MADHAV M. R. Analysis of the movements of battered piles. Proc. 1st Australia–New Zealand Conf. Geomechanics, Melbourne, 1971, 1, 268–275.

28. POULOS H. G. Behaviour of laterally loaded piles: II—pile groups. J. Soil Mech. Fdns Div. Am. Soc. Civ. Engrs, 1971, 97, No. SM5, 733–751.

29. POULOS H. G. Analysis of the settlement of pile groups. Géotechnique, 1968, 18, 449–471.

30. ABOUL-ELLA F. and NOVAK M. Dynamic analysis of turbine-generator foundations. Proc. Int. Symp. Foundation for Equipment and Machinery. American Concrete Institute, Houston, Texas, 1978, 26.

9. Mathematical modelling of piled foundations

D. D. LIOU (Bechtel Power Corporation) and J. PENZIEN (Earthquake Engineering Research Center, University of California)

An important step in the dynamic analysis of a pile-supported offshore structure is to derive the dynamic stiffness of its piled foundations. Current approaches for analysis and design of piled foundations use the subgrade reaction or static elasticity method for determining the stiffness of the soil–pile system. Since these approaches do not consider the dynamic interaction between structure, soil and piles that occurs during dynamic excitation, the subgrade stiffness obtained is frequency-independent and cannot properly account for the radiation damping effect. To evaluate the subgrade stiffness by rigorous mathematical method is a difficult task, because of the complicated boundary conditions that must be satisfied in the theoretical formulation. The method developed in this Paper provides an alternative approach to the solution of the problem. The method involves four steps. In the first, the soil resistance to the pile movement, obtained by numerical solution of the boundary value problem, is decomposed into the product of two orthogonal functions. This step yields the frequency-dependent subgrade stiffness. The second step is the computation of the dynamic stiffnesses of a single pile, using the subgrade stiffness obtained in step 1 and the pile's inertia and damping. The third step is the determination of the pile–group interaction factor using the soil resistance functions at adjacent piles when one pile in the pile group is excited. The fourth step is computation of the dynamic stiffnesses of a pile group from results of steps 2 and 3.

INTRODUCTION

Piled foundations are used for the support of a large number of structures in areas where the soil deposits exhibit low bearing capacity and medium-to-high compressibility. The behaviour of such structures subjected to dynamic loadings cannot be realistically predicted if the complete structure–foundation system has not been modelled accurately. However, present knowledge and understanding of the dynamic behaviour of piled foundations is far from complete, although it has been the subject of considerable interest and research in recent years. Therefore, it is important to give full consideration to the problem of defining an appropriate mathematical model for piled foundations.

2. Early analytical solutions for piles have been obtained along two principal lines: using a discrete model with lumped masses, springs and dashpots;[1-3] and using a continuous model and the two-dimensional theory of elasticity.[4] With a discrete model, the non-linearity of the surrounding soil deposit can be relatively easily introduced by the specification of arbitrary force–deformation characteristics for the spring. However, one encounters the difficulty of defining equivalent soil masses and fictitious dashpots to simulate radiation damping. The continuous models previously proposed treat the pile as a flexural bar buried in elastic, isotropic and homogeneous layers. They have the advantage that they can automatically incorporate in the foundation inertia effects and radiation damping. However, in the low frequency range, which is of special interest in earthquake engineering, they cannot predict the true dynamic behaviour of the pile.[3]

3. A more recent approach is the finite element method which can readily handle non-homogeneity and non-linearity. However, when directly applied to the pile problem, the solution is usually expensive due to excessive computer usage.

4. A simple model of the foundation is developed in this Paper which easily yields a solution. This new model is based on a discrete model for the pile and it uses three-dimensional theory of elasticity for the surrounding soil. Radiation damping is included in the new model. It may prove useful as an alternative to modelling the pile foundation by three-dimensional finite element methods.

MODELLING OF A SINGLE PILE

5. The classical theories of earth pressures are not reliable for determining lateral or vertical resistance of single piles. They assume mobilization of active and passive pressures, which do not occur except at complete failure. Satisfactory methods of determining lateral or vertical resistance of single piles or groups must be applicable to small deflexions. In some cases, the governing design criterion is the permissible lateral deflexion; in other cases, it is the maximum load that the pile can take without overstress. The governing equation for the lateral deflexion of a single pile with oscillating loads applied at its top is

$$EI\frac{\partial^4 y}{\partial z^4} + m\frac{\partial^2 y}{\partial t^2} + EIC_\mathrm{h}\frac{\partial^5 y}{\partial z^4 \partial t} + E_\mathrm{h}(z, y, \frac{\partial y}{\partial t})\, y = 0 \quad (1)$$

where y is the lateral displacement of the pile, EI is the flexural rigidity of the pile, m is the mass per unit length of the pile and C_h is the damping coefficient of the pile for lateral movement; E_h is the lateral soil resistance per unit length acting at height z on the pile and is called the lateral soil modulus or lateral subgrade stiffness.

6. The governing equation of a single pile pertinent to the harmonic vertical motion of the pile head is

$$m\frac{\partial^2 w}{\partial t^2} - AE\frac{\partial^2 w}{\partial z^2} - AEC_h\frac{\partial^3 w}{\partial z^2 \partial t} + E_v(z, t)w = 0 \qquad (2)$$

in which w is the vertical displacement of the pile and A is the cross-sectional area of the pile; E_v, the counterpart of E_h, is the vertical soil resistance per unit length acting at height z on the pile and is called the vertical soil modulus or vertical subgrade stiffness.

7. After determining the lateral and vertical soil moduli, which define the soil resistance forces to the pile movements, one must solve equations (1) and (2) in order to evaluate the impedance functions at the head of a single pile. The impedance functions of a pile can be defined as the transfer functions describing the ratios between the dynamic complex response displacements on the head of the pile and its surface harmonic exciting force.

8. The steady-state solution to equation (1) can be written as

$$y(z, t) = y(z)e^{i\omega t} \qquad (3)$$

where the complex amplitude is given by

$$y(z) = y_1(z) + iy_2(z) \qquad (4)$$

Substituting equation (3) into equation (1) yields the ordinary differential equation

$$EI(1 + iC_h\omega)\frac{d^4y(z)}{dz^4} + y(z)(E_h - m\omega^2) = 0 \qquad (5)$$

In the same way, the steady-state solution to equation (2) can be written as

$$w(z, t) = w(z)e^{i\omega t} \qquad (6)$$

Substituting equation (6) into equation (2) yields

$$-AE(1+iC_v\omega)\frac{d^2w(z)}{dz^2} + w(z)(E_v - m\omega^2) = 0 \qquad (7)$$

Solutions of equations (5) and (7) are straightforward if the soil moduli are constant with respect to z. The general solution of equation (7) is the sum of sine and cosine functions, while for equation (5) the solution is the sum of sine, cosine, hyperbolic sine and hyperbolic cosine functions. The dynamic stiffness of the pile can be determined as the complex end force producing unit displacement of the pile head. This unit displacement and the other end conditions represent the boundary conditions from which the integration constants can be established.

9. If the soil moduli are not constant with respect to z, equations (5) and (7) can be solved by a simple two-dimensional discretization method for each discrete frequency ω. The pile itself can be idealized as a lumped-mass system of multiple degrees of freedom with the displacements at n points along the length of the pile as the unknown parameters. The stiffnesses of a pile segment between two neighbouring control points can be formulated by the standard finite element method for beam elements. The resistances of the soil to the movements of the pile can be assumed to be lumped at the n control points and to be simulated by the soil springs. After the total stiffness, damping, and mass matrices are assembled, they can be combined together to form the total dynamic stiffness matrix. Finally, a well established static-condensation subroutine can be used to obtain the dynamic stiffness matrix at the head of a single pile.

APPROXIMATION OF SOIL MODULI

10. The problem of a single force acting at an interior point of a homogeneous half-space is called Lamb's problem. The solution of Lamb's problem can be used to approximate the effect of soil resistance to the pile movement. For the static case, a closed-form solution is possible as shown by Mindlin.[5]

11. The Mindlin equation, which gives the x component of displacement as produced by a single concentrated force P located at any arbitrary point $(0, 0, c)$ within an isotropic half-space and acting in the x direction, is

$$u_x(x, y, z) = \frac{P(0,0,c)}{16\pi(1-\nu)G}\left(\frac{3-4\nu}{R_1} + \frac{1}{R_2} + \frac{2cz}{R_2^3} + \frac{4(1-\nu)(1-2\nu)}{R_2+z+c} + x^2\left\{\frac{1}{R_1^3} + \frac{3-4\nu}{R_2^3} - \frac{6cz}{R_2^5} - \frac{4(1-\nu)(1-2\nu)}{R_2(R_2+c+z)^2}\right\}\right) \qquad (8)$$

in which ν is Poisson's ratio, c is the z distance of the load below the surface x–y boundary plane, and

$$R_1^2 = x^2 + y^2 + (z-c)^2$$
$$R_2^2 = x^2 + y^2 + (z+c)^2$$

The Mindlin equation permits one to characterize completely a static elastic half-space. This equation can be used to obtain a three-dimensional approximate static soil modulus. The horizontal interaction force between soil and pile can be assumed to be uniformly distributed along the length of the pile within each height interval $2h$, but the magnitude of the interaction force varies from one interval to the next. The general expression for the weighted average deflexion at the outside pile radius B, caused by a uniformly distributed interaction force over the height of one interval, is obtained by substituting the intensity of the interaction force $p(0, 0, \bar{c} \pm h)$ between points $(\bar{c} - h)$ and $(\bar{c} + h)$ for the concentrated load $P(0, 0, c)$ in equation (8) and integrating with respect to c over this interval. The static modulus is then taken as the ratio of intensity p and the weighted average deflexion at the outside pile radius. For Poisson's ratio ν equal to 0.5, the lateral soil modulus is

$$E_{\rm h}(B,z) = \frac{8\pi E}{3}\left(\sinh^{-1}\frac{\bar{c}+h-z}{B} - \sinh^{-1}\frac{\bar{c}-h-z}{B} + \sinh^{-1}\frac{\bar{c}+h+z}{B} - \sinh^{-1}\frac{\bar{c}-h+z}{B} + \frac{2}{3B^2}\left\{\frac{B^2(\bar{c}+h)-2B^2z+(\bar{c}+h)z^2+z^3}{[B^2+(\bar{c}+h+z)^2]^{1/2}} - \frac{B^2(\bar{c}-h)-2B^2z+(\bar{c}-h)z^2+z^3}{[B^2+(\bar{c}+h+z)^2]^{1/2}}\right\} - \frac{2}{3}\left\{\frac{z-(\bar{c}+h)}{[B^2+(\bar{c}+h-z)^2]^{1/2}} - \frac{z-(\bar{c}-h)}{[B^2+(\bar{c}-h-z)^2]^{1/2}}\right\} + \frac{4}{3}\left\{\frac{B^2z+(\bar{c}+h)z^2+z^3}{[B^2+(\bar{c}+h+z)^2]^{3/2}} - \frac{B^2z+(\bar{c}-h)z^2+z^3}{[B^2+(\bar{c}-h+z)^2]^{3/2}}\right\}\right)^{-1} \tag{9}$$

Fig. 1 shows the variation of average static lateral soil modulus with the slenderness ratio L/B of the pile.

12. The vertical soil modulus can be obtained in the same way as the lateral soil modulus. For Poisson's ratio equal to 0.5, the vertical soil modulus for a pile segment between the elevations $(\bar{c} - h)$ and $(\bar{c} + h)$ is

$$E_{\rm v} = \frac{8\pi E}{3}\left(2\sinh^{-1}\frac{\bar{c}+h-z}{B} - 2\sinh^{-1}\frac{\bar{c}-h-z}{B} + 2\sinh^{-1}\frac{\bar{c}+h+z}{B} - 2\sinh^{-1}\frac{\bar{c}-h+z}{B} + \frac{\bar{c}-h-z}{[B^2+(\bar{c}-h-z)^2]^{1/2}} - \frac{\bar{c}+h-z}{[B^2+(\bar{c}+h-z)^2]^{1/2}} - \frac{\bar{c}+h+z}{[B^2+(\bar{c}+h+z)^2]^{1/2}} + \frac{\bar{c}-h+z}{[B^2+(\bar{c}-h+z)^2]^{1/2}} - 4z\left\{\frac{1}{[B^2+(\bar{c}+h+z)^2]^{1/2}} - \frac{1}{[B^2+(\bar{c}-h+z)^2]^{1/2}}\right\} + 2z\left\{\frac{B^2+z^2+z(\bar{c}+h)}{[B^2+(\bar{c}+h+z)^2]^{3/2}} - \frac{B^2+z^2+z(\bar{c}-h)}{[B^2+(\bar{c}-h+z)^2]^{3/2}}\right\}\right)^{-1} \tag{10}$$

Fig. 2 shows the variation of average static vertical soil modulus with the slenderness ratio of the pile.

13. In general, the dynamic lateral soil modulus can be written as

$$E_{\rm h}(z, B, G, \nu, \omega) = E_{\rm h1}(z, B, G, \nu, \omega) + {\rm i}E_{\rm h2}(z, B, G, \nu, \omega) \tag{11}$$

in which z is the vertical distance from the surface of the soil, B is the outside radius of the pile, G is the shear modulus of the soil, ν is the Poisson's ratio of the soil, ω is the angular frequency at which the foundation is excited, and $E_{\rm h1}$ and $E_{\rm h2}$ are the real and imaginary parts of the lateral soil modulus, respectively.

14. Similarly, the dynamic vertical soil modulus can be written as

$$E_{\rm v}(z, B, G, \nu, \omega) = E_{\rm v1}(z, B, G, \nu, \omega) + {\rm i}E_{\rm v2}(z, B, G, \nu, \omega) \tag{12}$$

The calculation of the dynamic soil moduli would be just an extension of the static case if the closed-form solution of the dynamic Lamb's problem were available. Unfortunately, such a solution does not exist. The solution of the dynamic Lamb's problem involves complicated infinite integrals which can only be solved by numerical methods.[6, 7] In cylindrical co-ordinates (r, θ, z), the horizontal displacement component which is produced by a single concentrated, horizontal, harmonic force $Pe^{{\rm i}\omega t}$ acting at any arbitrary point $(0, 0, z_{\rm f})$ within an isotropic half-space can be expressed as

$$u(r, \theta, z) = \frac{P\omega e^{{\rm i}\omega t}}{GV_{\rm s}} \sum_{n=0}^{\infty} (f_{1n} + {\rm i}f_{2n})\cos n\theta \tag{13}$$

in which f_{1n} and f_{2n} are frequency-dependent displacement functions which are obtained from a numerical solution of the dynamic Lamb's problem. Terms in the series at the right hand side of equation (13) that are symmetric about the plane $\theta = \pm 90°$ will vanish automatically, as the loading condition is antisymmetric about that plane.

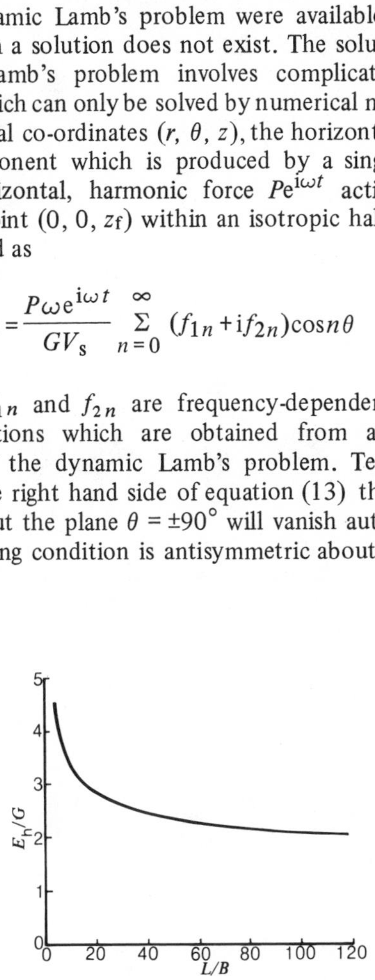

Fig. 1. Average lateral soil modulus versus slenderness ratio L/B of pile at zero frequency

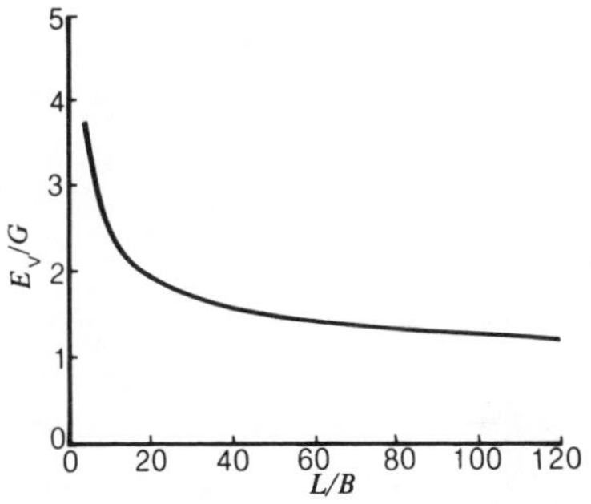

Fig. 2. Average vertical soil modulus versus slenderness ratio L/B of pile at zero frequency

Taking only the first two significant terms, equation (13) can be reduced to

$$u(r,\theta,z) = \frac{P\omega e^{i\omega t}}{GV_s}\,[f_{11} + if_{21} + (f_{12} + if_{22})\cos 2\theta] \quad (14)$$

For most of the cases, the displacement function f_{22} is negligible when compared with the values of the other three displacement functions. The displacement functions are more easily expressed in terms of dimensionless parameters, c, c_f and a, which are defined as

$$c = \omega z/V_s \quad (15)$$

$$c_f = \omega z_f/V_s \quad (16)$$

$$a = \omega r/V_s \quad (17)$$

in which V_s is the shear wave velocity of the soil.

15. When the value of a is small and the value of c is reasonably large, the horizontal displacement component for cases where $c = c_f$ can be adequately approximated by

$$u(r,\theta,c=c_f) = \frac{Pe^{i\omega t}}{32\pi Gr}\,[(6 + 2\cos 2\theta) - i5a] \quad (18)$$

Similarly, the vertical displacement component produced by a single concentrated vertical harmonic force $Qe^{i\omega t}$ acting at an arbitrary point $(0, 0, z_f)$ within an isotropic half-space can be expressed as

$$w(r,\theta,z) = \frac{Q\omega e^{i\omega t}}{GV_s}\,(F_1 + iF_2) \quad (19)$$

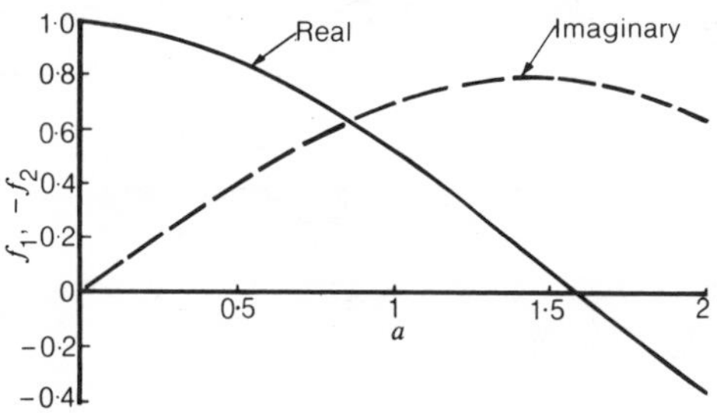

Fig. 3. Horizontal shape function $F_h(\omega)$ for $\nu = 0.5$

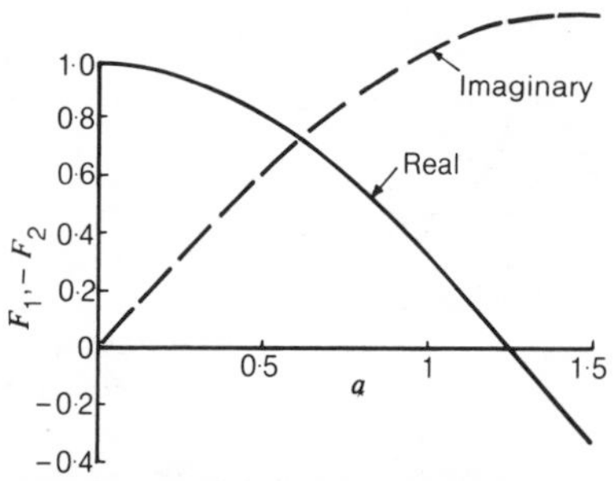

Fig. 4. Vertical shape function $F_v(\omega)$ for $\nu = 0.5$

16. Equations (13) and (19) can be combined with the static soil moduli as expressed by equations (9) and (10) to yield approximate dynamic soil moduli for engineering application, if the dynamic soil moduli are separated in the following forms:

$$E_h(z,B,G,\nu,\omega) = E_h(z,B,G,\nu,0)F_h(\omega) \quad (20)$$

$$E_v(z,B,G,\nu,\omega) = E_v(z,B,G,\nu,0)F_v(\omega) \quad (21)$$

in which $F_h(\omega)$ and $F_v(\omega)$ are frequency-dependent shape functions.

17. The separation of the dynamic soil moduli is similar to the method of separation of variables commonly employed in the solution of partial differential equations. All it means is that the complex-valued surface can be constructed from two orthogonal functions of a single variable. Functions $F_h(\omega)$ and $F_v(\omega)$ can be obtained by normalizing the inverse functions of $f_{11} + if_{21}$ and $F_1 + iF_2$. Figs 3 and 4 present the shape functions obtained for Poisson's ratio of 0.5 and values of a in the range 0–2.

TIP CONDITIONS

18. End conditions theoretically constitute part of the boundary conditions in the determination of dynamic pile stiffnesses. At the embedded end, the movement of the pile generates reactions from the soil lying below the level of the tip. It is necessary to determine the degree of fixity at the tip in order to evaluate correctly the responses of the pile.

19. For the static case, the degree of fixity at the tip can be approximated by again using the Mindlin equation. The equation is singular at the loading point, but this difficulty can be avoided by assuming that the vertical reaction of the soil below the level of the pile tip is uniformly distributed over the tip area. The vertical displacement at the centre of the pile tip caused by a uniform loading q is

$$w(0,0,L) = \frac{B^2 q}{8G(1-\nu)}\Bigg\{(3-4\nu)B + [8(1-\nu)^2 - (3-4\nu)](B^2+4L^2)^{1/2} - \frac{2L^2(3-4\nu)}{(B^2+4L^2)^{1/2}} - \frac{8L^4}{(B^2+4L^2)^{3/2}}\Bigg\} \quad (22)$$

where L is the length of the pile. The stiffness constant used to simulate the tip condition can be taken as the ratio of the total reaction force over the tip area to the approximate deflexion at the centre of the pile tip as expressed by equation (22).

MODELLING OF PILE GROUP

20. Both theory and tests have shown that the total bearing value of a group of friction piles, particularly in clay, may be less than the product of the bearing value of an individual pile multiplied by the number of piles in the

group.[8, 9] The reduction in value per pile depends on the size and shape of the pile group and on the size, shape, spacing and length of the piles. No reduction due to grouping occurs when the piles are end-bearing piles; however, for groups which partake of both actions, that portion taken by friction is reduced.

21. To calculate the total dynamic stiffness matrix of the pile group where there is no reduction of carrying capacity due to grouping, it is advantageous to choose the centroid of the pile group in the horizontal plane as the reference point. Then the dynamic stiffness coefficients are defined as forces that must act at the centroid to produce a sole unit displacement at the reference point. From this definition, the dynamic stiffness coefficients of the pile group are

$$\left.\begin{array}{ll} K_{\mathrm{hh}} = \sum\limits_i k^{\mathrm{e}}_{\mathrm{hh}} & K_{\mathrm{hr}} = \sum\limits_i k^{\mathrm{e}}_{\mathrm{hr}} \\ K_{\mathrm{vv}} = \sum\limits_i k^{\mathrm{e}}_{\mathrm{vv}} & \\ K_{\mathrm{rr}} = \sum\limits_i k^{\mathrm{e}}_{\mathrm{rr}} + \sum\limits_i k^{\mathrm{e}}_{\mathrm{vv}} x_i^2 & \end{array}\right\} \quad (23)$$

in which $k^{\mathrm{e}}_{\mathrm{hh}}$, $k^{\mathrm{e}}_{\mathrm{vv}}$ and $k^{\mathrm{e}}_{\mathrm{rr}}$ are the dynamic stiffness coefficients defined at the head of an element pile for the horizontal, vertical and rocking motions, respectively; $k^{\mathrm{e}}_{\mathrm{hr}}$ is the coupling term between horizontal and rocking motions; x_i is the horizontal distance parallel to the plane of action from the centre of the cross-section of ith element pile to the centroid of the pile group; and k_{hh}, k_{vv} and k_{rr} are the dynamic stiffness coefficients of the pile group for the horizontal, vertical and rocking motions. The summations are to be taken over all piles in the group.

GROUP EFFECT OF LATERALLY LOADED PILES

22. When piles within a group are spaced less than five diameters apart, like those in most onshore piled foundations, interaction effects between individual piles become prominent and can no longer be neglected. Because of these interaction effects between piles within a group, the actual dynamic stiffness coefficients at the reference point of the pile group will be less than those calculated by equation (23). Each dynamic stiffness coefficient of a pile group as expressed by equation (23) has to be multiplied by a reduction factor in order to obtain the actual dynamic stiffness coefficient. The reduction factors are, in general, not unique.

23. For laterally loaded pile groups, the reduction factor, which is normally called the group efficiency, can be obtained by using equation (18). Equation (18) gives, approximately, the horizontal component of displacement which is parallel and is produced by a single, concentrated, horizontal harmonic force applied at a certain depth below the surface of a uniform, elastic half-space. When the frequency of the applied concentrated force is small (i.e., when the applied force is close to the static force) the imaginary part of equation (18) can be neglected, yielding

$$u_{\mathrm{x}}(r,\theta,c{=}c_{\mathrm{f}}) = \lim_{\substack{\omega\to 0\\ a\to 0}} \frac{P(0,0,c_{\mathrm{f}})\mathrm{e}^{\mathrm{i}\omega t}}{32\pi G r}[(6{+}2\cos 2\theta){-}\mathrm{i}5a]$$

$$\approx \frac{P(0,0,c_{\mathrm{f}})}{32\pi G r}(6{+}2\cos 2\theta)$$

$$= \frac{P(0,0,c_{\mathrm{f}})}{32\pi G r}(8\cos^2\theta{+}4\sin^2\theta) \quad (24)$$

The function $(8\cos^2\theta + 4\sin^2\theta)$ expresses the trace of an ellipse whose major axis is twice as long as its minor axis. Therefore, the displacement field as expressed by equation (24) depends on the direction as well as the distance from the loading position.

24. Consider now the determination of the group efficiency for a laterally loaded pile group consisting of M individual piles. By applying a single, concentrated horizontal force at a certain depth of one of the intended pile axes within the pile group, and using equation (24), one can obtain the horizontal displacements of all the other unloaded pile axes at the same level as the applied force. The displacements are denoted by u_{rs}, where r refers to the pile axis being loaded, $s = 1, 2, \ldots M$, and $r \neq s$. The horizontal displacement at the loaded pile axis, u_{rr}, can be approximated by taking the weighted average of the displacements around the outside radius of the pile. By moving the concentrated horizontal load to each of the M intended pile axes within the pile group, one can obtain $M \times M$ displacement functions. The group efficiency for the laterally loaded pile group can now be approximated by

$$\zeta = \frac{\sum\limits_{r=1}^{M} u_{rr}}{\sum\limits_{r=1}^{M}\sum\limits_{s=1}^{M} u_{rs}} \quad (25)$$

To calculate the total dynamic stiffness matrix of a pile group with a pile cap, the stiffness and damping contributions from the pile cap should be properly considered. After they have been determined, the remaining procedure does not differ from that required for the case of a pile group without a pile cap. The dynamic stiffnesses of the pile cap can be simply taken as the same as for a rigid, massless plate resting on an elastic half-space.

CONCLUSION

25. Theory, formulae and a straightforward procedure for modelling pile foundations have been presented. The modelling method suggested bypasses the mathematical difficulty encountered in the rigorous solution of three-dimensional elasticity problems but still gives results within engineering accuracy.

ACKNOWLEDGEMENT

26. The Authors would like to express their deep gratitude to Messrs Kida, Takagi, Fugitani and Saito for generous permission to use their computer programs.

REFERENCES

1. PENZIEN J. (WIEGEL R. L., ed.). Soil–pile foundation interaction. In: Earthquake engineering. Prentice-Hall, Englewood Cliffs, NJ, 1970, Chapter 14, 349–381.

2. SUGIMURA Y. Dynamic behavior of a long pile foundation extended through soft soil layers during an earthquake. Waseda University, 1972.

3. YAMAMOTO S. and SEKI T. Earthquake responses of multi-story building supported on piles. Proc. 3rd Japanese Earthquake Engineering Symposium. Architectural Institute of Japan, Tokyo, 1970.

4. NOVAK M. Dynamic stiffness and damping of piles. Canadian Geotech. J., 1974, 11, No. 4, 574–598.

5. MINDLIN R. Displacements and stress due to nuclei of strain in the elastic half-space. Department of Civil Engineering and Engineering Mechanics, Columbia University, New York, 1961.

6. LIOU D. and PENZIEN J. Seismic analysis of offshore structures supported on pile foundation. University of California, Berkeley, 1977, EERC 77–25.

7. KIDA Y. et al. Coupling analysis between embedded wall and soil medium by wave propagation theory. Architecture Institute of Japan, Tokyo, 1976. Collection of lecture notes, Papers 2005 and 2056, 545–548.

8. WAGNER A. A. Lateral load tests on piles for design information. American Society for Testing and Materials, 1953, technical publication 154, 59–74.

9. ESASHI Y. et al. Pile spacing effects on lateral behavior of piles. Proc. 8th annual meeting Japanese Society for Soil Mechanics and Foundation Engineering, Tokyo, 1973, 451–454 (in Japanese).

10. Dynamic stiffness of piles

J. M. ROESSET (Professor of Civil Engineering, The University of Texas at Austin) and D. ANGELIDES (Supervising Structural Engineer, Brian Watt Associates, Houston, Texas)

The objective of the Paper is to present an overview of some of the results available on the dynamic stiffness of piles. In the linear elastic range it appears that a variety of procedures (Novak's, Nogami's, Blaney's) will give very similar results. Blaney's approach is particularly interesting because of its flexibility. More work needs to be done, however, in relation to non-linear soil behaviour and pile group effects.

INTRODUCTION

The response of piles to dynamic excitation, whether resulting from forces applied at their top or from seismic waves propagating through the surrounding soil, has been the subject of considerable interest and research in recent years. While present knowledge and understanding of the dynamic behaviour of pile foundations is still more limited than for rigid mats, and the number of approximate formulae for preliminary design estimates is still small, much has been learned through analytical and experimental work.

2. The dynamic stiffness of an isolated pile has been determined by two main types of procedure. The first of these is substitution of the surrounding soil by a Winkler foundation with distributed springs and dashpots, constant or dependent on frequency, or with lumped springs concentrated at a finite number of nodes (the spring constants being obtained from analytical considerations or from experimental data).

3. Penzien et al.[1] determined the spring constants in the linear elastic range from a static solution based on the use of Mindlin's equation. To account for non-linear soil behaviour they used a bilinear force deformation characteristic. Matlock[2] determined the non-linear spring characteristics from full-scale tests on a pile embedded in a clay stratum and suggested rules to construct the corresponding curves for any soil as a function of its undrained shear strength, the strain at one half of the maximum stress in a laboratory test, the pile diameter and the depth. These are the P–y and T–z curves, which are extensively used in the design of piles for offshore structures. While these rules were initially intended to estimate the load capacity of the pile and the behaviour under large static (monotonic or cyclic) loads, they have been also used for dynamic analyses.

4. Novak[3] derived the spring constants as a function of frequency using Baranov's equations. By opposition to the two previous solutions this is basically a dynamic formulation including implicitly inertia effects (through the frequency-dependence of the stiffnesses) and the radiation of waves away from the pile into the surrounding soil (reflected in the imaginary term of the stiffnesses). It assumes a plane strain condition and it is only valid for relatively high frequencies (the resulting static stiffness of the soil springs is zero). It is only applicable in addition to small levels of strain.

5. The second procedure for determining the dynamic stiffness of an isolated pile is to use a solution based on the theory of elasticity (elastodynamics) including the coupling between forces and displacements in the soil along the pile. This can be achieved by deriving an appropriate Green's function for the problem (the dynamic equivalent of Mindlin's equation) and applying collocation at discrete points along the pile or the boundary integral equation method, or by discretizing the soil with finite elements in conjunction with an adequate set of boundary conditions at a given radius.

6. Blaney[4] applied the latter approach to the determination of the coupled lateral and rocking stiffnesses of a pile embedded in a stratum of finite depth (underlain by rigid rock). By using a consistent boundary matrix in cylindrical co-ordinates derived by Kausel[5] for the analysis of circular foundations and assuming a soil layer whose properties were constant in the radial direction, Blaney was able to avoid the use of finite elements, modelling the pile as a series of regular beam elements and adding directly to the dynamic stiffness of the pile the boundary matrix at each particular frequency (after condensing the degrees of freedom of the nodal rings relating them to the degrees of freedom of the pile's nodes). The resulting procedure is not only particularly economical but it provides also an excellent and complete solution in the linear elastic range (small strains) for a point bearing pile. The soil properties can change with depth in any arbitrary fashion without any extra cost in computation. Blaney pointed out that the approach could be extended to include variation of soil

properties in the radial direction by placing several cylinders of soil (discretized with toroidal finite elements) between the pile and the far field (represented by the consistent boundary matrix) as shown in Fig. 1 and suggested that this extension could be used to approximate non-linear soil behaviour in the neighbourhood of the pile.

7. Angelides[6] has extended recently Blaney's work by studying the vertical stiffness of piles, by simulating the condition of a floating pile (making the depth of the soil deposit substantially greater than that of the pile and adding a cylinder with soil properties at the bottom of the physical pile) and by simulating in an approximate way non-linear soil behaviour.

8. Nogami and Novak[7–9] determined the pile stiffnesses for point bearing piles (finite soil stratum on a rigid base) by expressing the pile deformations in terms of the natural modes of the stratum. Coupling along the height of the pile was therefore included although vertical displacements of the soil were neglected under horizontal excitation and vice versa. This approach is conceptually similar to the principle underlying the derivation of the consistent boundary matrix and allows closed form solutions for the case of a uniform, homogeneous soil stratum in terms of Bessel functions.

9. All the above work refers to the dynamic stiffnesses of an isolated pile. To determine the stiffnesses of a complete pile foundation it is usual to consider a rigid slab connecting the pile caps and to neglect the interaction between the various piles through the soil. Very little work has been published to date on the dynamic stiffness of pile groups. Wolf,[10] using a finite element formulation, has recently presented some results which indicate important pile group effects when dealing with a layer of finite depth.

10. The purpose of this Paper is to summarize some of the results available at present, particularly those obtained by Blaney and Angelides, comparing them with other solutions.

SOIL STIFFNESSES

11. While models based on the assumption of a Winkler foundation neglect the coupling between forces and displacements in the soil along the height of the pile, their use is widely spread and they seem to provide results for the pile stiffnesses which are in good agreement with those of more complete solutions. It is therefore interesting to examine the equivalent soil stiffnesses resulting from a solution such as Blaney's and to compare them with those of other methods. Since the boundary matrix is a full matrix and not a diagonal one, in order to obtain the spring constant of an equivalent Winkler foundation it is assumed that a uniformly distributed (horizontal or vertical) force is applied along the cylindrical cavity which maintains a rigid cross-section. The resulting displacements along the height of the cavity are then inverted to obtain approximate stiffnesses. As was pointed out by Blaney these stiffnesses will vary with depth depending on the frequency of excitation (Fig. 2). In order to obtain a single value (corresponding to a uniform elastic foundation) the average stiffness over the top half of the cavity is computed. This selection is made because near the bottom the existence of a rigid base increases very much the soil stiffness.

12. It is convenient to represent the resulting soil stiffness (horizontal and vertical) in the form

$$K = (K_1 + iK_2) = K_0\,(k_1 + i a_0 c_1)$$

where $a_0 = \omega r/v_s$ is a dimensionless frequency, ω is the frequency of excitation in rad/s, r is the radius of the pile (and cavity), $v_s = G_s/\rho_s$ is the shear wave velocity of the soil, G_s its shear modulus and ρ_s its mass density.

13. In this form K_1 and K_2 represent the real and imaginary parts of the soil stiffnesses; K_0 is the static stiffness and k_1 and c_1 are dynamic coefficients (functions of frequency). One can interpret $K_1 = K_0 k_1$ as the stiffness of a typical spring, and $(1/\omega)K_2 = (r/v_s)\,c_1 K_0$ as the constant of a viscous dashpot. This dashpot reproduces the loss of energy in each cycle of vibration by radiation of waves from the pile into the far field (often referred to as radiation damping). Alternatively the quotient $K_2/2K_1$ can be interpreted as the effective soil damping (increasing with frequency). This soil damping is not the same as the effective damping for the pile: the latter will be a weighted average of the soil damping and the material damping of the pile itself.

14. When the soil is not purely elastic but it is assumed to have some internal dissipation of energy of a hysteretic nature (frequency-independent) to account for inelastic deformations one can approximately write

$$K = (K_1 + iK_2)(1 + 2iD) = K_0\,(1 + 2iD)(k_1 + i a_0 c_1)$$

where D is the internal damping ratio for the soil.

15. While this expression is only approximate (the approximation deteriorating for very large values of a_0) it provides a good insight into the effect of the internal soil

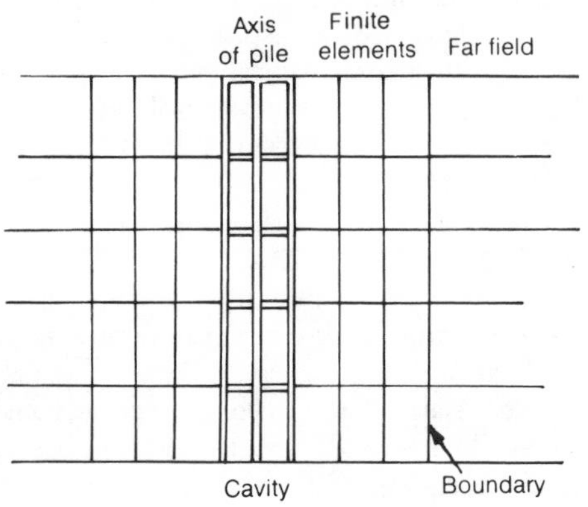

Fig. 1. Mathematical model

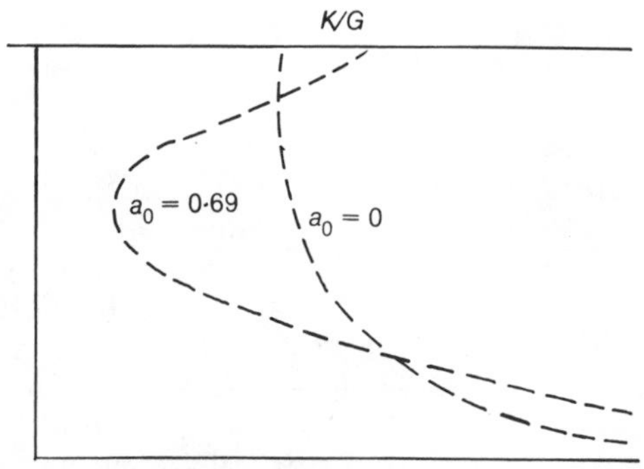

Fig. 2. Variation of soil stiffness

damping. The real and imaginary parts of the soil stiffnesses become then

$$K_{real} = K_1 - 2DK_2 = K_0(k_1 - 2Da_0c_1)$$

$$K_{imag} = K_2 + 2DK_1 = K_0(a_0c_1 + 2Dk_1)$$

The effect is thus particularly important for the K_{real} which will decrease with increasing frequency a_0.

16. Figure 3 shows the variation of the static horizontal soil stiffness K_0 divided by G_s as a function of the aspect ratio of the pile H/r and Poisson's ratio. It can be seen that for piles with an aspect ratio larger than 40 the values are almost constant. They are also practically the same for a point bearing and a floating pile in this range. Values obtained using Mindlin's solution and Penzien's formulation (as reported by Blaney) follow the same trend but are slightly smaller. From these results it has been suggested that values of K_0 between $2G_s$ and $4G_s$ (or $3G_s$ in the average) can be used for preliminary design purposes. It appears that a smaller variation is obtained expressing K_0 in terms of Young's modulus of elasticity E_s rather than the shear modulus G_s. For slender piles then $K_0 \approx E_s$. Fig. 4 shows similar results for vertical vibration. In this case, even at an aspect ratio H/r of 80 there is a small difference between the values for a point bearing and a floating pile. It appears, however, that a value of $K_0 \approx 0.6\ E_s$ would provide reasonable results.

17. Figure 5 shows the variation of the real part of the soil stiffness $K_1 = K_0k_1$ normalized with respect to the shear modulus, versus the dimensionless frequency a_0. For values of a_0 of the order of 0.5 the results are almost independent of the aspect ratio of the pile and they agree well with Novak's solution. In the low frequency range the values are not only affected by the aspect ratio but also by the existence of a rigid bottom. The results presented were obtained with 5% hysteretic damping in the soil. For a purely elastic soil deposit the real stiffness would in fact become zero at the natural frequency of the layer (which could go into resonance). This is the reason for the dip in the value of the stiffness for $H/r = 10$. For the more slender piles the soil stiffness tends to increase with frequency, reaching a value of approximately $4G_s$ or $1.4E_s$.

18. The corresponding results for vertical excitation are shown in Fig. 6 for a pile with an aspect ratio of 80. One should notice again the decrease in the stiffness approaching the natural dilatational frequency of the stratum in this case and the increase in stiffness with frequency reaching values of $2.3G_s$ or $0.8E_s$.

19. In order to account for the frequency variation of the soil stiffness it has been suggested some times to use added soil masses in combination with the springs and dashpots. This solution would be correct if the term K_1 varied as a second degree parabola with ω. It can be seen, however, that this is not the case. The concept of an added mass is thus incorrect and should be abandoned. Physically there is no such thing as a lumped mass of soil participating in the motion of the pile.

20. In Fig. 7, part (a) shows the variation of the imaginary part of the horizontal stiffness K_2 divided by G_s versus the dimensionless frequency a_0. The same quantity divided by a_0 is plotted in part (b). This latter value can be directly associated with the constants of equivalent viscous dashpots

$$c = \rho_s v_s r K_2 / G_s a_0$$

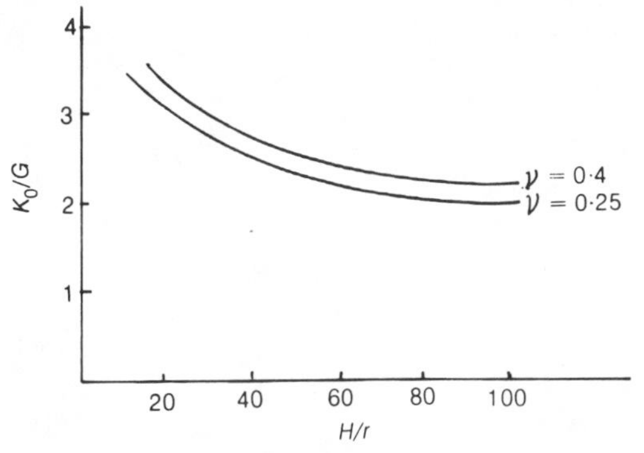

Fig. 3. Variation of horizontal K_0

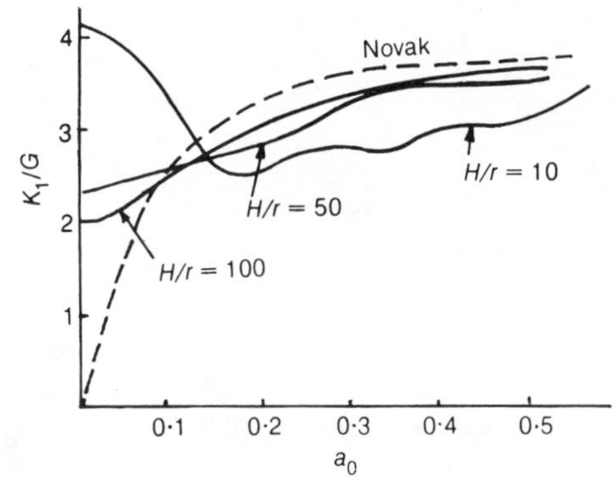

Fig. 5. Variation of horizontal K_1

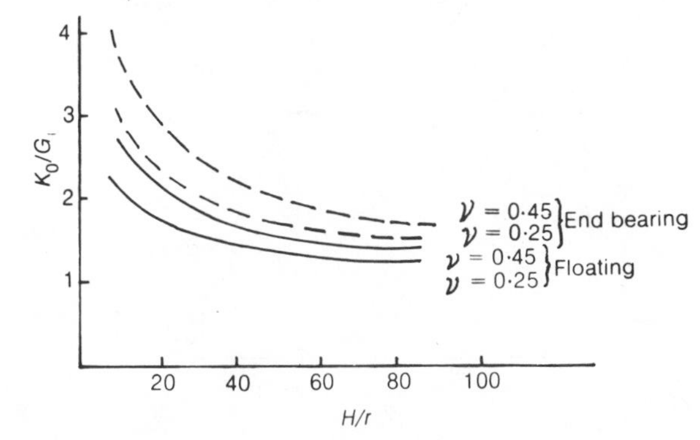

Fig. 4. Variation of vertical K_0

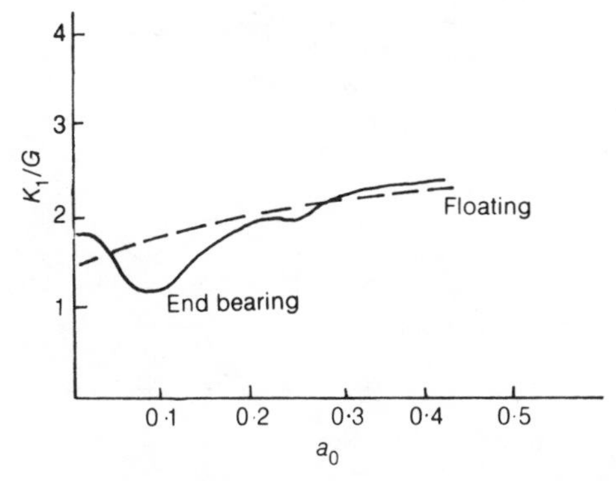

Fig. 6. Variation of vertical K_1; $H/r = 80$

It can be seen that for dimensionless frequencies a_0 larger than 0.3 the results are almost independent of the aspect ratio of the pile and are very similar to Novak's solution. In the low frequency range they depend on the ratio H/r and the stratum depth. For a layer of finite depth below the fundamental frequency of the stratum there cannot be any lateral radiation and the imaginary term would be zero if there were no internal soil damping. In the range of frequencies $a_0 = 0.5$ the dashpot constants would be approximately $c = 10\rho_s v_s r$ (limiting values $c \approx 8\rho_s v_s r$ for $\nu = 0.25$ and $9.75\rho_s v_s r$ for $\nu = 0.4$).

21. Analogous results are presented in Fig. 8 for the case of vertical vibration. The same trend is noticed with a value of the equivalent dashpot constant at $a_0 = 0.5$ of about $c = 7.5\rho_s v_s r$. The limiting value given by the slope of the nearly straight line in the curve of K_2/G_s versus a_0 corresponds to $c = 6.3\rho_s v_s r$ approximately.

22. In the selection of appropriate values for the spring and dashpot constants of an equivalent Winkler foundation some consideration must be given to the range of frequencies of interest. Of particular importance is the case of a finite soil layer over much stiffer rock-like material. Then

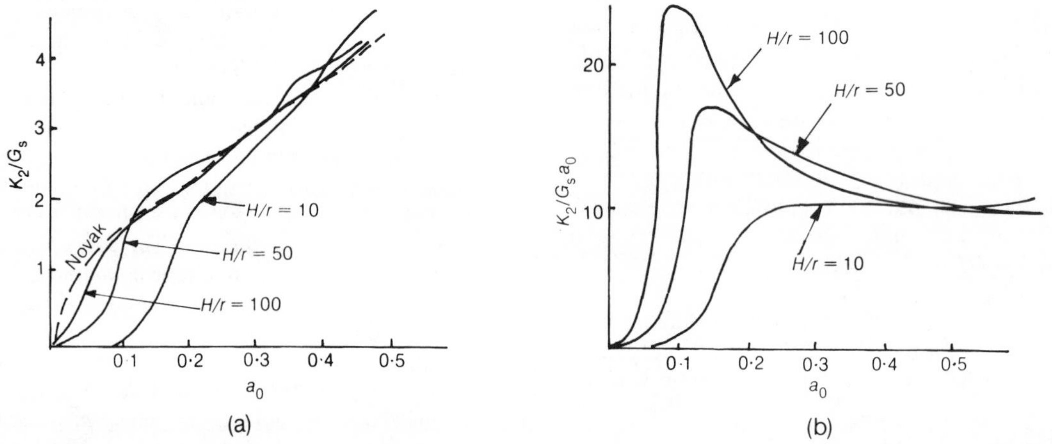

Fig. 7. (a) Variation of horizontal K_2; (b) horizontal K_2/a_0

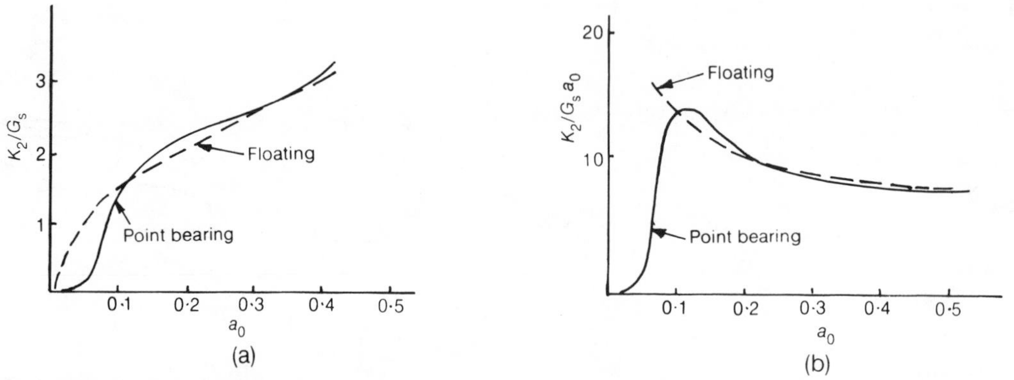

Fig. 8. (a) Variation of vertical K_2; (b) vertical K_2/a_0

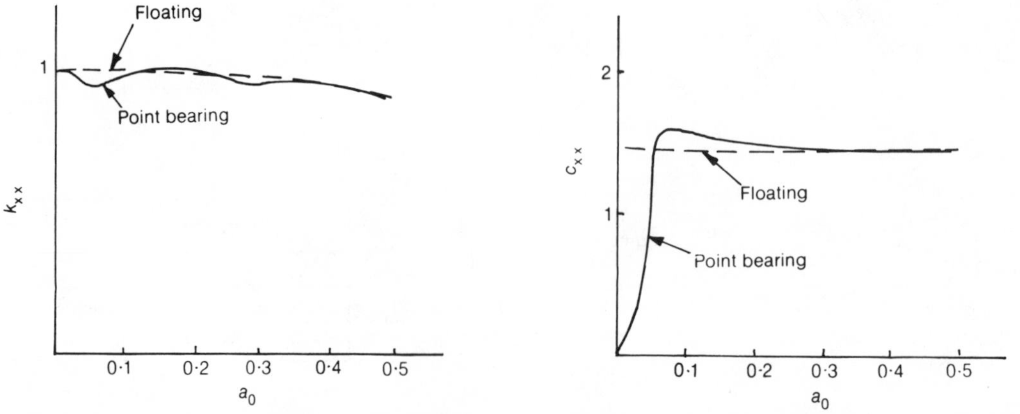

Fig. 9. Horizontal stiffness coefficients; $H/r = 50$

for frequencies smaller than the natural frequency of the stratum there is no radiation and the real parts of the soil stiffnesses may decrease considerably in the neighbourhood of this frequency (depending on the amount of internal soil damping).

23. The above solutions are only appropriate in the linear elastic range (i.e., under the assumption of very small levels of strain). For larger excitations and particularly near the top of the pile under lateral loading the non-linear behaviour of the soil must be accounted for. The best data available in this range are still those provided by Matlock[2] through the *P–y* curves. A comparison of analytical solutions with these curves is difficult because of the difference in the soil parameters used (Young's modulus or shear modulus versus undrained shear strength and critical strain). Yegian and Wright[11] used plane stress and plane strain two-dimensional finite element solutions with a non-linear soil model and joint elements around the pile to determine inelastic spring constants. Their solution indicated that the *P–y* curves for points at sufficient depth were between the plane stress and the plane strain predictions. Angelides, using the model of Fig. 1 and an equivalent linearization (iterative procedure adjusting the shear modulus of each element on the basis of the maximum shear strain), found also a reasonable agreement after a certain depth but a variation by a factor of 2–3 near the surface. Since the pile stiffness under horizontal excitation is controlled by the soil properties near the top the value of the horizontal stiffness predicted by the finite element model was 3–5 times larger than the result using the *P–y* curves. It appears that considerably more work, both analytical and experimental, is needed in this area. Angelides' results seem also to indicate that as the level of excitation increases the internal soil damping (of a hysteretic nature) increases up to a certain point but the radiation damping tends to decrease. Due to the variation of soil properties in the radial direction (the modulus increasing with increasing distance from the pile) some reflection of the waves seems to take place.

PILE STIFFNESSES

24. The dynamic stiffnesses of the pile can be represented by a stiffness matrix of the form

$$\mathbf{K} = \begin{vmatrix} K_{xx} & K_{x\phi} & 0 \\ K_{\phi x} & K_{\phi\phi} & 0 \\ 0 & 0 & K_{zz} \end{vmatrix}$$

where the terms K_{ij} are again complex functions of frequency and can be written as

$$K_{ij} = K_{ij\text{o}}\,(k_{ij} + \mathrm{i}a_\text{o}c_{ij}) = K_{ij1} + \mathrm{i}\,K_{ij2}$$

$K_{ij\text{o}}$ is the static stiffness and k_{ij} and c_{ij} are dynamic coefficients (functions of frequency); $K_{ij\text{o}}K_{ij} = K_{ij1}$ represents the real part of the stiffness (constant of an equivalent spring element) and $K_{ij\text{o}}a_\text{o}c_{ij} = K_{ij2}$ the imaginary part ($(r/v_s)\,K_{ij\text{o}}c_{ij}$ would be the constant of an equivalent dashpot). For a hysteretic soil it is no longer possible to multiply the previous expression by the factor $(1 + 2\mathrm{i}D)$ unless the pile itself has the same material damping.

25. Based on his studies for piles with an aspect ratio larger than 30, Blaney had suggested the approximate formulae for the terms $K_{ij\text{o}}$ under lateral and rocking excitation

$$K_{xx\text{o}} \approx 2\,\frac{E_\text{p}I}{r^3}\left\{\frac{E_\text{s}}{E_\text{p}}\right\}^{0\cdot75}$$

$$K_{x\phi\text{o}} \approx -1\cdot2\,\frac{E_\text{p}I}{r^2}\left\{\frac{E_\text{s}}{E_\text{p}}\right\}^{0\cdot5}$$

$$K_{\phi\phi\text{o}} \approx 1\cdot6\,\frac{E_\text{p}I}{r}\left\{\frac{E_\text{s}}{E_\text{p}}\right\}^{0\cdot25}$$

The actual exponents fitted to the numerical results were closer to 0.80, 0.54 and 0.27 respectively but they were rounded off. It is interesting to notice that using the analogy of the Winkler foundation with horizontal springs $K_\text{o} = \alpha E_\text{s}$ the results for an infinitely long pile would be

$$K_{xx\text{o}} \approx 1\cdot7\alpha^{0\cdot75}\,\frac{E_\text{p}I}{r^3}\left\{\frac{E_\text{s}}{E_\text{p}}\right\}^{0\cdot75}$$

$$K_{x\phi\text{o}} \approx -1\cdot1\alpha^{0\cdot5}\,\frac{E_\text{p}I}{r}\left\{\frac{E_\text{s}}{E_\text{p}}\right\}^{0\cdot5}$$

$$K_{\phi\phi\text{o}} \approx 1\cdot5\alpha^{0\cdot25}\,\frac{E_\text{p}I}{r}\left\{\frac{E_\text{s}}{E_\text{p}}\right\}^{0\cdot25}$$

These values would agree well with those suggested by Blaney for $\alpha \approx 1.2$. In a similar way for the vertical case and a very long pile

$$K_{zz\text{o}} \approx 0\cdot56\,\frac{EA}{r}\,\alpha^{0\cdot5}\left\{\frac{E_\text{s}}{E_\text{p}}\right\}^{0\cdot5}$$

which agrees well with the results of Angelides' work for $\alpha \approx 0.75$.

26. Figure 9 shows the variation of the real and imaginary parts of the K_{xx} term (k_{xx} and c_{xx}) for a typical pile with $H/r = 50$ as a function of frequency. For values of a_o less than 0.5 it can be assumed that both k_{xx} and c_{xx} are practically constant, independent of frequency. The exception is the term c_{xx} for a point bearing pile, on a soil layer of finite depth, which would be zero below the fundamental frequency of the stratum. The value of c_{xx} is a function of the ratio E_s/E_p and varies typically between 1 and 2 (for most practical cases it would be between 1.5 and 2).

27. Figures 10 and 11 show the corresponding results for the terms $x\phi$ and $\phi\phi$. The value of $c_{x\phi}$ varies typically (depending on the E_s/E_p ratio) between 1 and 1.5. The value of $c_{\phi\phi}$ is much smaller than those of c_{xx} and $c_{x\phi}$, starting from zero and increasing to values of 0.2–0.5 at $a_\text{o} = 0.5$.

28. The results for the vertical stiffness are presented

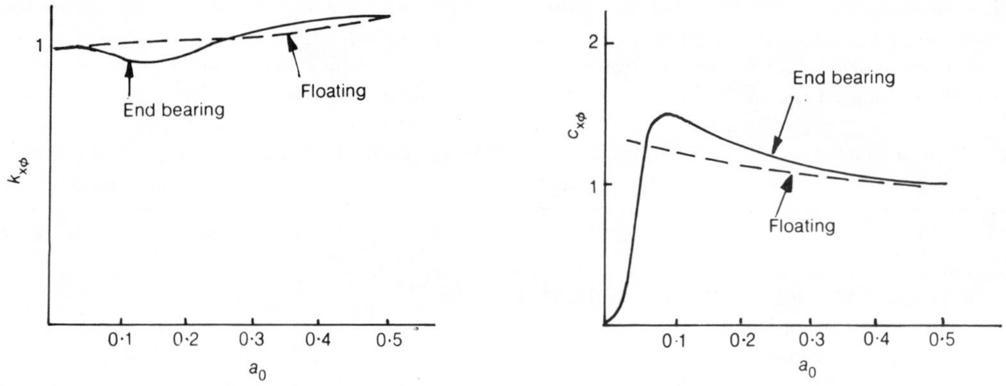

Fig. 10. Coupling horizontal–rocking stiffness; H/r = 50

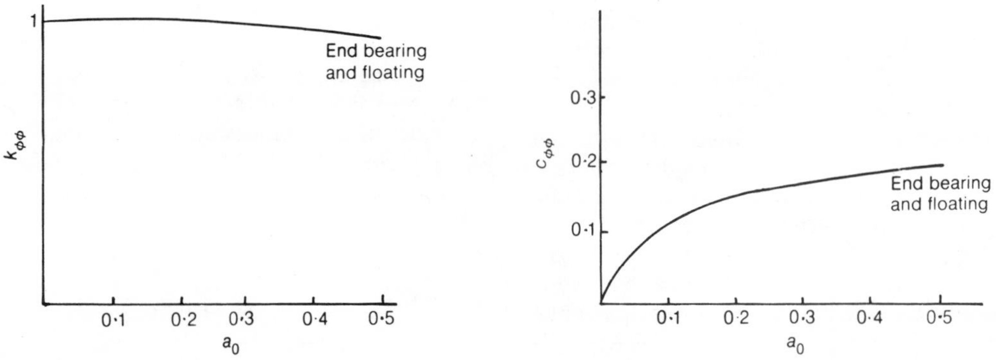

Fig. 11. Rocking stiffness coefficients; H/r 50

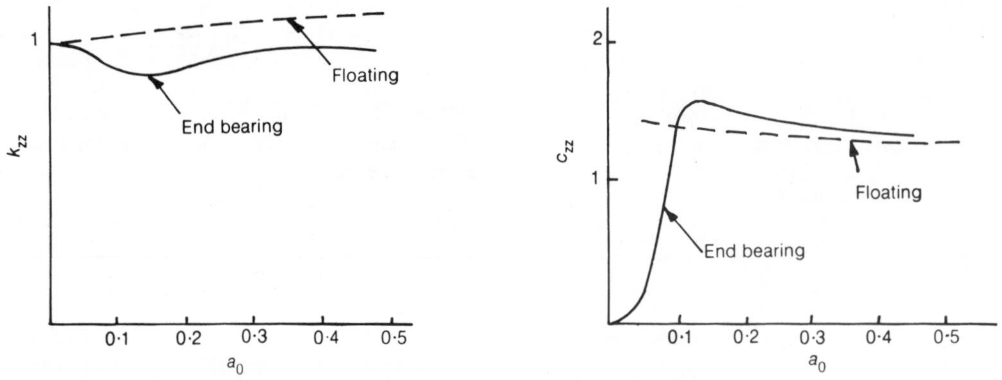

Fig. 12. Vertical stiffness coefficients

in Fig. 12. The difference between the end bearing and the floating pile is now more marked, as could be expected. The static stiffness of the floating pile is smaller than that of the end bearing pile so in fact the real stiffnesses become almost equal at $a_0 = 0.5$ (one increases and the other decreases). For shorter piles (H/r of the order of 20) the differences are more significant and affect also importantly the imaginary term.

PILE FOUNDATIONS

29. It is customary to assume that the piles are sufficiently distant from each other to ignore any interaction through the soil. Under these conditions if a rigid slab connects the pile heads the total stiffnesses of the pile foundation are given by

$$K_{zz} = \Sigma K_{zz}$$

$$K_{xx} = \Sigma K_{xx}$$

$$K_{x\phi} = K_{\phi x} = \Sigma K_{x\phi}$$

$$K_{\phi\phi} = \Sigma(K_{\phi\phi} + x_i^{\,2} K_{zz})$$

where x_i is the distance from each pile to the axis of rotation of the slab.

30. The studies by Wolf indicate, however, that due to pile group effects the foundation stiffness may be substantially reduced from the value given by the straight summation of each pile's contribution, that the variation with frequency may be much more pronounced (particularly in the case of a finite soil stratum) and that the radiation damping may increase considerably (these last two observations would

indicate a much larger effective radius of the foundation). More work on the dynamic stiffness of pile groups is clearly necessary.

CONCLUSIONS

31. In the linear elastic range it appears that a variety of procedures (Novak's, Nogami's, Blaney's) will give very similar results. Blaney's approach is particularly interesting because of its flexibility. More work needs to be done, however, in relation to non-linear soil behaviour and pile group effects.

REFERENCES

1. PENZIEN J. et al. Seismic analysis of bridges on long piles. J. Engng Mech. Div. Am. Soc. Civ. Engrs, 1964, June, No. EM 3.
2. MATLOCK H. Correlations for design of laterally loaded piles in soft clay. 2nd Offshore Technology Conf., Houston, 1970.
3. NOVAK M. Dynamic stiffness and damping of piles. Canadian Geotech. J., 1974, 11, No. 4, 574–598.
4. BLANEY G. W. et al. Dynamic stiffness of piles. 2nd int. conf. Numerical Methods in Geomechanics, Blacksburg, Virginia, 1976.
5. KAUSEL E. et al. Dynamic analysis of footings in layered media. J. Engng Mech. Div. Am. Soc. Civ. Engrs, 1975, Oct., No. EM5.
6. ANGELIDES D. et al. Dynamic stiffness of piles. Research report, Civil Engineering Department, Massachusetts Institute of Technology, 1979.
7. NOGAMI T. and NOVAK M. Soil–pile interaction in vertical vibration. Int. J. Earthq. Engng Struct. Dynamics, 1976, 4, Jan.–Mar., No. 3, 277–293.
8. NOGAMI T. and NOVAK M. Resistance of soil to a horizontally vibrating pile. Int. J. Earthq. Engng Struct. Dynamics, 1977, 5, July–Sept., No. 3, 249–261.
9. NOVAK M. and NOGAMI T. Soil pile interaction in horizontal vibration. Int. J. Earthq. Engng Struct. Dynamics, 1977, 5, July–Sept., No. 3, 263–282.
10. WOLF J. P. and VON ARX G. A., Impedance function of a group of vertical piles. Proc. Conf. Soil Dynamics and Earthquake Engineering, Pasadena, California, 1978. American Society of Civil Engineers.
11. YEGIAN M. and WRIGHT S. G. Lateral soil resistance–displacement relationships for pile foundations in soft clays. 5th Offshore Technology Conf., Houston, 1973.

BIBLIOGRAPHY

BERGER E. et al. Simplified method for evaluating soil–pile–structure interaction effects. 9th Offshore Technology Conf., Houston, 1977.

REESE L. et al. Analysis of laterally loaded piles in sand. 6th Offshore Technology Conf., Houston, 1974.

ROESSET J. M. and KAUSEL E. Dynamic soil structure interaction. 2nd Int. Conf. Numerical Methods in Geomechanics, Blacksburg, Virginia, 1976.

11. Theoretical studies of piles using the finite element method

F. BAGUELIN and R. FRANK, Dr-Ing (Laboratoire Central des Ponts et Chaussées at Nantes and Paris, France)

The aim of this Paper is to present the use of the finite element method for determining realistic models of behaviour of piles. The main results obtained so far concerning single piles under lateral or axial static loads are given. For axial loads, piles have been studied in ideal linear elastic (isotropic and cross-anisotropic) or dilatant media. The mechanism of shaft friction has been clarified and has been extended to non-linear media. The model of pile's behaviour proposed by Cambefort is proved to be valid by these studies. The mobilization of the shaft friction with the vertical displacements at a given depth is quantified, taking into account if necessary a remoulded zone of soil around the pile. The influence of the slenderness ratio of the pile is also studied. For lateral loads, the finite element method has been used to study the validity of the subgrade reaction modulus theory. The influence of the shape of the pile, of the remoulding and of the plastification of the soil is analysed by a plane strain model. The influence of the length of the pile, of the relative pile–soil stiffness and of the loading conditions is studied in linear isotropic elasticity with a three-dimensional model using Fourier series. Future research is examined and general conclusions are drawn concerning the use of the finite element method in foundation engineering.

INTRODUCTION

The finite element method (FEM) is increasingly widely used for calculating and designing all kinds of structure, including pile foundations. It is a very powerful tool which enables numerous factors that influence the behaviour of piles to be taken into account, even if there are still many other factors which cannot yet be included. By using it, we implicitly recognize that our usual and classical methods of determining pile foundations are too simplistic or not proven valid, if not erroneous. However, for various reasons, the FEM cannot be, at least for the moment, a current method of designing foundations. One reason for this is that it is a heavy and comparatively expensive tool (i.e. for serious studies). It is still a method that only a few specialists can use easily and it is out of reach of most practitioners. Another reason, probably the most important and embarrassing one, is that in general we do not know which rheological model or which values of soil characteristics should be used for the specific soil of the study: the FEM is in advance of our rheological knowledge of soils.

2. The efficiency of the FEM can be used in another way—without meaning, of course, that the use already mentioned should be abandoned. It can be used very successfully for pure research: for determining new realistic models of behaviour of piles. This includes not only parametric studies of piles (i.e., studies resulting in charts or formulas in terms of different factors) but also studies of the exact mode of behaviour of piles, of the intimate mechanisms involved in the soil masses (e.g., elementary stress systems induced).

3. The long term aim is to establish far more satisfactory current methods of designing piles than the actual ones, including the answer to the question about the characteristics to be used, especially by telling which soil tests are suitable for reproducing similar stress systems.

4. It is not necessary to use sophisticated rheological models for such studies. On the contrary, we must start with very simple models and develop them step by step, in order to know the influence of each characteristic. The only question is then to make sure that the qualitative results obtained are independent of the models or can reasonably be extended to more complicated ones.

5. We have been working in this direction for several years in the Laboratories of Ponts et Chaussées. Interesting results have been obtained for single piles under vertical or horizontal loads.[1–10] This Paper gives the most significant of these results, while trying to focus on the practical use of the FEM for such studies.

6. Throughout all these studies, the finite element program ROSALIE of our own laboratories has been used,[11] and thus the approximations and algorithms of the method are fully known, which is essential.

THE MECHANISM OF SHAFT FRICTION OF PILES

7. The theoretical analyses reported here, for the single pile under vertical loads, deal only with the initial phase of the loading itself. No mention is made either of the natural conditions and the influence of the installation of the pile, or of the failure mechanisms in the soil mass or at the soil–pile interface, which are still under study.

8. The FEM has been used[3, 6] to clarify the mechanism

of shaft friction in both linear isotropic elastic and linear dilatant media (which have the property of changing their relative volume not only under isotropic stresses but also under deviator stresses). This mechanism has been confirmed by Orsi[8] in cross-anisotropic elastic media of vertical axis, also using the FEM.

Mesh

9. Both geometry and deformation being symmetrical about the pile's vertical axis, the problem is reduced to a two-dimensional one.

10. The mechanism of shaft friction has been studied on piles of slenderness ratio $D/2r_0$ of 10 (where D is the length and r_0 the radius of the pile). The mesh of any radial plane is shown in Fig. 1.[6] The soil is completely fixed to the pile. The elements are isoparametric triangles with six nodes (i.e., the displacements are approximated by second order polynomials and the stresses after differentiation will be linear along one side). There are 660 elements and 1400 nodes. The band width of the rigidity matrix is 134.

11. This mesh is very dense in the vicinity of the pile, where the displacements and the stresses are expected to vary quickly. It is also the zone in which the stress system needs to be carefully studied. The density of elements has been checked by comparing the shear and normal stresses on the soil–pile interface, calculated in the soil and calculated in the pile. These agree to a high degree of accuracy.

12. The boundaries of the mesh were 15 diameters away from the pile and 11.5 diameters beneath it. The difference is always less than 5% between calculations with the two following fixing conditions: the two boundaries are completely fixed (friction boundaries) and only the normal displacements to each boundary are set to zero (slipping boundaries). The result of a subsequent study[8] of the influence of the distance of these boundaries is given below.

13. Between these boundaries the mesh is composed of horizontal and vertical lines. This allows remouded zones or heterogeneous layers to be introduced easily. It is also easy to study the different functions with radial distance or with depth independently.

Direct FEM results

14. The following synthesis can be made for the main part of the shaft, and for the adjacent soil, between the head and base disturbances.

15. The vertical loading of the pile produces hardly any increase in the normal stress σ_r compared with the vertical shear stress τ_{rz} in the vicinity of the pile. However, in the case of the dilatant media, σ_r exhibits a relative maximum within the soil mass at $r = 2r_0$. The stresses σ_z and σ_θ are also negligible in isotropic elasticity. With dilatancy they show more substantial increase, but remain much smaller than τ near the shaft.

16. The movements are practically only vertical in this region; the vertical loading produces almost no radial movement. These results show clearly the predominant mode of deformation near the pile's shaft: it is a pure shearing of vertical concentric annuli (Fig. 2).

17. The shear stress $\tau_{rz} = \tau$ decreases, in the soil mass, and is inversely proportional to the radial distance r, at least until $r = 5r_0$:

$$\tau = \frac{\tau_0 r_0}{r} \tag{1}$$

where τ_0 is the friction on the shaft. (This result is consistent with $\delta\sigma_z/\delta z \approx 0$ near the shaft.)

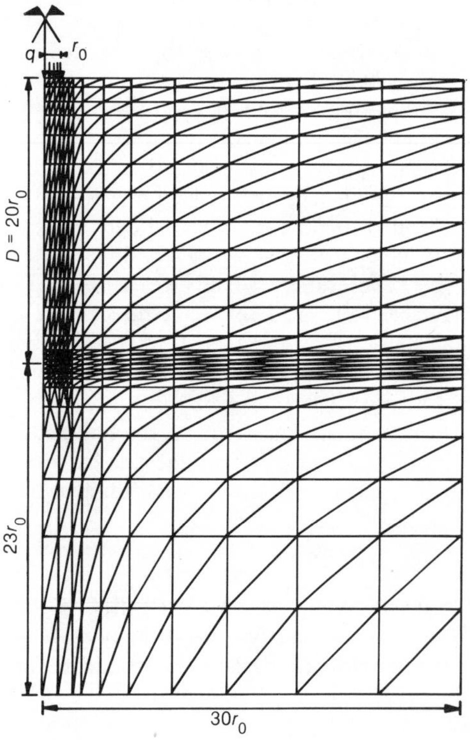

Fig. 1. Mesh of the radial plane for the study of the mechanism of shaft friction[1, 3, 6]

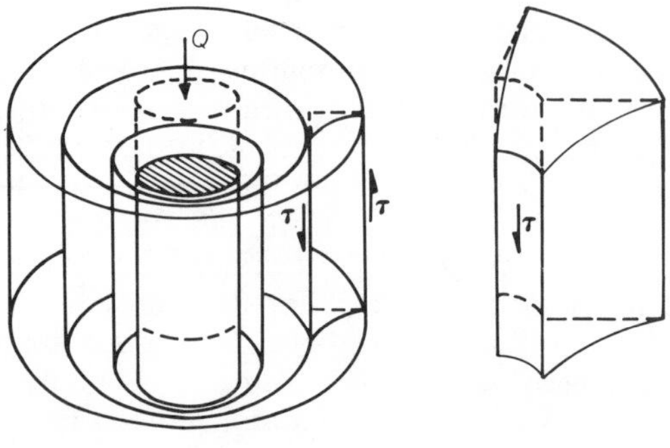

Fig. 2. Pure shearing of vertical concentric annuli[1, 6]

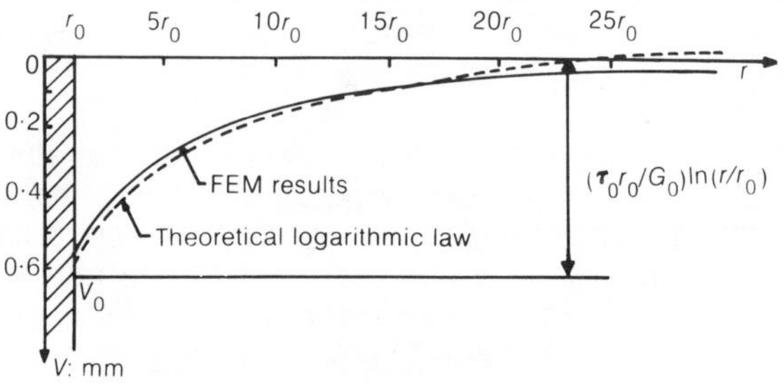

Fig. 3. Vertical displacements of the soil (E = 30 MPa, ν = 0.3) at mid-depth of the pile (r_0 =0.3 m, $E = 3 \times 10^4$ MPa, ν = 0.3)[1, 6]

18. $\delta u/\delta z$ is negligible in comparison with $\delta V/\delta r$ (u and V being the radial and vertical displacements) when calculating shear strains γ_{rz}. This, combined with equation (1), leads to the following logarithmic law for V with respect to r:

$$V(r) = V_0 - \frac{\tau_0 r_0}{G_0} \ln\frac{r}{r_0} \qquad (2)$$

where G_0 is the shear modulus of the soil (independent of r, when there is no remoulding), and V_0 is the downward movement of the pile. Figure 3 shows that this law has been found valid up to a radius of $15r_0$ or $20r_0$ at mid-depth of a fairly rigid pile. Cooke[12] postulated independently identical laws for τ and γ and derived the same logarithmic variation for V, which he compared, with reasonably good success, to experimental measurements. Randolph[13] checked numerically this variation and used it to derive his solution for pile groups.

19. This model is very well suited to take into account a remoulded zone of soil around the pile shaft, due to the installation of the pile. If the modulus G becomes $G(r) < G_0$ in this zone, then the vertical displacement V_0 of the pile becomes V_r:

$$V_r = V_0 - \frac{\tau_0 r_0}{G_0} \ln\frac{r_1}{r_0} + \tau_0 r_0 \int_{r_0}^{r_1} \frac{dr}{G(r)r} \qquad (3)$$

where r_1 is the radius of the remoulded zone. The finite element calculations with remoulded zones of Waschkowski[10] and Randolph[13] confirm this equation.

20. The ratio τ_0/V_0 is very sensibly the same at different depths and for compressible or rigid piles (Fig. 4). This gives an interesting support to Cambefort's model of behaviour of piles under vertical loads.[14] Moreover, this mobilization parameter is found to be directly proportional to Young's modulus of the soil. In the case of vertical heterogeneity (different soil layers) it is proportional to the modulus of the adjacent soil. This can also be written

$$V_0 = \frac{k\tau_0 r_0}{G_0} \qquad (4)$$

where V_0, τ_0 and G_0 are relative to the depth under consideration and where k is a constant depending only on the slenderness ratio of the pile and on Poisson's ratio of the soil.

21. The mean value of 3 was proposed for k from these calculations with piles of slenderness ratio 10. This value is expected to be rather small because of the proximity of the horizontal boundary. Orsi[8] studied the influence of the boundaries, by imposing on them the analytical isotropic elastic displacements obtained when replacing the pile by a vertical segment under the same load. He found that for these piles ($D/2r_0 = 10$) the horizontal boundary should be at least $80r_0$ beneath the pile's base, if we wish to estimate k with 5% precision when imposing $V = 0$ on this boundary. His study, which concerns, more widely, self-boring friction probes of different slenderness ratios inserted in the ground, yields the values of k shown in Fig. 5 for the friction probe at the ground surface in isotropic elastic media. These values hold true for cross-anisotropic media of vertical axis. G_0 or $G(r)$, in equations (2)–(4), then represents the vertical shear modulus. Randolph[13] had estimated the general value

$$k = \ln[2.5\,D\,(1-\nu)/r_0] \qquad (5)$$

from integral equation analyses in isotropic elastic media. These values are also shown in Fig. 5 for $\nu = 0.33$ and $\nu = 0.5$ and are quite close to Orsi's in the range of study common to both. Values from Poulos and Davis[28] also compare well.

Extension of the model

22. Because of its intrinsic nature, it was possible to extend the mechanism of shaft friction, demonstrated for linear elastic or dilatant media, to non-linear media.[15]

23. Let $\gamma_{rz}(\tau)$ be a non-linear shear law of the soil:

$$\gamma_{rz} = \frac{\tau}{G_0} + \chi(\tau) \qquad (6)$$

The integration of the linear term gives V_0 of equation (4) and it can be shown that the integration of $\chi(\tau)$, which represents the additional displacement ΔV_0 due to non-linearity or curvature effect, can be approximated by

$$\Delta V_0 \approx \tau_0 r_0 \int_0^{\tau_0} \frac{\chi(\tau)d\tau}{\tau^2} \qquad (7)$$

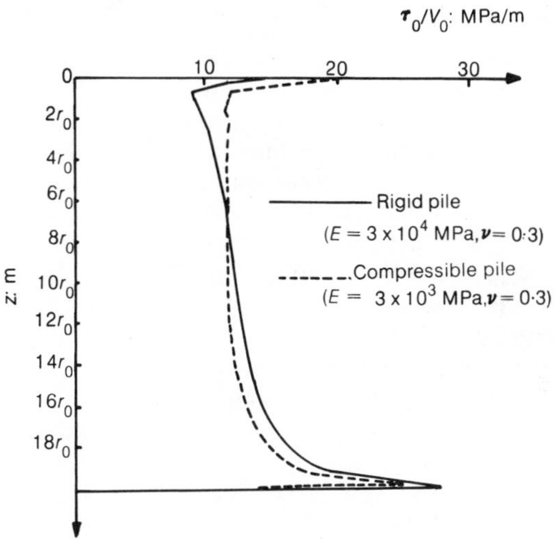

Fig. 4. Variation of Cambefort's parameter τ_0/V_0 with depth in a homogeneous soil (E = 30 MPa, ν = 0.3)[1,6]

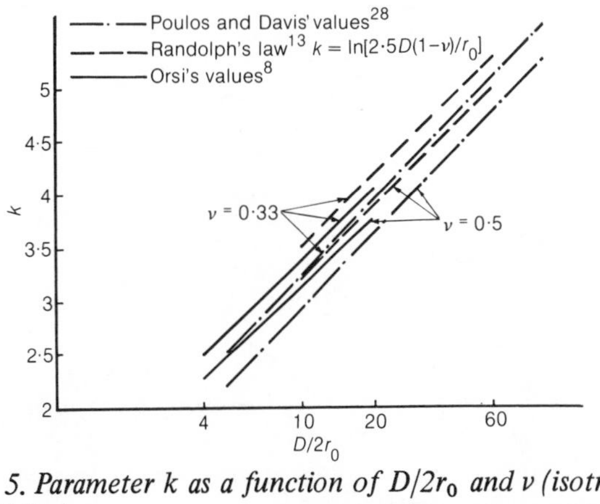

Fig. 5. Parameter k as a function of $D/2r_0$ and ν (isotropic elasticity)

If, for instance, we write $\chi(\tau)$ in a polynomial form

$$\chi(\tau) = A_2\tau^2 + A_3\tau^3 + \ldots + A_n\tau^n + \ldots \quad (8)$$

we get

$$V_0 = kr_0 \left[\frac{\tau_0}{G_0} + \frac{A_2\tau_0^2}{k} + \frac{A_3\tau_0^3}{2k} + \ldots + \frac{A_n\tau_0^n}{k(n-1)} + \ldots \right] \quad (9)$$

showing that non-linear terms are less important on V_0 than on γ_{rz}. In other words the shaft friction curve $\tau_0(V_0/r_0)$ will tend to stay nearer to its initial tangent (of slope G_0/k) than the shear curve $\tau(\gamma_{rz})$ to its own initial tangent (of slope G_0) (Fig. 6).

Practical applications

24. A few selected practical applications are described in the following paragraphs.

25. The fact that σ_r does not increase significantly at the soil/pile interface is of great value for the self-boring pressure-meter tests[16, 17] as it allows for measurement of the natural horizontal pressures 'at rest' in dilative soils such as sands. One could have feared, indeed, that dilatancy caused by the insertion of the probe would increase this pressure.

26. The extension of the model to non-linear behaviour provides a method of determining in situ the vertical shear law of soils from the shaft friction curves given by the self-boring friction probe.[18–20]

27. About piles themselves, the first striking result is that the parameter τ_0/V_0 of Cambefort's model[14] is reasonably constant with depth in a homogeneous soil. We can also now calculate it from the elementary elastic characteristics of the soil and take into account a remoulded annulus of soil around the shaft if necessary (using equations (3) and (4)). Gambin[21] and Cassan[22] applied to pile tests the same ideas as Cambefort's, together with pressure-meter characteristics. By combining his elementary mobilization laws of shaft friction and base load, Cambefort calculated analytically the displacements and the stresses at any point of the pile. Baguelin and Venon[23] extended this work by giving the most general load–settlement curves until failure for compressible and rigid piles. These two studies are supported by experimental data of pile tests. The method used by the French Laboratories of Ponts et Chaussées for determining experimentally the necessary mobilization laws are given by Baguelin et al.[1] Randolph,[13] similarly, combined equation (4) for the shaft with a Boussinesq type law for the base, in order to estimate the load–settlement curves of piles in linear elasticity. He also gave another interesting application of some of these results by studying pile groups.

THE LATERAL REACTION MECHANISM OF PILES

28. It is now common practice to study piles under horizontal loads using Winkler's coefficient of subgrade reaction or, more generally, non-linear reaction curves relating at each level the force density P to the deflexion u_0. It is a very adaptable method for both experimental and theoretical studies and it is capable of describing the entire process until failure. The problem is to know how to determine these reaction curves. The main difficulty lies in the fact that they depend not only on soil properties after installation of the pile, but also on the pile itself (shape, length, relative soil–pile stiffness) and on the loading conditions (nature of the load, fixing conditions). Poulos[24] proposed a rather general solution in linear elasticity from integral equation analyses on a vertical rectangular strip. His study showed quite clearly the influence of the characteristics of the piles and of the mode of loading.

29. The work reported here, using the FEM, is divided into two distinct parts. The first part deals with factors that affect the direct vicinity of the pile: remoulding or plasticity of the soil, and shape of the cross-section of the pile. It is assumed that their influence can be studied by a two-dimensional model (plane strain in the horizontal plane). The second part takes into account the length and the relative stiffness of the pile as well as the loading conditions. These are studied on a cylindrical pile in linear elasticity. In this case the problem, which is a three-dimensional one, can be simplified by using Fourier series (cf. Zienkiewicz[25]).

30. Again the initial state of stress in the soil prior to the loading does not come into account. It would have an influence when plasticity is considered. In the present study Tresca's criterion in the horizontal plane has been examined (for cohesive undrained soils), and we need only assume that the initial stress system is isotropic in the horizontal plane.

Bidimensional study

31. *Effect of remoulding around a circular section in linear elasticity.* The effect of the remoulding of the soil due to the installation of the pile has been studied by Saïd[9] for a circular section, by reducing the elastic modulus in a zone of radius r_1 around the pile section (Fig. 7). The analytical solution has been given using Airy functions for a uniform translation of the soil/pile interface. This solution depends on E and ν of the intact soil, on the ratios

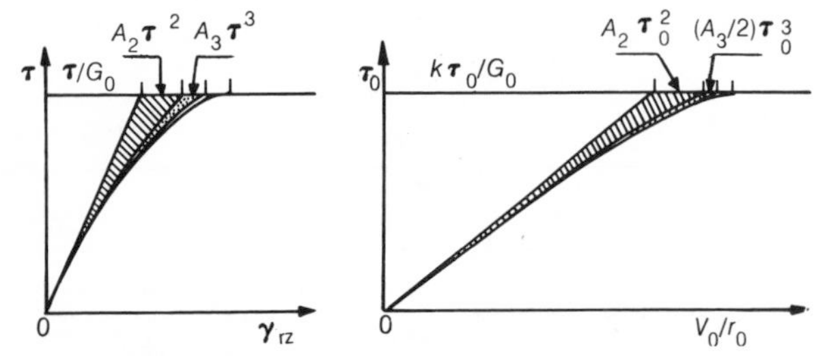

Fig. 6. Polynomial development of the shear law and incidence on the shaft friction law[15]

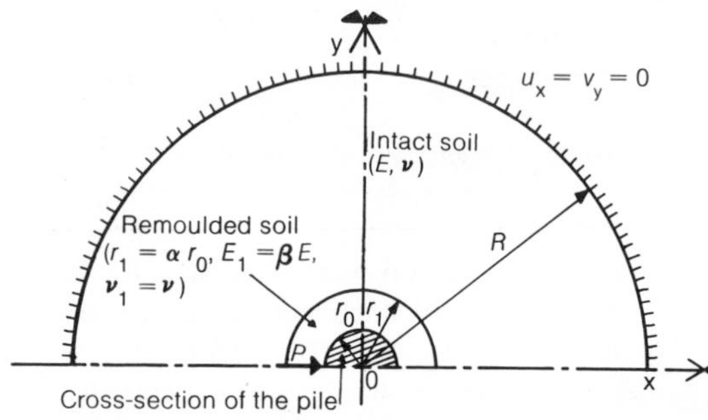

Fig. 7. Bidimensional model for a circular pile with a remoulded annulus of soil[4, 9]

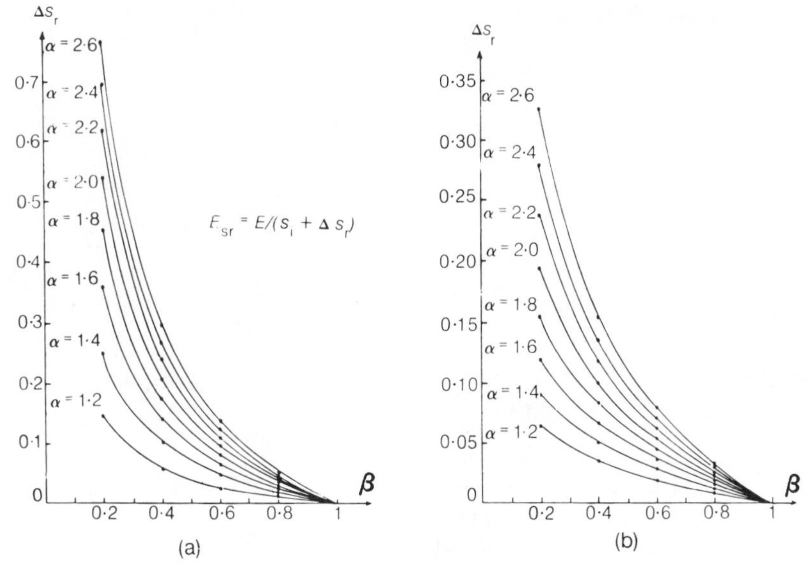

Fig. 8. Effect of disturbance Δs_r (circular pile):[2] *(a) $\nu = 0.33$; (b) $\nu = 0.5$*

$\alpha = r_1/r_0$ and $\beta = E_1/E$ ($\alpha \geqslant 1$ and $\beta \leqslant 1$) and also on the radius R of the outside boundary of the model which is fixed. When R tends to infinity, the displacement of the section of the pile also tends to infinity. This means that the bidimensional assumption does not hold at large distances. This is not annoying because we wish only to know the influence of the disturbed ring, which is proved to affect only the vicinity of the pile.[4]

32. The influence of remoulding on the dimensionless displacement $s = u_0E/P$ can be summed up and quickly computed by the charts of Fig. 8 giving the increases Δs_r as functions of α, β and ν only (once R is large enough, of course).

33. Thus the modulus of reaction of the soil is given by

$$E_{sr} = \frac{P}{u_0} = \frac{E}{s_i + \Delta s_r} \tag{10}$$

where $s_i = E/E_{si}$ is the dimensionless displacement of the circular section in an intact elastic medium, which can only be calculated by three-dimensional analyses (see below).

34. *Effect of the shape of the section in linear elasticity.* The effect of the shape of different rectangular cross-sections has been studied using the FEM by Carayannacou-Trezos.[5] A typical mesh is shown in Fig. 9. The outside boundary is at $R = 25r_0 = 12.5B$, where the shape of the section or the remoulding have no more effect. There is perfect pile–soil adhesion. Sections of slenderness ratio L/B of 1 (square pile), 2, 3 and 5, as well as 1/2, 1/3 and 1/5, have been studied. The shape effect has been examined in an intact soil for $\nu = 0.33$ and $\nu = 0.45$ and also in a disturbed soil for $\nu = 0.33$. The results are synthesized in Fig. 10 giving the shape correction Δs_{fr} as a function of the remoulding characteristics α and β.[2] These calculations were in fact made with the same width e of remoulded soil in front and on the side of the rectangular sections. α is then defined as $1 + e/r_0$.

35. This correction has to be added to the previous correction Δs_r of Fig. 8 when the rectangular section is in a disturbed soil. Thus, the general relation for the dimensionless displacement of the section is

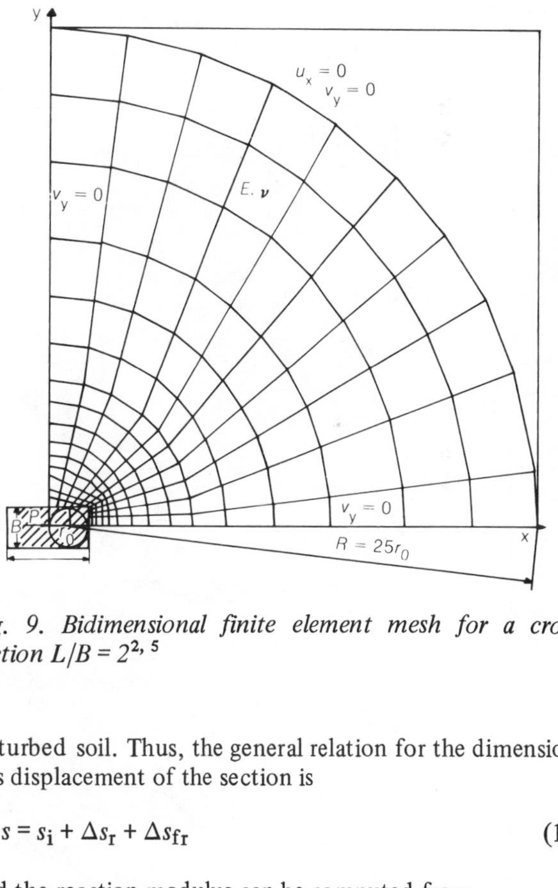

Fig. 9. Bidimensional finite element mesh for a cross-section $L/B = 2$[2, 5]

$$s = s_i + \Delta s_r + \Delta s_{fr} \tag{11}$$

and the reaction modulus can be computed from

$$E_{sfr} = \frac{E}{s_i + \Delta s_r + \Delta s_{fr}} \tag{12}$$

In the case of an intact soil $\Delta s_r = 0$ and the values for Δs_{fr} are obtained for $\beta = 1$ in Fig. 10. These corrections are the total corrections and E_{sf} is an overall reaction modulus: i.e., no separation has been made between normal (front or rear) reactions and tangential reactions. This is made elsewhere.[2] This separation may allow calculation of elongated section piles by coupling of the two different elementary stress systems—pressure-meter type expansion for the normal reactions and side friction for the tangential reactions.

36. *Effects of plasticity around a circular section.* As already mentioned, the case of a cohesive soil in undrained behaviour only has been considered so far.[4] The criterion used is Tresca's:

$$|\sigma_1 - \sigma_2| \leqslant 2c_u$$

where c_u denotes the undrained cohesion. In the case of a disturbed zone around the pile, the cohesion of this zone has been reduced by the same degree as the Young's modulus: $c_{u2} = \beta c_u$, assuming the same plastic yield strain. The circular section of Fig. 7 has thus been studied using the FEM with elasto-plastic behaviour of the soil by Saïd.[9] No relative slippage or soil–pile separation has been introduced. The influence of plastic yield is again proved to take place only near the pile. Fig. 11[4] gives an example of plasticity expansion in front of the section (or in the back, the problem being antisymmetrical about the y axis). The analogy with a cylindrical pressure-meter expansion is obvious.

37. The additional dimensionless displacement Δs_p due to plasticity can be computed from Fig. 12.

38. In the case of an intact soil

$$\Delta s_{pi} = \frac{r_o c_u}{P} \mathrm{f}\left\{\frac{P - P_{ei}}{r_o c_u}\right\} \tag{13}$$

and

$$E_{sp} = \frac{E}{s_i + \Delta s_{pi}} \tag{14}$$

where P_{ei} is the elastic limit load for an intact soil:

$$P_{ei} = 2\pi r_o c_u$$

39. In the case of a remoulded zone around the pile

$$\Delta s_{pr} = \frac{s_i + \Delta s_r}{s_i} \frac{r_o c_u}{E} \mathrm{f}\left\{\frac{P - P_{er}}{r_o c_u}\right\} \tag{15}$$

where P_{er} is the elastic limit load with a remoulded zone:

$$P_{er} = \frac{2\pi r_o \beta c_u}{c_2}$$

where c_2 is a function of α, β and ν.[4]

Three-dimensional study

40. The full problem of piles under horizontal loads is a three-dimensional problem. Though there is an axisymmetrical geometry in the case of circular piles, the loading does not have the same symmetry, and the deformations thereof are three-dimensional. Nevertheless great advantage of the geometry can be taken by expressing the loads and the displacements in Fourier series[25] avoiding full three-dimensional finite element analysis.

41. We can even study our problem by one single Fourier harmonic, if we wish to have uniform translations

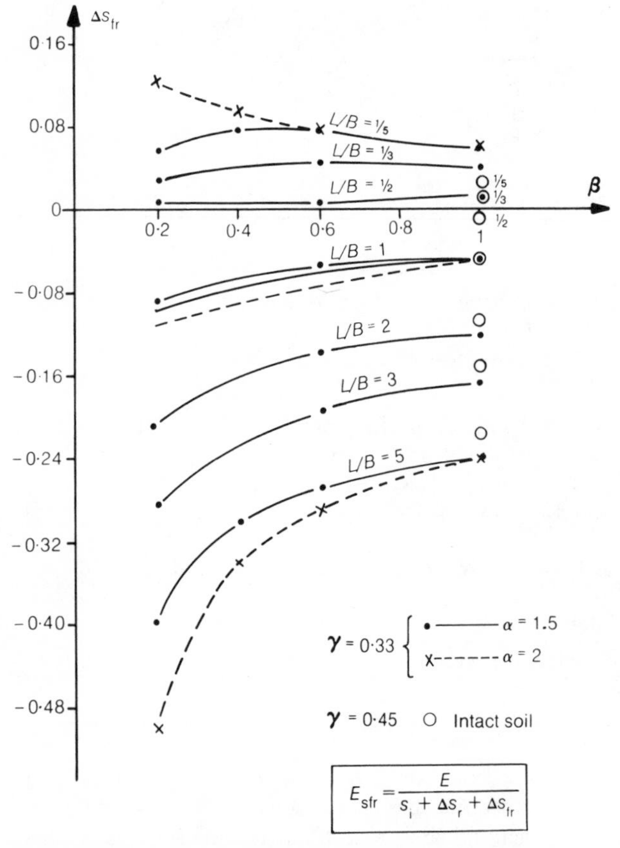

Fig. 10. Shape effect Δs_{fr} for rectangular piles[2]

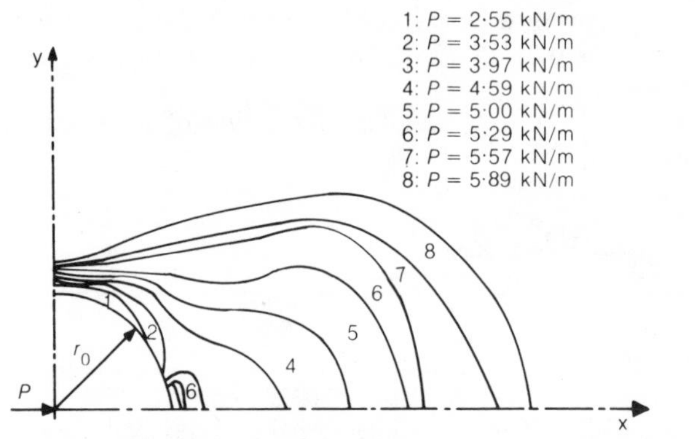

Fig. 11. Expansion of plastic zones in the case of an intact soil:[4, 9] *$E = 13$ MPa, $c_u = 30$ kPa, $r_0 = 10^{-2}$ m*

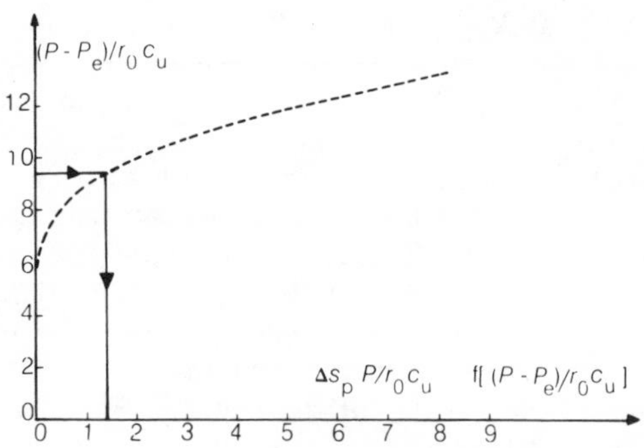

Fig. 12. Additional displacement due to plasticity (intact model)[4]

of the cross-section of the piles, thus assumed to be infinitely rigid in the r–θ plane. In this case displacements u_r and V_θ of the section are necessarily in the form $u_0 \cos\theta$ (u_0 being the translation) and the method requires radial or tangential loads of the same form. Carayannacou-Trezos[5] thus chose a radial density $g(r) = (H/\pi r_0)\cos\theta$, leading to a total horizontal force H. Force or moment resulting loads at the pile head can be easily obtained this way.

42. To date, piles of slenderness ratio $D/2r_0 = 10$ and $D/2r_0 = 25$ have been studied with different relative stiffnesses in homogenous linear elastic soils with $\nu = 0.33$. (Poisson's ratio is assumed to have little influence according to various authors.) Only results concerning the reaction modulus E_{si} (or s_i) are quoted here. Full results of the analyses are given in the literature.[2, 5, 7] Fig. 13 shows the mesh of the radial plane for piles with $D/2r_0 = 10$. The bottom boundary is situated $0.5D$ beneath the pile base and the side boundary at $1.75D$ distance. All three displacements u, v and w were set at zero on the first one, only u and v on the second one. A model with an outside radius of $2D$ instead of $1.75D$ has been tested. It led to no significant difference for the displacements of the pile.

43. The soil is in perfect adhesion to the pile (i.e., all three displacements are continuous between the pile and the soil).

44. Figures 14 and 15 give the variations of $E_{si}/E = 1/s_i$ (s_i to be put in equation (10)) with depth for a force load and a moment load for the $D/2r_0 = 10$ piles. The dimensionless relative soil–pile stiffness factor is D/l_0^* where $l_0^* = (4E_pI_p/E)^{1/4}$, E_pI_p being the bending rigidity of the pile and E Young's modulus of the soil. (The length l_0^* is similar to the relative stiffness factor used in analytical solutions for Winkler soils. Only the elastic modulus E of the continuum is used instead of the reaction modulus E_{si}.) These results clearly show that, even in a homogeneous elastic continuum, the reaction modulus is not constant with depth. This is only approximately true for very rigid piles (low values of D/l_0^*).

45. The problem is to determine a realistic unique value for E_{si} which can be used in simple solutions. We have calculated the value $E_{si}(u)$ which, used in the analytical solutions of the Winkler soil either for an infinitely rigid

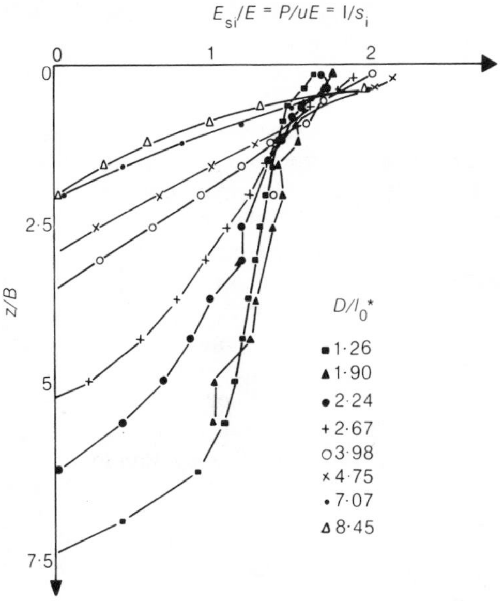

Fig. 14. Variation of E_{si}/E with depth for a force load (pile: $D/2r_0 = 10$, soil: $\nu = 0.33$)[2]

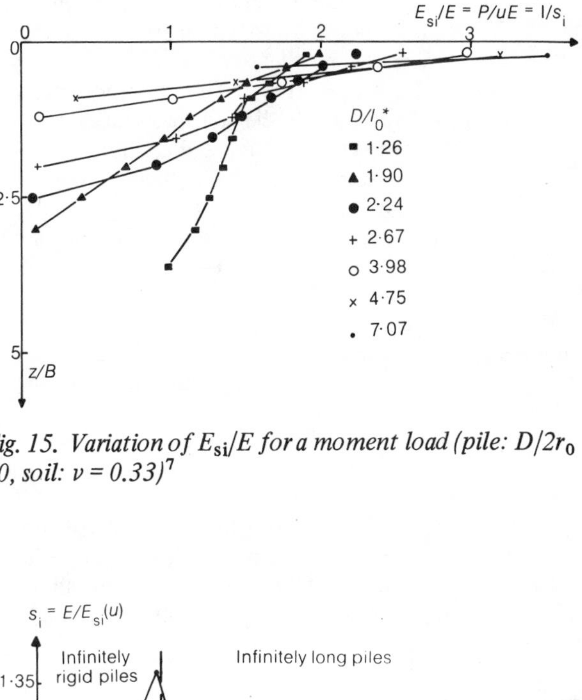

Fig. 15. Variation of E_{si}/E for a moment load (pile: $D/2r_0 = 10$, soil: $\nu = 0.33$)[7]

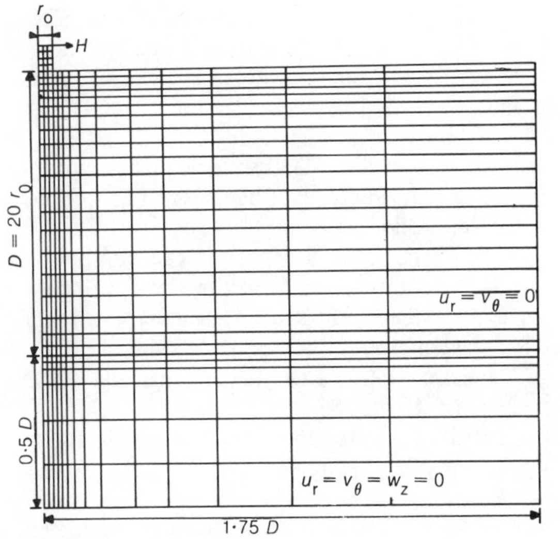

Fig. 13. Mesh of the radial plane for three-dimensional study of a pile $D/2r_0 = 10$[2, 5, 7]

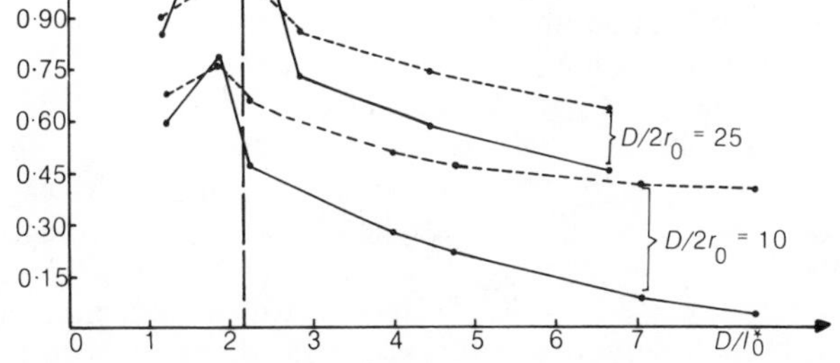

Fig. 16. Variations of equivalent reaction moduli $E_{si}(u)$ with length, rigidity and nature of loading[7]

pile or an infinitely long pile, would give the same head displacements as those obtained by the FEM.

46. These values are given in Fig. 16. They show clearly the influence of the relative soil–pile stiffness, the nature of the loading and the slenderness ratio of the pile.

47. The maximum bending moment for a force load at the head of the pile, or the maximum shear force for a moment load, if they are calculated by the corresponding analytical solutions with these reaction moduli, are larger than the values given by the FEM or very close to them. These values are, with this respect, on the safe side.

48. However, our finite element calculations give head deflexions usually smaller than those of Poulos[24] (whose results are given for $\nu = 0.5$). They match fairly well (except for the very flexible piles under moment load) the general expression given for flexible piles by Randolph[13] who also used the FEM.

FUTURE RESEARCH

49. There is still a lot more to do for such theoretical studies of piles. Even for single piles the research is far from complete. Some more analyses are already possible. Other analyses need preliminary theoretical or experimental studies.

50. This is the case, for instance, of the problem of the initial state of stress in the ground after installation of the pile but before it is loaded. These initial stresses are wanted for any calculation beyond the elastic phase and they play an important role. This is true not only for plasticity or other non-linear behaviour of the soil mass but also for the soil–pile contact problems (traction and friction).

51. For piles under vertical loads, the mechanism of shaft friction is now quite clear. Future research should look at the problem of the propagation of the full mobilization of the side friction and at the punching mechanism at the tip.

52. For piles under horizontal loads, complementary two-dimensional calculations should take into account soil–pile separation at the back of the section or relative slippage in granular or cohesive media. For this kind of study, the problem of matching realistically the contact plasticity criterion with the soil mass criterion usually arises. Here again the theoretical studies cannot be separated from experimental observations. As for three-dimensional studies, different particular points remain to be examined. When it is possible, it will also be interesting to check the validity of some hypotheses or results of the two-dimensional study.

CONCLUSION

53. By giving a synthesis of the results already available for single piles under horizontal or vertical loads, we have tried to show how the FEM can be used for improving our theoretical knowledge of the behaviour of piles. This way of using the FEM applies, of course, to any kind of structure. We think of diaphragm or sheet pile walls for which common design practice is not fully convincing and which can easily be studied with two-dimensional models.

54. For axial loads, the mechanism of shaft friction has been clearly demonstrated by the FEM analyses in linear media (elastic-isotropic or cross-anisotropic and dilatant media). The model of a pile's behaviour proposed by Cambefort[14] is proved to be valid by these studies. The mobilization of the shaft friction with the vertical displacements, at a given depth, has been quantified taking into account, if necessary, a remoulded zone of soil around the pile. The influence of the slenderness ratio of the pile has also been studied.

55. For lateral loads, the FEM has been used to study the validity of the subgrade reaction modulus. The influence of the shape of the cross-section of the pile, of the remoulding and of the plastification of the soil has been quantified by plane strain studies in the horizontal plane. The influence of the length of the pile, of the relative pile–soil stiffness and of the loading conditions (force load or moment load) has been studied in linear isotropic elasticity with a three-dimensional model using Fourier series. Equivalent reaction moduli have been computed, which can be used in the analytical solutions for the homogeneous Winkler soil.

56. This shows how the FEM can also be used to evaluate existing design methods. This can be the case too, of methods, such as the pressure-meter method,[26] which have an experimental basis but no precise theoretical support, and also the case of methods of a theoretical nature but requiring simplistic hypotheses (e.g., bearing capacity theories).

57. Furthermore, the FEM is a hope for refining these existing methods or for developing new ones. This must be accompanied by a corresponding geotechnical progress in two fields: the modelling of soil behaviour and the measurement of soil properties. On the first point, progress has already been made with the strain-hardening elasto-plastic models such as the Cam-Clay model of the University of Cambridge and the Ylight model of the Université Laval of Québec. On the second point, the self-boring devices for in situ testing, such as those developed by the Laboratories of Ponts et Chaussées (self-boring pressure-meter, friction probe, shear meter, permeameter and lateral penetrometer[16,18,27]) and the Camkometer developed by the University of Cambridge[17] are the most significant advances of the past few years.

REFERENCES

1. BAGUELIN F. et al. La capacité portante des pieux. Annls Inst. Tech. Bâtim., 1975, July–Aug., No. 330, SF/116, 1–22. (English summary: Bearing capacity of piles. Materials and Building Research, No. 41, Annls Inst. Tech. Bâtim., 1978, July–Aug., No. 363, 111–119.)

2. BAGUELIN F. et al. Effets de forme et effets tridimensionnels de la réaction latérale des pieux. Bull. Liaison Laboratoires des Ponts et Chaussées, 1979, Nov.–Dec., No. 104, Ref. 2384.

3. BAGUELIN F. et al. Etude de la capacité portante des pieux par la méthode des éléments finis: influence de la dilatance du milieu. Anais do 1° seminário Brasileiro do método dos elementos finitos aplicado à mecânica dos solos, Rio de Janeiro, 1974. COPPE, Universidade Federal do Rio de Janeiro, 1974, 287–302.

4. BAGUELIN F. et al. Theoretical study of lateral reaction mechanism of piles. Géotechnique, 1977, 27, Sept., No. 3, 405–434.

5. CARAYANNACOU-TREZOS S. Comportement des pieux sollicités horizontalement. Dr-Ing thesis, Université Pierre et Marie Curie (Paris VI), 1977.

6. FRANK R. Etude théorique du comportement des pieux sous charge verticale: introduction de la dilatance. Dr-Ing thesis, Université Pierre et Marie Curie (Paris VI), 1974, CNRS No. A0 10766. Laboratoires des Ponts et Chaussées, 1975, rapport de recherche No. 46.

7. FRANK R. and GERMAIN P. Comportement des pieux soumis à un moment en tête. Laboratoire Central des Ponts et Chaussées, Paris, 1978, rapport du Dépt Sols et Fond.

8. ORSI J. P. L'autoforage et le frottement latéral des pieux: étude théorique de l'essai à la sonde frottante. Dr-Ing thesis, Ecole Nat. Ponts et Chaussées, Paris, 1978.

9. SAID Y. H. Etude théorique des pieux sollicités horizontalement. Dr-Ing thesis, Université Pierre et Marie Curie (Paris VI), 1975, CNRS No. A0 11 518.

10. WASCHKOWSKI E. Contribution à l'étude du comportement du sol autour d'une sonde pressiométrique et d'un pieu chargé verticalement. Ing thesis, Conservatoire Nationale des Arts et Métiers, Orléans, 1976.

11. GUELLEC P. et al. La méthode des éléments finis et le système ROSALIE. Bull. Liaison Laboratoires des Ponts et Chaussées, 1976, Jan.–Feb., No. 81, Ref. 1801, 152–178.

12. COOKE R. W. The settlement of friction pile foundations. Proc. conf. tall buildings, Kuala Lumpur, 1974, 7–19.

13. RANDOLPH M. F. A theoretical study of the performance of piles. PhD thesis, University of Cambridge, 1977.

14. CAMBEFORT H. Essai sur le comportement en terrain homogène des pieux isolés et des groupes de pieux. Annls Inst. Tech. Bâtim., 1964, Dec., No. 204, SF/44, 1477–1518.

15. BAGUELIN F. et al. Quelques résultats théoriques sur l'essai d'expansion dans les sols et sur le frottement latéral des pieux. Bull. Liaison Laboratoires des Ponts et Chaussées, 1975, July–Aug., No. 78, Ref. 1687, 131–136.

16. BAGUELIN F. et al. Expansion of cylindrical probes in cohesive soils. J. Soil Mech. Fdn Div. Am. Soc. Civ. Engrs, 1972, 98, Nov., No. SM 11, Proc. Paper 9377, 1129–1142.

17. WROTH C. P. and HUGHES J. M. O. An instrument for in-situ measurement of the properties of soft clays. Proc. 8th Int. Conf. Soil Mech., Moscow, 1973, Vol. 1.2, 1/75, 487–494.

18. AMAR S. et al. The self-boring placement method and soft clay investigation. In: Geotechnical aspects of soft clays. Asian Institute of Technology, Bangkok, 1977, 337–357.

19. BAGUELIN F. and JEZEQUEL J. F. Further insights on the self-boring technique developed in France. Proc. Am. Soc. Civ. Engrs specialty conf. In situ measurement of soil properties, Raleigh, North Carolina, 1975. American Society of Civil Engineers, New York, 1976, II, 231–243.

20. BAGUELIN F. et al. L'autoforage et le frottement latéral des pieux. Proc. 6th Eur. conf. soil mech. fdn engng, Vienna, 1976. International Society Soil Mechanics and Foundation Engineering, Österreichischer National komitee im ÖIAV, Vienna, 1976, Vol. 1.2, III/1–2, 339–344.

21. GAMBIN M. Calcul du tassement d'une fondation profonde en fonction des résultats pressiométriques. Sols-Soils, 1963, Dec., No. 7, 11–31.

22. CASSAN M. Le tassement des pieux; synthèse des recherches récentes et essais comparatifs. Sols-Soils, 1966, No. 18–19, 43–58 and 1968, Mar., No. 20, 23–40.

23. BAGUELIN F. and VENON J. P. Influence de la compressibilité des pieux sur la mobilisation des efforts résistants. Journées nationales sur le comportement des sols avant la rupture, Paris, 1971. Proc. Bull. Liaison Laboratoire des Ponts et Chaussées, 1972, June, Spécial, 308–322.

24. POULOS H. G. Behavior of laterally loaded piles: I–Single piles. J. Soil Mech. Fdn Div. Am. Soc. Civ. Engrs, 1971, 97, May, No. SM5, Proc. Paper 8092, 711–731.

25. ZIENKIEWICZ O. C. The finite element method in engineering science. McGraw Hill, London, 1971.

26. BAGUELIN F. et al. The pressuremeter and foundation engineering. Trans Tech Publications, Clausthal-Zellerfeld, W. Germany, 1978.

27. BAGUELIN F. et al. Le pénétromètre latéral autoforeur, Proc. 9th Int. Conf. Soil Mech., Tokyo, 1977, Vol. 1, 1–7, 27–30.

28. POULOS H. G. and DAVIS E. H. The settlement behaviour of single axially loaded incompressible piles and piers. Géotechnique, 1968, 18, Sept., No. 3, 351–371.

12. Application of a numerical procedure for laterally loaded structures

C. S. DESAI, PhD, MIStructE, MASCE (Professor) and T. KUPPUSAMY, PhD (Assistant Professor; Department of Civil Engineering, Virginia Polytechnic Institute and State University, Blacksburg, Virginia, USA)

It is recognized that laterally loaded structures such as piles and retaining walls generally require three-dimensional analysis. However, it can be cumbersome to perform a three-dimensional analysis incorporating all the major factors such as material and geometric non-linearities. A simple one-dimensional analysis based on the finite element method is presented. The procedure uses a beam column element for the structure and non-linear springs for the soil. The non-linear behaviour of the soil is incorporated by using a generalized Ramberg–Osgood model.

INTRODUCTION

Laterally loaded structures such as piles and retaining walls generally require three-dimensional idealizations. For realistic computations of stress–deformation and ultimate behaviour of these structures it becomes necessary to consider other factors like non-linear behaviour of soil, interaction effects between soil and structure and nature of loading. Fig. 1 shows the three-dimensional nature of axially and laterally loaded structures. If a long pile (in an offshore environment) and its foundation experience large displacements, it may become necessary to consider geometric non-linearity in addition to the material non-linearity of the problem.

2. It is difficult to incorporate all the (major) factors in a solution procedure, analytical or closed form. Numerical procedures such as the finite element method can be used to handle many of the complexities. These complexities can make the solutions difficult from the viewpoints of mathematical derivation and computational effort. Some investigations have been performed or are under way toward solution of the general three-dimensional problem using the finite element method.[1,2] For instance, the Authors and co-workers have been engaged in obtaining solutions for two- and three-dimensional idealizations including geometric and material non-linearities.[1,3,4] The results are, however, in research stages and not yet relevant for practical application.

3. On the other hand, it is possible to adopt simplifying assumptions and devise solutions based on one-dimensional idealizations. The purpose of this Paper is to present one such procedure based on the finite element method. The main emphasis will be on the application potential of the proposed procedure.

BACKGROUND

4. A number of previous investigations have solved the problem of a laterally loaded pile as a one-dimensional idealization in conjunction with the beam-on-elastic-foundation approach by using the finite difference method.[5–8] This procedure has proved to be useful for a number of practical problems. However, often the procedure has been used for loading in one direction and essentially for a single pile. Furthermore, the soil behaviour is simulated by interpolation for material moduli based on a number of data points on the stress–strain or p–y curves.[6,7]

NUMERICAL PROCEDURE USED

5. The procedure used in this Paper is based on a general finite element formulation that includes both lateral and axial behaviour. It also, in an approximate manner, includes simulation of some of the construction sequences, which allows its use for retaining and tie-back walls. The stress–strain response in the form of p–y (or in general terms resistance response[9,10]) curves is simulated by using the idea of a modified form of Ramberg–Osgood model.[11]

Idealization

6. It is assumed that the problem of a laterally loaded structure in a soil foundation can be replaced by an approximate but equivalent one-dimensional system (Fig. 2). The three translational responses in the x, y and z directions respectively are assumed to be uncoupled, thereby permitting superimposition of their effects. The rotational modes of deformation can also be included; however, in the results presented here they are not considered. The main

idea here is to develop a general one-dimensional finite element procedure that can be used for a number of problems.

7. The three-dimensional soil response is replaced by non-linear springs in the three co-ordinate directions (Fig. 2).

Details of finite element procedure[12]

8. The lateral displacements u and v and the axial displacement w are approximated respectively by using cubic and linear approximation functions:

$$\begin{Bmatrix} u \\ v \\ w \end{Bmatrix} = \begin{bmatrix} [N_b] & 0 & 0 \\ 0 & [N_b] & 0 \\ 0 & 0 & [N_a] \end{bmatrix} \begin{Bmatrix} \{q_{b1}\} \\ \{q_{b2}\} \\ \{q_a\} \end{Bmatrix} \qquad (1)$$

where

$$[N_b] = [1-3s^2+2s^3 \quad ls(1-2s+s^2) \quad s^2(3-2s) \quad ls^2(s-1)]$$

$$[N_a] = [1-s \quad s]$$

$$\{q_{b1}\}^T = [u_1 \quad \theta_{x1} \quad u_2 \quad \theta_{x2}]$$

$$\{q_{b2}\}^T = [v_1 \quad \theta_{y1} \quad v_2 \quad \theta_{y2}]$$

$$\{q_a\} = [w_1 \quad w_2]$$

and u_i $(i = 1, 2)$, v_i $(i = 1, 2)$ and w_i $(i = 1, 2)$ are element nodal displacements, θ_{xi} $(i = 1, 2)$, θ_{yi} $(i = 1, 2)$ are the nodal rotations of the structural elements (Fig. 3) and s is a local co-ordinate.

9. Use of a variational procedure leads to the following element equations:

$$[k]\{q\} = \{Q\} \qquad (2)$$

where

$$[k] = \begin{bmatrix} \alpha_x[k_x] & 0 & 0 \\ 0 & \alpha_y[k_y] & 0 \\ 0 & 0 & [k_w] \end{bmatrix} \qquad (2b)$$

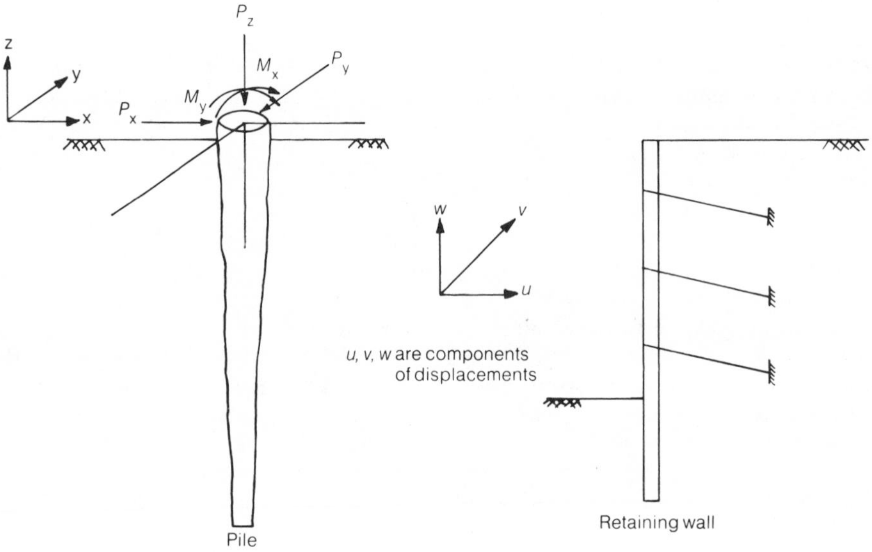

Fig. 1. Axially and laterally loaded structures

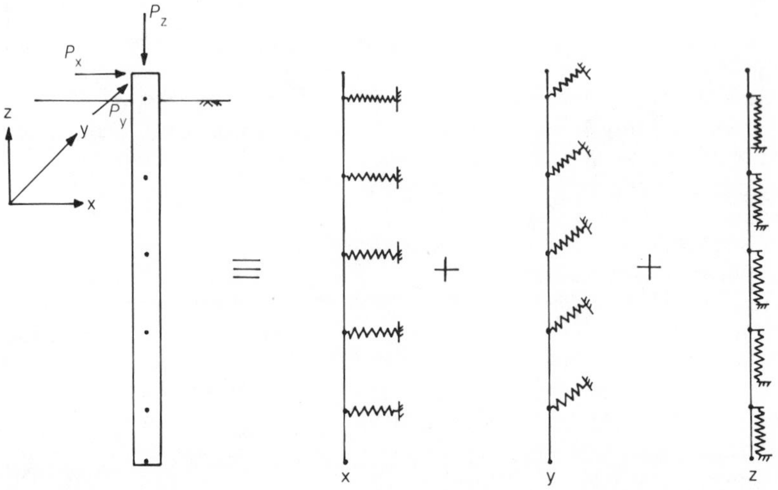

Fig. 2. Responses in x, y and z directions

$\alpha_x = EI_x/l^3$

$\alpha_y = EI_y/l^3$

$$[k_x] = \begin{bmatrix} 12 & -6l & -12 & -6l \\ & 4l & 6l & 2l^2 \\ \text{symm.} & & 12 & 6l \\ & & & 4l^2 \end{bmatrix} \quad (2c)$$

$[k_y]$ is as $[k_x]$ but with appropriate sign changes

$$[k_w] = \frac{AE}{l}\begin{bmatrix} 1 & -1 \\ -1 & 1 \end{bmatrix} \quad (2d)$$

$\{q\}$ is the vector of nodal unknowns and $\{Q\}$ is the vector of nodal forces:

$$\{Q\} = {}_V\!\int [N]^T \{\bar{X}\}\, dV + {}_S\!\int [N]^T \{\bar{T}\}\, dS + P_{i1}\ (i = 1, 2, \ldots, N) \quad (3)$$

Here $\{\bar{X}\}$ and $\{\bar{T}\}$ denote body forces and surface tractions and P_{i1} denotes (local) load at a point and N = number of load points.

Variable geometry

10. It is possible to derive equation (2d) with variable (linear) cross-sectional area and the modulus E within an element as

$$A = [1-s \quad s]\begin{Bmatrix} A_1 \\ A_2 \end{Bmatrix}$$
$$E = [1-s \quad s]\begin{Bmatrix} E_1 \\ E_2 \end{Bmatrix} \quad (4)$$

where A_1, A_2 and E_1, E_2 are the cross-sectional areas and moduli at nodes 1 and 2 respectively. Then the matrix in equation (2d) for axial behaviour becomes

$$[k_w] = \lambda \begin{bmatrix} 1 & -1 \\ -1 & 1 \end{bmatrix}$$

where

$$\lambda = \frac{1}{6l}[A_1\ E_1\ A_2\ E_2]\begin{bmatrix} 0 & 1 & 0 & \frac{1}{2} \\ 1 & 0 & \frac{1}{2} & 0 \\ 0 & \frac{1}{2} & 0 & 1 \\ \frac{1}{2} & 0 & 1 & 0 \end{bmatrix}\begin{Bmatrix} A_1 \\ E_1 \\ A_2 \\ E_2 \end{Bmatrix}$$

For the case of variable A and E, average (mid-section) values of E and I can be used for the bending modes, equations (2b) and (2c).

11. The equations for the elements are assembled and the boundary conditions in terms of displacements or their derivatives are introduced. The solution of the equations gives nodal displacements which are then used to find moments, M, and shear forces, V, as

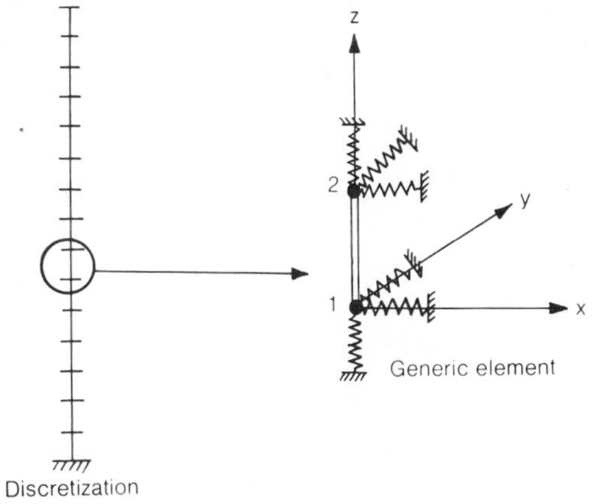

Fig. 3. One-dimensional beam column element

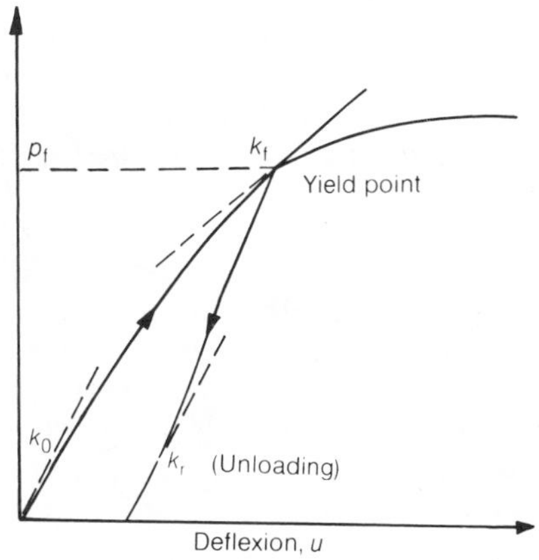

Fig. 4. Ramberg–Osgood model

$$\begin{Bmatrix} M_x \\ M_y \end{Bmatrix} = \begin{Bmatrix} EI_x \dfrac{d^2u}{dz^2} \\ EI_y \dfrac{d^2v}{dz^2} \end{Bmatrix} = \begin{bmatrix} EI_x[B_b] & 0 \\ 0 & EI_y[B_b] \end{bmatrix}\begin{Bmatrix} \{q_{b1}\} \\ \{q_{b2}\} \end{Bmatrix} \quad (5)$$

$$\begin{Bmatrix} V_x \\ V_y \end{Bmatrix} = \begin{Bmatrix} EI_x \dfrac{d^3u}{dz^3} \\ EI_y \dfrac{d^3v}{dz^3} \end{Bmatrix} \quad (6)$$

Here $[B_b]$ = transformation matrix.

Non-linear soil behaviour

12. A general resistance-response that can include the p–y concept as a special case has been recently proposed.[9,10] For the purpose here, the p–y approach used previously[6,8] is adopted. However, the p–y curves are simulated by using a different and somewhat general procedure.

13. Figure 4 shows a modified form of the Ramberg–Osgood model[11] used. Accordingly, the relation between

the resistance p (p_x, p_y or p_z) and the displacement (u, v or w) is assumed; for instance p_x is expressed as

$$p_x = \frac{(k_0 - k_f)u}{\left[1 + \left\{\dfrac{(k_0 - k_f)u}{p_f}\right\}^m\right]^{1/m}} + k_f u \qquad (7)$$

where k_0 is the initial stiffness, k_f the 'final' stiffness, p_f the yield resistance and m the order of the curve. (For $m = 1$, $k_f = 0$ and $p_f = p_0$, equation (7) reduces to a hyperbola.)

14. An incremental iterative procedure is used to incorporate the non-linear behaviour of soil. The non-linearity is assumed as piecewise linear and at each step of loading the parameters are updated by using equation (7).

Simulation of excavation and tie-backs

15. The foregoing procedure is modified to permit incorporation of some construction sequences such as excavation and tie-backs. This capability can be used to analyse approximately, perhaps as a preliminary step in analysis and design, retaining structures with tie-backs idealized as one-dimensional.

16. Excavation is simulated by applying to the structure surface tractions equivalent to the lateral earth pressure caused by in situ stresses. Then the anchor load is introduced at a given point on the structure. The procedure involves a number of iterative cycles until equilibrium is established at every stage. For example, in Fig. 5, after stage 1 of excavation up to node 3, the first anchor is installed. Loads opposite in sign to the surface traction are then applied at nodes 1 and 2 and the springs at nodes 1 and 2 are deleted. The anchor load is applied in a number of increments; for each increment, iterations are performed to reach approximately the equilibrium state. The procedure is continued for subsequent stages.

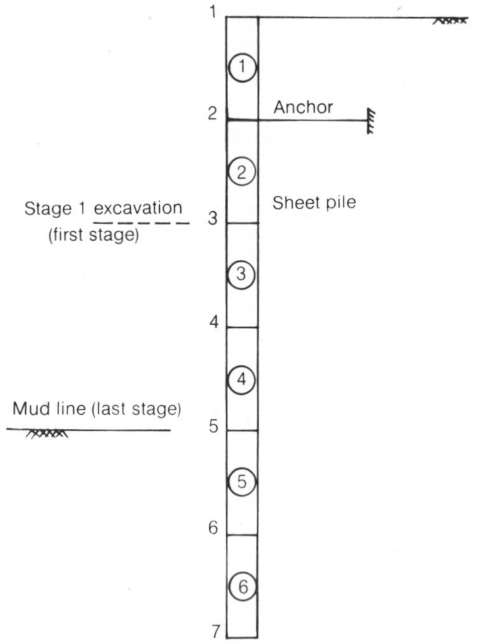

Fig. 5. Simulation of excavation

APPLICATIONS

17. In order to establish the validity of the procedure, a number of problems have been solved.[12] They involve comparisons of numerical prediction with closed form solutions, laboratory and/or field observations. Here, three examples are presented.

Example 1: laterally loaded pile—a field problem

18. A wooden pile with the following properties was tested in the field at the site of the Arkansas Lock and Dam No. 4:[13]

length of the pile = 40 ft
diameter = 14 in
E (pile) = 1.6×10^6 lbf/in^2
I (pile) = 1950 in^4

The pile was driven in a dense sand with angle of friction $\phi = 40°$, coefficient of earth pressure at rest $K_0 = 0.4$, and coefficient of active earth pressure $K_a = 0.22$. Table 1 shows the values of p–y curves at various depths obtained by using the criteria proposed by Reese and Matlock.[6–8] This data was used to develop the parameters in the Ramberg–Osgood model, equation (7). In the numerical procedure, increments of load equal to 2 kips were applied. Fig. 6 shows the comparisons between the numerical predictions from the computer code SSTIN-1DFE and the field observation; the code is part of a general purpose computer series CANDICE. Fig. 6 also includes numerical predictions obtained by using a program, COM52, based on the finite difference method.[6]

Example 2: pile with general loading

19. The formulation and code SSTIN-1DFE can be applied for general loading conditions. In order to illustrate this, the problem shown in Fig. 7 is solved. The pile is subjected to loads F_x, F_y and F_z and moments M_x and M_y at the top. Linearly varying surface tractions are also applied along the top portion of the pile.

Table 1. p–y curves at various depths

Depth = 0		Depth = 8 ft		Depth = 10 ft		Depth = 24 ft		Depth = 32 ft	
y, in	p, lbf/in	y, in	p, lbf/in	y, in	p, lbf/in	y, in	p, lbf/in	y, in	p, lbf/in
0.000	0.00	0.000	0	0.000	0	0.000	0	0.000	0
0.045	29.46	0.240	1292	0.330	4051	0.330	6076	0.330	8102
0.106	58.93	0.606	2585	0.819	8102	0.819	12153	0.819	16204
0.195	88.39	1.090	3877	1.470	12153	1.470	18229	1.470	26306
0.352	117.85	1.930	5170	2.620	16204	2.620	24306	2.620	32408
0.700	117.85	3.860	5170	5.240	16204	5.240	24306	5.240	32408

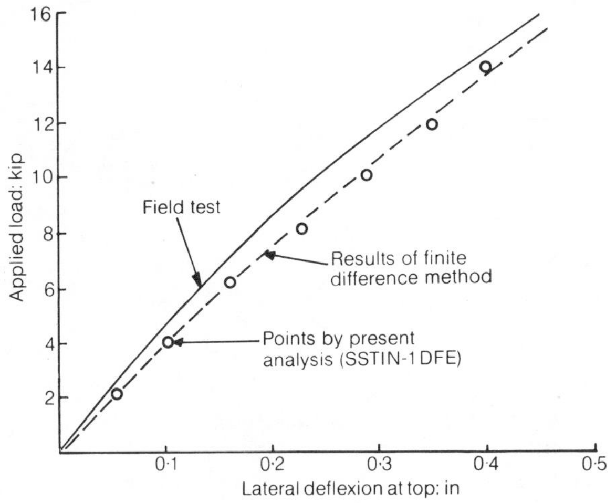

Fig. 6. Wooden pile; Arkansas River, Lock and Dam No. 4

20. The soil is replaced by non-linear springs in the x, y and z directions. It is assumed that the properties are the same in both lateral directions. The following parameters were chosen for the soil response:

k_0 at the top = 0
k_0 at the bottom = 980.7 kN/m^2
p_0 at the top = 0
p_0 at the bottom = 1961.4 kN/m^2
k_f for all depths = 0
$m = 1$

The other properties of the system were assumed as follows:

length of pile = 2.80 m
E (pile) = 19.61 × 10^6 kN/m^2
area = 0.1 m^2
$I_x = I_y = 1.0 \times 10^{-4}$ m^4

21. The applied loads (concentrated loads, surface traction and moments) were divided into three increments. For the concentrated loads and moments the increments were

$\Delta F_x = 4.9$ kN
$\Delta F_y = 4.9$ kN
$\Delta F_z = 19.6$ kN
$\Delta M_x = \Delta M_y = 0.1$ kN m

22. Surface traction forces may be acting due to external pulls by mooring forces in offshore piles. Such forces can be distributed over the elements. Linearly varying surface tractions can be specified. In this problem linearly varying surface traction was applied along the top 0.40 m of the pile above the ground level. The surface traction was applied in both x and y directions, varying linearly from 980 kN/m^2 at the top to 4900 kN/m^2 at the bottom (Fig. 7).

23. The results of the problem obtained by using the code SSTIN-1DFE are shown in Fig. 8. Part (a) shows a plot of M_x and M_y with depth. Part (b) shows the deflexions of typical nodes 1, 3 and 5. Each node experiences u, v and w components of displacement; however, only the components u and v of typical nodes are shown. The resultant displaced position of each node in plan is shown by a dotted line.

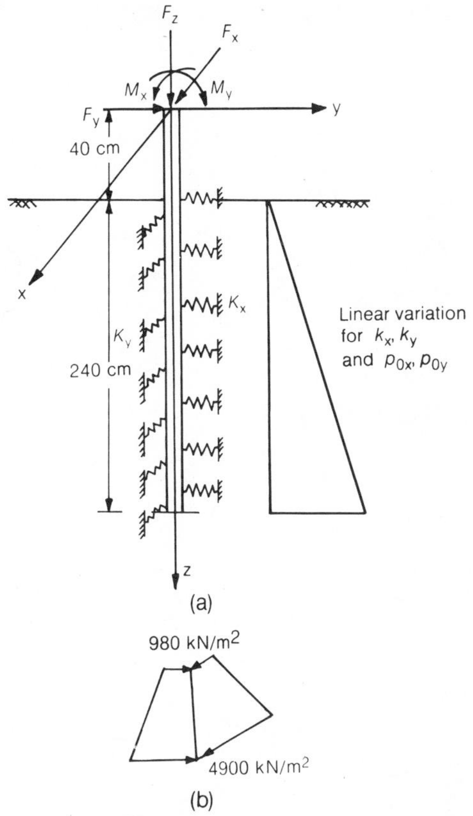

Fig. 7. General pile problem: (a) pile; (b) surface traction

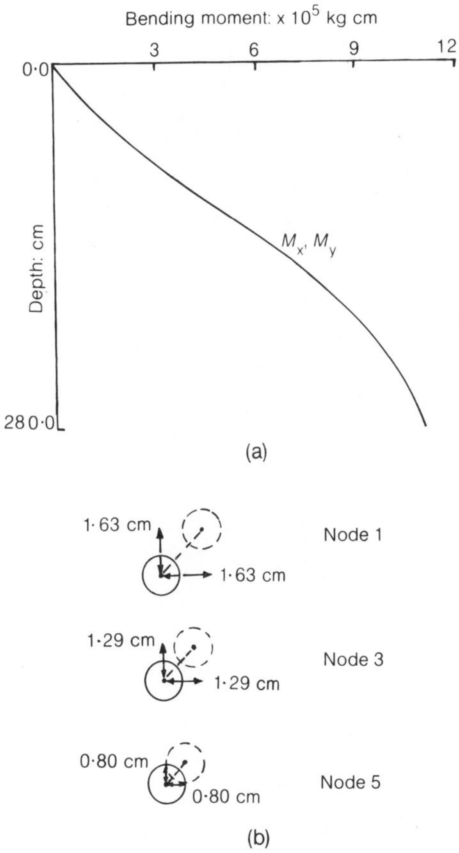

Fig. 8. Pile with general loading: (a) bending moments versus depth; (b) displacement of nodes in plan

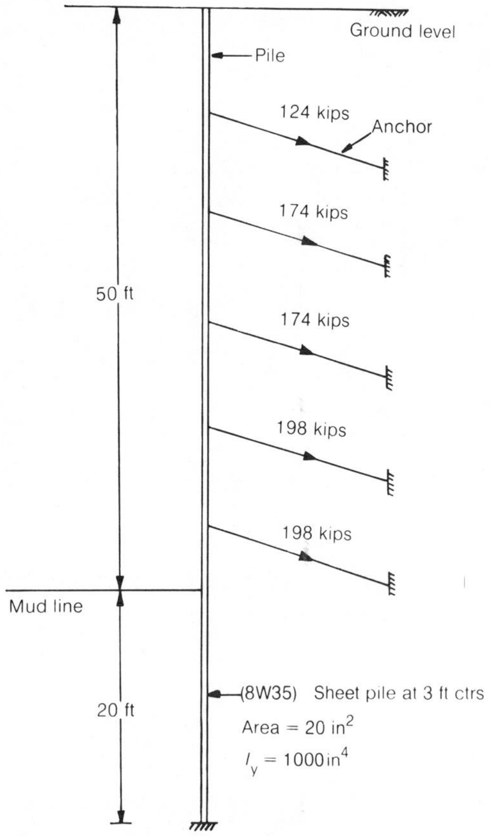

Fig. 9. Sheet pile wall[14]

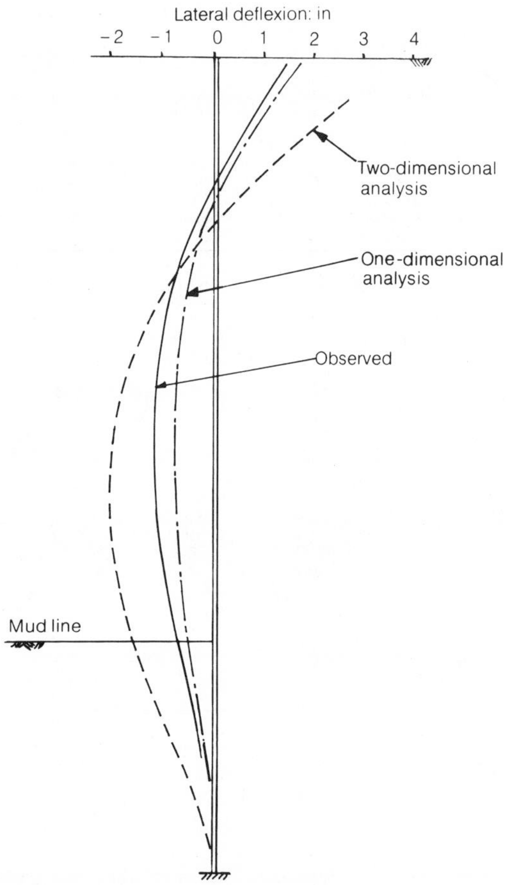

Fig. 10. Lateral deflexion of sheet pile wall

Table 2. p–y curve

y, in	p, lbf/in
0.000	000.0
0.075	623.6
0.150	741.7
0.225	820.8
0.300	882.0
0.600	1048.9
1.200	1247.3
5.700	1764.0

Example 3: tied back sheet pile wall including simulation of excavation and installation of tie-backs

24. A feature of the code SSTIN-2DFE lies in its capacity to solve problems of retaining walls with tie-backs, idealized as one-dimensional. (Furthermore, the simulation of construction sequences is possible, as already explained.)

25. The problem considered here has been solved previously using a two-dimensional finite element procedure.[14]

26. The sheet pile wall was used for supporting (open) deep excavation for a bank building in Seattle, Washington.[14] Fig. 9 shows the dimensions and properties of the structure–foundation system. The tie-backs or anchors were installed in stages with their load capacity in the range 120–200 kips.

27. The foundation and back soil was a cohesive material with an average undrained shear strength of 2.1 UStonf/ft^2. The p–y curves were evaluated on the basis of this value. Table 2 shows the points on the computed curve. Using this data the parameters for the Ramberg–Osgood model were found as

$$k_0 = 8314.7 \text{ lbf/in}^2$$
$$p_f = 1764.0 \text{ lbf/in}$$
$$k_f = 0$$
$$m = 1$$

28. The process of excavation is simulated approximately by applying a surface traction equal to the lateral earth pressure due to the initial (in situ) stresses caused by overburden. For this problem, the amount of such force during a stage of excavation is computed from the value of undrained strength, 2.1 UStonf/ft^2. This force was equal to 32.70 lbf/in^2 and was applied in increments up to the depth excavated at each stage. For each excavation stage five load increments were applied.

29. The anchors were installed at 3 ft intervals (in the longitudinal direction). The applied anchor load per unit length was obtained by dividing the total anchor load by 3. This load was applied in five equal increments. A total of eleven steps of construction sequence, comprising six excavation stages and five anchors, were adopted.

30. Figure 10 shows the comparison between the results of the deflexions of the wall at the end of construction up to the mud line and anchoring all the tie-backs, obtained from the present one-dimensional analysis using the SSTIN-1DFE code, observed field data, and the two-dimensional analysis. Better agreement with the field data is shown by the results from the present analysis than by those from the two-dimensional analysis. Such an agreement may indeed be fortunate; however, the trends of the

results are encouraging. It is felt that the present procedure can provide preliminary results for anchored sheet pile walls. At the same time the cost of such computation is much smaller than that for the two-dimensional procedure. For example, the sheet pile wall problem described above with eleven construction sequences and five iterations at each step was solved in about 3 min of CPU time by an IBM 370/176 computer.

Comments on cost

31. Formulation by the finite element method involves an increase in the derivations and the amount of data to be handled. This is partly due to the general provisions in the program. A preliminary comparison for a simple pile problem with load in one direction indicates that the finite difference procedure is less time-consuming; it takes about 4 s of CPU time per increment of load against 7 s for the present code. However, the cost ratio will change in favour of the finite element method with inclusion of the foregoing additional factors. Moreover, the CPU time and the associated cost with both procedures is quite small, of the order of about $5 for the entire incremental solution for a given problem; hence, the cost factor may not have significant relevance.

CONCLUSIONS

32. A simple one-dimensional analysis based on the finite element method has been developed. A number of problems have been solved satisfactorily by use of the procedure. It is believed that in view of the simplicity and reduced computer time, the procedure has significant potential for use in analysis and design of axially and laterally loaded structures.

REFERENCES

1. DESAI C. S. and APPEL G. C. Three-dimensional analysis of laterally loaded structures. Proc. 2nd Int. Conf. Numerical Methods in Geomechanics, Blacksburg, Va, 1976.
2. WITTKE W. et al. Three-dimensional calculation of the stability of caverns, tunnels, slopes and foundations in anisotropic jointed rock by means of the finite element method. Schriftenr. Dtsch Geo. Erd- Grundbau Dtsch Beitr Geotech., 1972, No. 1.
3. DESAI C. S. and PHAN H. V. Three-dimensional finite element analysis including material and geometric nonlinearities. Second Int. Conf. Computational Methods in Nonlinear Mechanics, Austin, Texas, 1979.
4. DESAI C. S. et al. Geometric and material nonlinear three-dimensional finite element analysis for structures moving in ground. Dept Civil Engng, Virginia Polytechnic Institute and State University, Blacksburg, Va, 1978, report VPI-E-78-11.
5. GLESSER S. M. Lateral load tests on vertical fixed head and free head piles. Symp. Lateral Load Test Piles. American Society for Metals, 1953, spec. pap. 154.
6. REESE L. C. Laterally loaded piles: program documentation. J. Geotech. Engng Div. Am. Soc. Civ. Engrs, 1977, 103, Apr., GT4.
7. REESE L. C. and MATLOCK H. Non-dimensional solutions for laterally loaded piles with soil modulus assumed proportional to depth. Proc. 8th Texas Conf. Soil Mechanics Foundation Engineering, University of Texas, 1956.
8. REESE L. C. and DESAI C. S. (DESAI C. S. and CHRISTIAN J. T. (eds)). Laterally loaded piles. In: Numerical methods in geotechnical engineering. McGraw-Hill, 1977, chapter 9.
9. DESAI C. S. Analysis of laterally loaded structures. Proc. Conf. Finite Elements in Nonlinear Solid and Structtural Mechanics, Geilo, Norway, 1977.
10. DESAI C. S. et al. Interaction and load transfer in vehicle guide way systems. Dept Civil Engng, Virginia Polytechnic Institute and State University, Blacksburg, Va, 1978. Report to Dept Transportation, University Research, Washington, DC.
11. DESAI C. S. and WU T. H. A general function for stress–strain curves. Proc. 2nd Int. Conf. Numerical Methods in Geomechanics, Blacksburg, Va, 1977.
12. DESAI C. S. and KUPPUSAMY T. User's manual and background for a computer code for axially and laterally loaded piles and retaining walls (SSTIN-1DFE). Dept Civil Engng, Virginia Polytechnic Institute and State University, Blacksburg, Va, 1978, report 78–2.
13. FRUCO AND ASSOC. Results of tests on foundation materials, Lock and Dam No. 4. US Army Engr. Div. Lab., Southwestern, Corps of Engrs, SWDGL, Dallas, Texas, 1962, Reports 7920 and 7923.
14. CLOUGH G. W. et al. Design and observations of tied back wall. Proc. Sp. Conf. Performance of Earth and Earth Supported Structures, Purdue University, 1972.

13. Analysis of some reported case histories of laterally loaded pile groups

P. K. BANERJEE and T. G. DAVIES (Department of Civil and Structural Engineering, University of Wales, University College, Cardiff)

A brief account of recent developments of the boundary element method in pile group analysis is given. The general formulation developed by using the reciprocal work theorem is applied to some classes of pile group problems involving inhomogeneous soils. The computational efforts involved are greatly reduced by the selection of the most advantageous point force solutions in each case. The analysis presented here includes features such as slip at the pile–soil interface as well as yielding of the soil. A systematic study of some reported case histories of laterally loaded piles is described to demonstrate the practical utility of these analyses.

INTRODUCTION

During the last decade remarkable progress has been made in the development of the boundary element method as a problem-solving tool. For bulky three-dimensional solids the method has distinct advantages over the finite element method. This development has reached such a state that it is now possible to carry out routine analysis of two- and three-dimensional elastic and elasto-plastic problems involving any number of piece-wise homogeneous zones.[1–5]

2. For certain classes of problems, the computational work can be substantially reduced by the development of appropriate point force solutions. In addition, in the present problem the efficiency of the algorithm may be greatly enhanced by the adoption of a hybrid procedure in which the finite size components of the problem (e.g., piles) may be modelled by using the finite element or the finite difference method and the exterior region by the boundary element method. Examples of such applications may be found in works of Butterfield and Banerjee,[6, 7] Poulos,[8–10] Banerjee and Driscoll[11, 12] and Banerjee and Davies,[13–15] in which the working load (elastic) responses of piles and pile groups are explored.

3. In this Paper the Authors present these analyses in a unified manner by stressing the fundamental energy principles on which they are based, and demonstrate their extension to non-linear behaviour such as slipping and yielding. The analysis is applied to some reported case histories to illustrate its usefulness for routine analysis of pile group problems.

METHODS OF ANALYSIS FOR ELASTIC RESPONSE

The general boundary element formulation

4. The basic requirements for developing a boundary element formulation for elasto-statics are a reciprocal identity such as the reciprocal work theorem, and a point source solution of the governing differential equation (e.g., the Kelvin[2] solution for a point load within an infinite solid).

5. If we consider two elastic equilibrium states (one real and the other virtual) of displacements (u_i, u_i^*), tractions (t_i, t_i^*) and body forces (f_i, f_i^*), then by virtue of the reciprocal theorem we have

$$\int_S t_i(x)\,u_i^*(x)\,\mathrm{d}S + \int_V f_i(z)\,u_i^*(z)\,\mathrm{d}V = \int_S t_i^*(x)\,u_i(x)\,\mathrm{d}S + \int_V f_i^*(z)\,u_i(z)\,\mathrm{d}V \quad (1)$$

where S is the surface enclosing the volume V, x is a point on the surface S and z is a point within V.

6. If we choose the virtual state as that arising out of the point force solution, it can be shown[16] that

$$\alpha u_j(\xi) = \int_S [G_{ij}(x,\xi)t_i(x) - F_{ij}(x,\xi)u_i(x)]\,\mathrm{d}S + \int_V f_i(z)\,G_{ij}(z,\xi)\,\mathrm{d}V \quad (2)$$

where

$\alpha = 0$ if ξ is outside S

$\alpha = \frac{1}{2}$ if ξ is on S

$\alpha = 1$ if ξ is inside S

and $G_{ij}(x,\xi)$ is a function which gives the displacements at a point x due to a unit force vector at a point ξ, and $F_{ij}(x,\xi)$ is a function which gives the corresponding surface tractions.

7. In the absence of any body forces we can recast equation (2) in matrix notation as

$$\alpha \mathbf{I} \boldsymbol{u} = \int_S [\mathbf{G}\boldsymbol{t} - \mathbf{F}\boldsymbol{u}]\,\mathrm{d}S \quad (3)$$

where, for three-dimensional problems, **G** and **F** are 3 × 3 matrices, $\boldsymbol{t}$ and $\boldsymbol{u}$ are 3 × 1 vectors, and **I** is the unit matrix.

8. If we divide the surface into a number (N) of boundary elements and replace the integral by a piece-wise summation we can write equation (3) for the pth node on the surface as

$$\alpha \mathbf{I} \boldsymbol{u}^p = \sum_{q=1}^{N} \left[\left(\int_{\Delta S^q} \mathbf{G}^{pq} \mathbf{M}^q \, dS \right) \boldsymbol{t}^q - \left(\int_{\Delta S^q} \mathbf{F}^{pq} \mathbf{N}^q \, dS \right) \boldsymbol{u}^q \right] \quad (4)$$

where $\mathbf{M}^q$ and $\mathbf{N}^q$ are the shape functions for the variations of the tractions and displacements respectively over the qth boundary element, and $\boldsymbol{t}^q$ and $\boldsymbol{u}^q$ are the nodal values of tractions and displacements respectively for the qth boundary element.

9. By using equation (4) for all nodal points on the surface we can form the following system of equations:

$$\mathbf{A}\boldsymbol{t} - \mathbf{B}\boldsymbol{u} = 0 \quad (5)$$

NOTATION

A, B	final system matrices obtained from the general boundary element method of analysis
$\boldsymbol{b}$	pile cap boundary condition vector
C_u	undrained cohesion of soil
D	shaft diameter
D	finite difference coefficient matrix for the pile compressibility and the pile flexibility
E, E_0, E_1	Young's modulus of soil
E_pI_p	flexural rigidity of pile
f_i, f_i^*	real and virtual body forces per unit volume
f_{ij}	global flexibility coefficients for the pile cap
F_{ij}, **F**	traction kernel functions due to a unit force vector
G_{ij}, **G**	displacement kernel functions due to a unit force vector
H	strain kernel function due to a unit force vector
K	final system matrix for the soil domain
L	embedded length of the pile
m	rate of increase of Young's modulus with depth
N	number of boundary elements
P, H, M	loading system on the pile cap (cf. Fig. 1)
S, V	surface and volume respectively
S/D	spacing to diameter ratio
t_i, t_i^*	real and virtual surface tractions
$\boldsymbol{t}'$	surface traction on the surface of the complementary region
$\boldsymbol{t}_0$	initial surface traction generated by slip
u_i, u_i^*	real and virtual displacements
$\boldsymbol{u}'$	displacement on the surface of the complementary region
$\boldsymbol{u}, \boldsymbol{t}$	vector equivalent of u_i, t_i
U	displacement vector at the pile–soil interface
u, w	horizontal and vertical pile cap displacements
α	a constant
ϕ	traction at the pile–soil interface elements
Φ	traction vector at the pile–soil interface
ν	Poisson's ratio
σ_0	initial stress vector within the volume cells

10. In any well posed problem exactly half the nodal values of $\boldsymbol{u}$ and $\boldsymbol{t}$ will be specified, and equation (5) may be solved to obtain the remaining values.[1, 5, 16] For a piece-wise homogeneous solid, equation (5) may be formed for each zone; the equations are then assembled and solved by use of a procedure similar to that used in the finite element method.[2–4]

Simplification of the analysis for pile group problems

11. If we consider the complementary region V^+ which is outside S and apply surface tractions $\boldsymbol{t}'$ and displacements $\boldsymbol{u}'$ on S it can be shown[17] that for any point within V we have

$$\boldsymbol{u} = \int_S [\mathbf{G}(\boldsymbol{t} + \boldsymbol{t}') - \mathbf{F}(\boldsymbol{u} - \boldsymbol{u}')] \, dS \quad (6)$$

Since $\boldsymbol{u}'$ may be chosen such that $\boldsymbol{u} - \boldsymbol{u}' = 0$ we have

$$\boldsymbol{u} = \int_S \mathbf{G}\phi \, dS \quad (7)$$

where ϕ is a fictitious traction equal to $\boldsymbol{t} + \boldsymbol{t}'$.

12. In pile group problems it has been shown by Butterfield and Banerjee[18] that the fictitious traction ϕ is nearly identical to the real surface traction at the pile–soil interface for pile slenderness ratios greater than 5. We can visualize this physically by considering the region V to be the soil domain with the pile–soil interface as the boundary S. If a surface traction $\boldsymbol{t}$ is applied to the region V the complementary region V^+ (i.e., soil-filled pile domain) suffers almost a rigid body displacement, hence $\boldsymbol{t}' \approx 0$.

13. If we divide the pile–soil interface into a number of surface elements and assume a variation of ϕ over these elements, we can write equation (7) for all nodal points of the pile–soil interface as

$$U = \mathbf{K}\Phi \quad (8)$$

14. Assuming that the function **G** satisfies the zero traction boundary condition on the ground surface we can couple equation (8) with a corresponding set obtained from the considerations of the compressibility and the flexural rigidity of the pile domain:[11]

$$U = \mathbf{D}\Phi + \boldsymbol{b} \quad (9)$$

where **D** is a matrix of finite difference coefficients and $\boldsymbol{b}$ is the boundary condition vector.

15. By satisfying the interface equilibrium and compatibility conditions we obtain

$$(\mathbf{K} - \mathbf{D})\Phi = \boldsymbol{b} \quad (10)$$

which can be solved to obtain Φ.

16. The final solution relating the axial load P, the horizontal load H and the moment M at the pile head (Fig. 1) to the vertical displacement w, the horizontal displacement u and the rotation θ is given by the global pile head flexibility equations:[11]

$$\begin{Bmatrix} w \\ u \\ \theta \end{Bmatrix} = \begin{bmatrix} f_{11} & f_{12} & f_{13} \\ f_{21} & f_{22} & f_{23} \\ f_{31} & f_{32} & f_{33} \end{bmatrix} \begin{Bmatrix} P \\ H \\ M \end{Bmatrix} \quad (11)$$

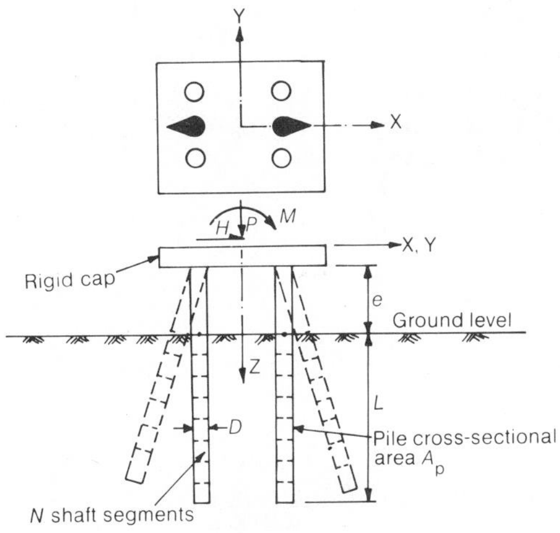

Fig. 1. A typical pile group problem

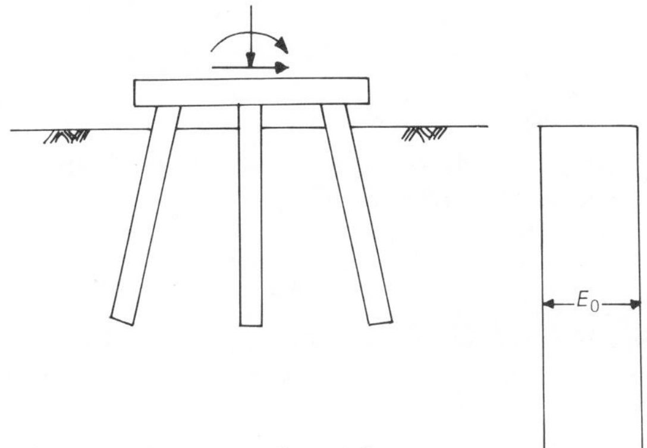

Fig. 2. Homogeneous soil model

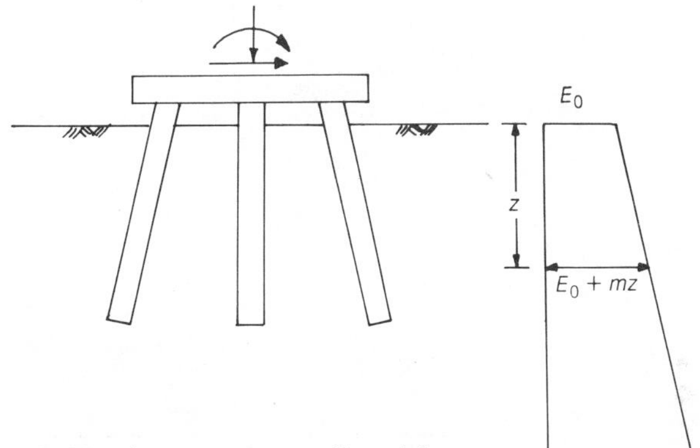

Fig. 3. Non-homogeneous soil model

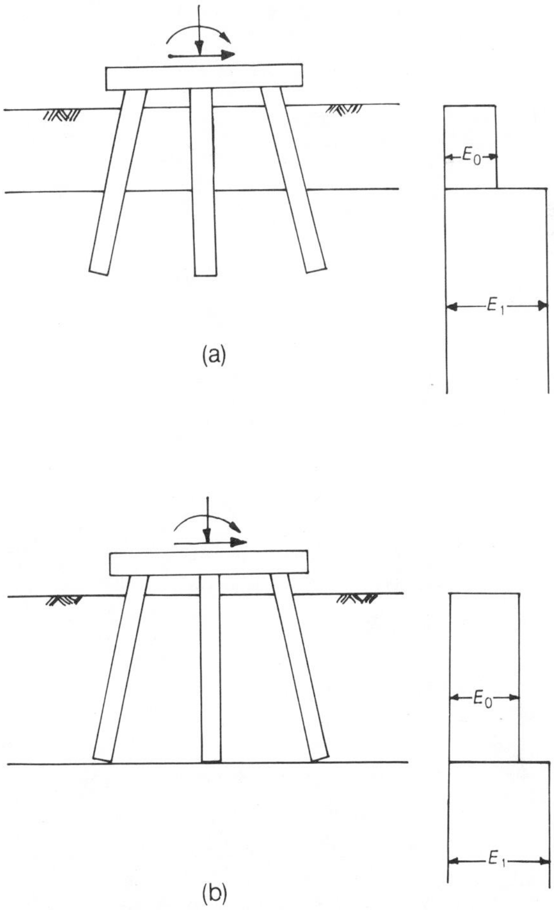

Fig. 4. Two-layer model

where $f_{ij} = f_{ji}$ and for symmetrical vertical pile groups $f_{12} = f_{13} = f_{21} = f_{31} = 0$.

17. The above analysis has been developed for completely general pile groups with or without ground contacting caps and, depending on the choice of the function **G**, can be applied to the following classes of problems:

(a) homogeneous soils of infinite depth by using Mindlin's solution;[19]
(b) two-layer soils by using an integral transform solution;[20]
(c) a linear increase of soil modulus with depth.[15, 21]

Figs 2–4 show typical problems that can be idealized by use of the present analysis.

METHODS OF ANALYSIS FOR NON-LINEAR PROBLEMS

18. Recently the Authors and co-workers have shown[16, 22] how the general boundary element method of analysis can be extended to elasto-plasticity by using an initial stress, an initial strain or a modified body force algorithm. Essentially, these algorithms use an incremental and iterative procedure in which the effects of yielding and slipping are introduced by distributing initial stresses over volume cells and distributing initial surface tractions over slip-surfaces, respectively.

19. In the presence of any initial stresses and initial surface tractions (both are unknown a priori) we can write equation (7) as

$$\dot{u} = \int_S \mathbf{G}\dot{\underset{\sim}{\Phi}}\,\mathrm{d}S + \int_{S_0} \mathbf{G}\dot{t}_0\,\mathrm{d}S + \int_V \mathbf{H}\dot{\underset{\sim}{\sigma}}_0\,\mathrm{d}V \qquad (12)$$

where S_0 is a part of the surface S where slipping is likely to occur, $\dot{t}_0$ and $\dot{\underset{\sim}{\sigma}}_0$ are initial surface tractions and initial stresses respectively arising from any non-linear event, **H** is a function which gives the strains at a point x due to a unit force at a point ξ, and the superior dots indicate incremental quantities.

20. Equation (12), which is valid for an infinitesimal increment of loading, may be discretized by dividing the pile surface into surface elements and the plastic region into volume cells in a manner similar to that described above;

i.e.,

$$\dot{U} = \mathbf{K}\dot{\underset{\sim}{\Phi}} + \mathbf{K}\dot{t}_0 + \underset{\sim}{\mathbf{H}}\dot{\underset{\sim}{\sigma}}_0 \tag{13}$$

This equation can be combined with the incremental form of the pile domain equation (9); i.e.,

$$\dot{U} = \mathbf{D}\dot{\underset{\sim}{\Phi}} + \dot{b} \tag{14}$$

to yield the final solution as before.

21. The initial stresses $\dot{\underset{\sim}{\sigma}}_0$ and the initial traction $\dot{t}_0$ generated, as a result of yielding and slipping, are determined by an interative solution similar to that used in the finite element method.[23]

22. It is interesting to note that the final system of equations may be written as

$$(\mathbf{K} - \mathbf{D})\dot{\underset{\sim}{\Phi}} = \dot{b} - \mathbf{K}\dot{t}_0 - \mathbf{H}\dot{\underset{\sim}{\sigma}}_0 \tag{15}$$

23. If we compare equations (10) and (15) we can see that the incremental analysis essentially involves the solution of the same system of equations with different right

Table 1. Theoretical analyses of McLelland and Focht's experimental data

E_0, lbf/in²	m, lbf/in² per in	E_1, lbf/in²	Lateral displacement, in	Maximum bending moment, 10^3 lbf ft	Comments
–	–	–	1.2	360	Experimental
400	–	–	1.15	210	Homogeneous
170	1.7	–	1.25	270	Inhomogeneous
170	2.3	–	1.07	245	Inhomogeneous
–	2.3	–	1.55	300	Inhomogeneous
–	3.4	–	1.17	290	Inhomogeneous
300	–	3000	1.34	280	Two-layer
300	–	6000	1.23	300	Two-layer
225	–	3000	1.58	340	Two-layer

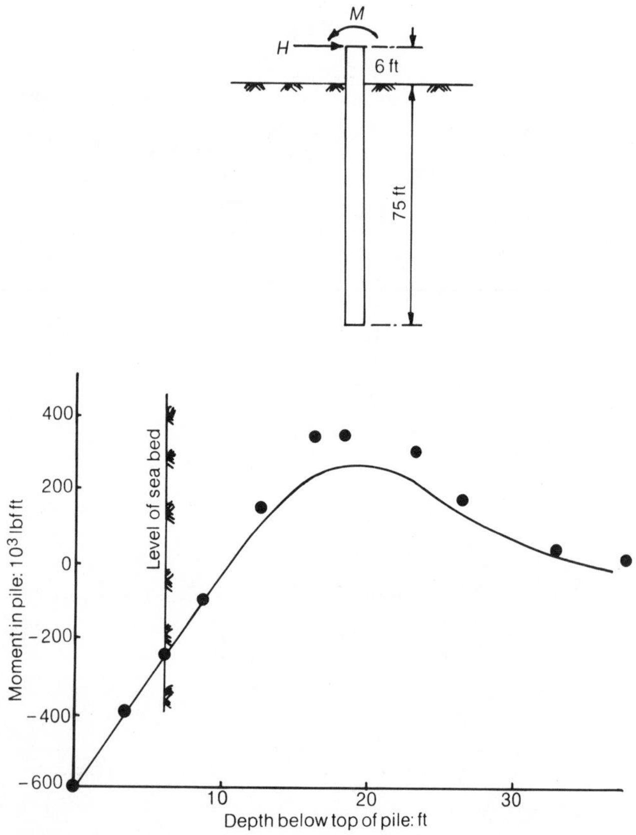

Fig. 5. (a) Full-scale test of McLelland and Focht;[24] (b) comparison of results; $E_0 = 0.25 \times 10^5$ lbf/ft², $m = 3.00 \times 10^3$ lbf/ft² per ft, horizontal displacement of head—experimental = 1.2 in, theoretical = 1.25 in

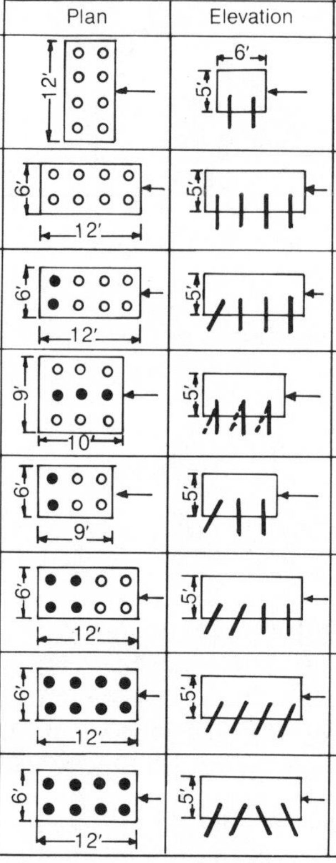

Test	Lateral load for ¼″ deflexion: Test	Lateral load for ¼″ deflexion: Theory	Lateral deflexion due to vertical load of 20 t per pile: Test	Lateral deflexion due to vertical load of 20 t per pile: Theory
1	4·8	4·8	0·0	0·0
2	5·8	5·3	0·0	0·0
3	7·0	7·3	0·04	0·06
4	7·1	6·8	0·06	0·08
5	7·3	8·1	0·05	0·07
6	9·0	8·4	0·07	0·11
7	9·0	8·2	0·21	0·27
8	15·8	11·7	0·0	0·0

○ = vertical ● = raked

Fig. 6. Comparison with the full-scale tests of Feagin[27]

hand sides. Therefore it is essential to store the results of matrix reductions so that subsequent solutions may be obtained at a fraction of the original cost.

24. Within the framework of the algorithm described above, for monotonic loading the isotropic strain hardening model base on the von Mises yield surface, and for cyclic loading a kinematic strain hardening model based on the same yield surface, may be used.[22]

COMPUTER PROGRAM

25. A suite of computer programs which implements these analyses is available. These programs have been written in a user-orientated form so that an engineer with little knowledge of numerical analysis can use them for routine analysis of practical problems. Comprehensive documentation and user guidance are also available.

26. The series of case histories studied below demonstrates the versatility of these programs.

ANALYSIS OF SOME REPORTED CASE HISTORIES

27. Despite the fact that there are numerous publications which describe in situ and model tests on pile groups, there are only a few case histories where comprehensive data regarding the loading and/or the soil properties are available. The major difficulty is that the instrumentation problems are formidable. In what follows the Authors analyse model and full-scale tests reported by McLelland and Focht,[24] Druery and Ferguson,[25] Davisson and Salley[26] and Feagin[27] by using a number of soil parameters (Figs 2–4), and attempt to establish the relationship between the soil properties and the deformation parameters used in the analysis. Since Poisson's ratio does not have significant effect on the lateral loading response, $\nu = 0.5$ is adopted throughout.

Comparison with McLelland and Focht[24]

28. McLelland and Focht carried out a full-scale test on a 24 in pipe pile driven into a sea bed. Details of the test were as follows:

total length of pile = 81 ft
embedded length of pile (L) = 75 ft
diameter of pile (D) = 2 ft
flexural rigidity (E_pI_p) = 1.62×10^{11} lbf in^2
horizontal load at the pile head (H) = 6×10^4 lbf
applied moment at the pile head (M) = -6×10^5 lbf ft

29. The site investigation showed a clay layer extending to a depth of 146 ft below the mud line which is submerged under 33 ft of water. Measurements of shear strength, determined from triaxial tests, indicated an approximately linear increase of strength with depth from a value of 1.6 lbf/in^2 at sea bed level to 10.1 lbf/in^2 at a depth of 40 ft.

30. The key results of several elastic analyses using the non-homogeneous soil model ($E(z) = E_0 + mz$) and the two-layer (E_0, E_1) model are shown in Table 1. The best fit results are shown in Fig. 5.

31. From theoretical considerations the behaviour of a flexible pile under lateral loads is largely governed by the soil modulus over the top third of the pile length. The representative value of the shear strength over this critical region is 4.3 lbf/in^2. The average value of elastic modulus over the same depth for the first five comparisons presented in Table 1 ranges approximately from 350 lbf/in^2 to 500 lbf/in^2 (i.e., $E/C_u \approx 80$–120).

32. Despite the fact that the measured shear strength increases linearly with the depth, a series of analyses was carried out by using the two-layer soil model, since the measured bending moment distribution indicated that very little lateral resistance was offered by the top 15 ft of the subsoil. It can be seen from Table 1 that the stiffness of this top 15 ft of soil has a significant effect on the horizontal displacement and the maximum bending moments. It is noteworthy that a 100% increase in the stiffness of the lower layer causes only a small effect on the overall response of the system. Although the solution with E_0 = 300 lbf/in^2 and E_1 = 6000 lbf/in^2 provides one of the best fits with the experimental results, the value of E_1 is probably too unrealistic for the lower layer.

Comparison with Feagin[27]

33. Feagin carried out a series of full-scale tests on groups of battered wooden piles fixed in concrete test monoliths subjected to lateral loads. Various configurations of the piles were constructed in order to investigate the effects of group geometry on lateral stiffness.

34. Details of the tests are as follows:

embedded length of piles (L) = 30 ft
mean diameter of the pile head = 13 in
mean diameter of the pile base = 9 in
angle of batter = 20°

35. Over the length of the piles the borehole log showed fine to coarse sand containing occasional gravel. Practically the same conditions existed below the pile tips with slight increase in coarseness up to a depth of 75 ft where the bedrock was encountered.

36. The Authors' programs cannot at present handle tapered piles although underreamed piles are permissible. In consequence an average pile diameter of 12 in was adopted, being representative over the critical part of the pile. The pile cap was assumed to be rigid and free-standing.

37. Although sand is not an elastic material it is assumed here that its behaviour can be characterized by a pseudo-elastic modulus whose magnitude is proportional to the in situ effective stresses. This assumption seems to be justified in the absence of a reasonable alternative. Accordingly a zero surface modulus was adopted and the parameter m in the non-homogeneous soil model was adjusted until a reasonable agreement with the experimental results for the first test monolith was obtained. For m = 40 lbf/in^2 per in the overall theoretical results, shown in Fig. 6, are in substantial agreement with the experimental data and reflect reasonably well the variations associated with changes in the group configurations. For the first test monolith with m = 30 lbf/in^2 per in and m = 50 lbf/in^2 per in the lateral loads required to produce 0.25 in lateral displacement are 3.7 tons and 5.3 tons per pile respectively.

38. These tests were carried out at relatively close pile spacings (S/D = 3) and in consequence the favourable effects produced by increasing the number of battered piles in the groups are magnified by the decreased interaction of the piles. In addition an increase in the lateral stiffness occurs in raked piles due to partial transfer of the lateral

Table 2. Druery and Ferguson tests

	Test 1	Test 2
Length of pile (L), mm	146	132
Diameter of pile (D), mm	6.4	9.53
Eccentricity of loading (e), mm	19.1	35.6
Young's modulus of piles (E_p), kN/mm^2	100	100

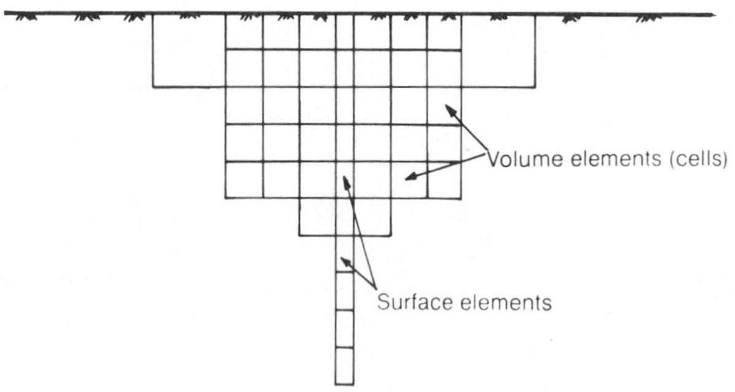

Fig. 7. Arrangement of internal cells

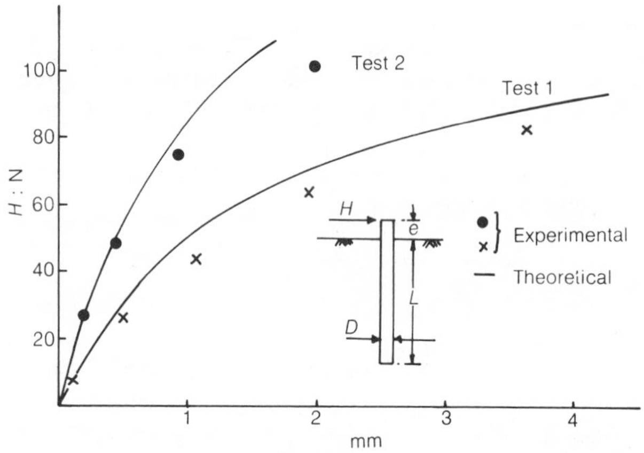

Fig. 8. Comparison with Druery and Ferguson[25]

Table 3. Comparison with model test results by Davisson and Salley; experimental horizontal displacement = 0.0074 in, theoretical horizontal displacement = 0.0078 in

Pile	Moment at pile head, lbf in		Shear at pile head, lbf		Axial load at pile head, lbf	
	Experiment	Theory	Experiment	Theory	Experiment	Theory
1	11.5	9.4	5.0	3.7	18.0	15.3
2	13.2	10.9	5.2	4.2	2.75	2.24
3	11.2	9.4	5.2	3.7	12.8	15.3
4	10.8	10.9	3.8	4.2	3.05	2.24
5	10.6	10.0	3.8	3.9	6.5	8.2
6	10.4	10.0	3.6	3.9	5.4	8.2

Table 4. Theoretical analyses of Davisson and Salley's experimental data

E_0, lbf/in^2	m, lbf/in^2 per in	Piles 1 and 3			Piles 2 and 4			Lateral displacement, in	Comments
		Moment, lbf in	Shear, lbf	Axial load, lbf	Moment, lbf in	Shear, lbf	Axial load, lbf		
–	–	11.4	5.1	15.4	12.0	4.5	2.9	0.0074	Experimental (average)
500	–	7.0	3.9	15.5	8.5	4.6	3.15	0.0056	Homogeneous
–	40	9.4	3.7	15.3	10.9	4.2	2.24	0.0078	Inhomogeneous
–	50	8.6	3.6	15.9	9.9	4.1	1.64	0.0066	Inhomogeneous

load by the development of the axial shear stress along the pile shaft.

39. The lateral deflexions due to vertical loads were not specified accurately for each group configuration by Feagin; therefore some of these test results had to be interpolated. Nevertheless the theoretical results are in very good agreement with the experimental data.

Comparison with Druery and Ferguson[25]

40. A series of model tests was carried out by Druery and Ferguson on solid brass piles embedded in kaolin clay in order to obtain the load–deflexion curves to failure. The details of two of these tests chosen from the extreme range of their investigation are shown in Table 2. The undrained cohesive strength of the soil C_u was measured, by means of an unconfined compression test, to be 0.36 N/mm^2.

41. The volume and the surface discretization used for carrying out the boundary element analysis is shown in Fig. 7. The volume cells are restricted to regions that are expected to yield.

42. The results of a non-linear analysis, assuming the homogeneous soil model with a value of Young's modulus for the soil of 170 N/mm^2 ($E/C_u \approx 500$), are depicted in Fig. 8. Since yielding of the soil takes place even at low load levels, the shape of the load–deflexion curve for flexible piles is largely governed by the shear strength of the soil. Consequently, much higher and more realistic values of soil modulus can be employed, as here, which correspond more closely to the values obtained for the modulus/strength ratio obtained elsewhere (e.g., plate-loading tests[28]). This result indicates the importance of selecting the secant modulus relevant to the particular load level in an elastic analysis. Unfortunately it is difficult to give proper guidance on this point since it is dependent not only on the flexibility of the pile but also on the pile head conditions and on the presence of any adjacent piles. The value of Young's modulus selected here is similar to that adopted by Poulos,[29] who analysed these data using a simplified analysis and demonstrated the relatively insignificant effect of changes in the elastic modulus. The results given here are in broad agreement with those obtained by Poulos using his simplified analysis.

Comparison with Davisson and Salley[26]

43. Davisson and Salley carried out a small-scale test on a group of raking piles subjected to combined vertical and horizontal loads and moments. They measured the load distribution between the piles and the horizontal displacement. The necessary details of their test are

length of piles (L) = 21 in
outer diameter of piles = 0.50 in
inner diameter of piles = 0.44 in
Young's modulus of piles (E_p) = 1.0×10^7 lbf/in^2
angle of rake = 18.5°

44. The test was carried out in a cylindrical tank of length and diameter of 48 in. These dimensions were thought to be sufficiently large to cause minimal interference to the pile group response. The sand used in the test was described as dried, fine and fairly uniform with about 7% passing the No. 200 sieve. The horizontal pile cap displacement and the moments, shears and axial loads at the pile heads were measured.

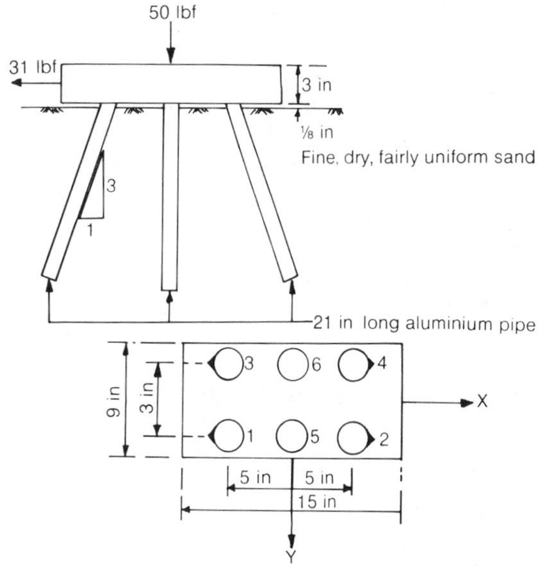

Fig. 9. Model test by Davisson and Salley[26]

45. Pipe piles can be analysed as a matter of course by the programs. The group is free-standing and is assumed to have a rigid pile cap. Details of the group configuration and the theoretical results are given in Fig. 9 and Table 3. Table 4 shows the sensitivity of the analysis to changes in the elastic parameters.

46. An examination of the test results reveals evidence of some experimental error, in particular a small violation of equilibrium and marked non-symmetry of moments and loads (Fig. 9). The theoretical results for moments, shears and axial loads always obey equilibrium, as it is an integral part of the analysis. The results for m = 40 lbf/in^2 per in are in very good agreement with the experimental data. The results for the bending moments at the pile heads are underestimated for all values of the chosen elastic parameters.

CONCLUSIONS

47. A complete method of analysis for obtaining the linear and non-linear response of pile groups has been outlined. The method of analysis has been applied to a number of reported case histories to demonstrate that it can be used to analyse practical pile group problems.

48. It has been shown that the analysis is capable of taking into account the effects of the geometry and loading conditions. (For example, one can compare the second and fourth case histories: a geometrical scale factor of about 20 exists, yet the same deformation parameters apply.)

49. A major advantage of the analysis is its considerable economic advantage over the more general methods of analysis using the general boundary element or the finite element algorithms.

ACKNOWLEDGEMENT

50. The Authors wish to express their sincere gratitude to the Department of Transport for providing financial support for the work described in this Paper.

REFERENCES

1. CRUSE T. A. and RIZZO F. J. (eds). Proc. Conf. Boundary Integral Method. American Society of Mechanical Engineers, New York, 1975, AMD vol. 11.
2. BANERJEE P. K. Integral equation methods for the analysis of three-dimensional piece-wise non-homogeneous elastic solids of arbitrary shape. Int. J. Mech. Sci., 1976, 18, 293–303.
3. LACHAT J. C. and WATSON J. O. Effective numerical treatment of boundary integral equations: a formulation for three-dimensional elastostatics. Int. J. Numer. Meth. Engng, 1976, 10, 991–1005.
4. BANERJEE P. K. and BUTTERFIELD R. (eds). Developments in boundary element methods—I. Applied Science Publishers, London, 1979.
5. CRUSE T. A. and WILSON R. B. Advanced applications of boundary integral equation methods. Nucl. Engng Des., 1978, 46, 223–234.
6. BUTTERFIELD R. and BANERJEE P. K. Analysis of axially loaded compressible piles and pile groups. Géotechnique, 1971, 21, No. 1, 43–60.
7. BUTTERFIELD R. and BANERJEE P. K. The problem of pile cap and pile group interaction. Géotechnique, 1971, 21, No. 2, 135–142.
8. POULOS H. G. Laterally loaded piles—single piles. J. Soil Mech. Fdns Div. Am. Soc. Civ. Engrs, 1971, 97, May, SM5, 711–731.
9. POULOS H. G. Laterally loaded piles – pile groups. J. Soil Mech. Fdns Div. Am. Soc. Civ. Engrs, 1971, 97, May, SM5, 733–751.
10. POULOS H. G. Analysis of the settlement of pile groups. Géotechnique, 1968, 18, 449–471.
11. BANERJEE P. K. and DRISCOLL R. M. C. A program for the analysis of pile groups of any geometry subjected to any loading conditions. Department of the Environment, London, 1975, HECB/B/7.
12. BANERJEE P. K. and DRISCOLL R. M. C. Three-dimensional analysis of raked pile groups. Proc. Instn Civ. Engrs, Part 2, 1976, 61, Dec., 653–671.
13. BANERJEE P. K. and DAVIES T. G. Analysis of pile groups embedded in Gibson soil. Proc. 9th Int. Conf. Soil Mech., Tokyo, 1977.
14. BANERJEE P. K. (SCOTT C. R. (ed.)). Analysis of axially and laterally loaded pile groups. In: Developments in soil mechanics. Applied Science Publishers, London, 1978, 317–346, chapter 9.
15. BANERJEE P. K. and DAVIES T. G. The behaviour of axially and laterally loaded piles embedded in nonhomogeneous soils. Géotechnique, 1978, 28, No. 3, 309–326.
16. BANERJEE P. K. and MUSTOE G. The boundary element method for two-dimensional problems of elasto-plasticity. Proc. int. conf. recent advances in boundary element method. Pentech Press, London, 1978, 283–300.
17. LAMB H. Hydro-dynamics. Pergamon, Oxford, 1932, 3rd edn.
18. BUTTERFIELD R. and BANERJEE P. K. A note on the problem of a pile reinforced half-space. Géotechnique, 1970, 20, 100–103.
19. MINDLIN R. D. A force in the interior of a semi-infinite solid. Physics, 1936, 7, May, 195–202.
20. DAVIES T. G. Linear and non-linear analysis of piles and pile groups. PhD thesis, University College, Cardiff, 1979.
21. DAVIES T. G. and BANERJEE P. K. The displacement field due to a point load at the interface of a two-layer elastic half-space. Géotechnique, 1978, 28, No. 1, 43–56.
22. BANERJEE P. K. et al. (BANERJEE P. K. and BUTTERFIELD R. (eds)). Two and three-dimensional problems of elasto-plasticity. In: Developments in boundary element method—I. Applied Science Publishers, London, 1979, chapter IV.
23. ZIENKIEWICZ O. C. The finite element method. McGraw Hill, London, 1977, 3rd edn.
24. McCLELLAND B. and FOCHT J. A. Soil modulus for laterally loaded pile. J. Soil Mech. Fdns Div. Am. Soc. Civ. Engrs, 1956, Oct., SM4, Paper 1081.
25. DRUERY B. M. and FERGUSON R. A. An experimental investigation of the behaviour of laterally loaded piles. BE thesis, University of Sydney, 1969.
26. DAVISSON M. T. and SALLEY J. R. Model study of laterally loaded piles. J. Soil Mech. Fdns Div. Am. Soc. Civ. Engrs, 1970, Sept., SM5, 1605–1627.
27. FEAGIN L. B. Lateral load tests on groups of battered and vertical piles. Symposium on lateral load tests on piles. American Society of Civil Engineers, New York, 1953, 12–20.
28. MARSLAND A. Laboratory and in situ measurements of the deformation moduli of London clay. Proc. symp. interaction of structures and foundations, Birmingham, 1971, 7–17. Midland Society Soil Mechanics and Foundation Engineering.
29. POULOS H. G. Load deflection prediction for laterally loaded piles. Aust. Geomech. J., 1973, G3, No. 1, 1–8.

14. A linear elastic interpretation of model tests on single piles and groups of piles in clay

R. BUTTERFIELD (Professor of Soil Mechanics, Southampton University) and
N. GHOSH (Senior Engineer, Halcrow Middle East, Dubai)

Although it has recently become quite common to carry out sophisticated computer analyses of piles and groups of piles, modelling them as elastic members embedded in some form of elastic half-space, no comprehensive systematic study has yet been published of the working load stiffness of a range of such systems in order to investigate whether the elastic analyses are reasonably valid. The fundamental questions are whether the working load response is linear elastic, and if so, whether the same elastic modulus interprets the behaviour of pile groups of any geometry. This Paper reports such a study carried out on a variety of model piles (single piles, and 2 × 2 and 3 × 3 groups), with floating and bedded caps (two sizes), loaded vertically (both centrically and eccentrically) in a bed of stiff, remoulded London Clay. The results provide strong support for the elastic stiffness analysis of the model systems; the load sharing between piles and cap; the validity of superposition of vertical load and moment effects; and the values of the group settlement ratios predicted by numerical analyses of the quality of the DoE PGROUP program.

INTRODUCTION

In recent years elaborate computer programs have been devised which analyse the load–displacement behaviour of single piles and groups of piles as elastic members embedded in a linear elastic 'soil', notably the work of Poulos[1] and the PGROUP program suite[2] stemming from the work of Butterfield and Banerjee.[3,4]

2. The key output from such analyses is the complete pile group stiffness matrix, and the associated settlement ratio, determined from a consistent and rigorous model within the assumptions of elastic linearity and homogeneity. This latter restriction has been relaxed in recent publications by Banerjee et al.[5,6] wherein the analysis has been extended to include layered and Gibson soil models. The inputs to such analyses, in addition to the group geometry and pile properties, are the psuedo-elastic parameters for the idealized soil, say Poisson's ratio (ν) and shear modulus (G). Since the analytical results are not very sensitive to ν, attention focuses on the magnitude of the modulus G.

3. Unfortunately, even within the assumptions of linear soil behaviour, there is very little experimental information available from which G might be assessed; the most common rule of thumb for clays being that $G \approx \beta c_u$ (Skempton and Henkel[7]). Early results pointed to β values of the order of 40 whereas more recent experimental evidence suggests that $\beta \approx 150$ provides a better assessment.[8,9] The principal research project summarized in this Paper set out to examine the following points, among others, using carefully controlled experiments on model pile systems installed in a large bed of quite stiff, remoulded London Clay.

(a) Is the response of single piles and groups of piles essentially linear under vertical centric and eccentric working loads?

(b) If so, what values of shear modulus (G), back-calculated from PGROUP, will best fit the single pile and group results and does a *unique* value of G interpret *all* such tests (with $\nu = ½$)?

(c) What value of β best correlates G and c_u via $G = \beta c_u$?

4. If reasonable linearity were established, with an approximately unique G value, then not only would the results support the use of PGROUP but also imply that, in practice, moduli determined from precise single-pile load tests could be used more confidently to predict pile group stiffnesses. The justification for using small-scale models rather than prototype piles was simply that large numbers of tests with a variety of group geometries could be carried out quickly, cheaply and repeatably, although the tests were not truly modelling full-scale piles. However, should the answers to points (a) and (b) not be affirmative from the model tests, then it is extremely unlikely that affirmative answers would be obtained from tests on prototype piles.

TESTS

5. The equipment and the single vertical pile axial load results are described elsewhere [8,10] and therefore only a short description is included here. The soil bed (1.5m × 1.5 m × 1.4 m) was a very uniform, hand-prepared bed of remoulded London Clay sealed and matured for over a

Table 1

	L/D	S/D	e = 0 (centric loads)		e = B/12		e = B/6		e = B/3	
			Q/w, MN/m	G, MN/m²	Q/w, MN/m	G, MN/m²	Q/w, MN/m	G, MN/m²	Q/w, MN/m	G, MN/m²
2 × 2 groups	30	2.5	20.2	19.6	18.6	19.7	19.6	20.7	18.4	19.4
	30	2.5	19.0	18.1	—	—	18.5	19.5	—	—
	30	2.5	21.9	21.5 *	13.4	14.2	15.4	16.2	14.1	14.9
	30	2.5	21.0	20.6 *	16.8	17.8	17.5	18.5	17.3	18.3
	40	2.5	16.2	15.5	16.5	14.4	15.4	13.5	15.7	13.8
	40	2.5	17.8	17.1	16.9	14.7	—	—	17.3	15.2
	20	5.0	13.6	15.9	15.6	15.8	15.3	15.4	15.8	15.9
	20	5.0	14.7	17.2	—	—	—	—	—	—
	30	5.0	—	—	24.0	20.9	15.3	17.6	19.7	17.2
	30	5.0	—	—	(27.8)	(24.2)	20.2	21.3	18.2	15.8
	30	5.0	—	—	21.5	18.7	24.5	18.6	21.0	18.3
	30	5.0	—	—	22.5	19.6	23.3	20.3	21.4	18.6
	40	5.0	—	—	20.3	15.5	21.5	16.5	19.8	15.2
	40	5.0	—	—	22.8	17.4	22.4	17.2	20.8	15.9
Mean	—	—	—	18.2	—	17.2	—	17.9	—	16.5
Standard deviation ($\bar{\sigma}$)	—	—	—	2.0	—	2.3	—	2.2	—	1.7
3 × 3 groups	30	2.5	17.1	14.7	23.9	22.0	23.3	21.4	22.9	21.0
	30	2.5	20.0	17.2	22.7	20.9	24.1	22.2	23.7	21.8
	40	2.5	22.1	16.2	20.0	16.8	21.0	17.6	24.2	20.2
	40	2.5	23.3	17.2	21.3	17.8	21.3	17.9	24.3	20.3
	20	5.0	15.5	12.4	21.4	15.7	19.0	13.9	22.5	16.5
	20	5.0	19.0	15.7	20.9	15.4	—	—	—	—
	30	5.0	21.0	13.8	20.6	14.1	22.4	15.3	24.5	16.7
	30	5.0	20.3	13.7	21.7	14.8	22.9	15.7	23.0	15.7
	40	5.0	25.1	14.1	26.3	16.7	27.3	17.3	30.0	18.6
	40	5.0	26.2	14.8	27.1	17.2	28.7	18.2	30.6	19.4
Mean values (mean † G 17.8)	—	—	—	15.0	—	17.1	—	17.7	—	18.9
Standard deviation (mean $\bar{\sigma}$ 3.0)	—	—	—	1.5	—	2.4	—	2.6	—	2.0

Overall grand mean $G = 17.5$ MN/m², standard deviation = 2.2 MN/m².

* From five tests.
† Including multiple tests.

year in a large steel tank (for the clay w = 25%, standard deviation $\bar{\sigma}$ = 0.58% from 77 tests; c_u = 100 kN/m^2, $\bar{\sigma}$ = 8 kN/m^2; w_L = 65%; w_p = 23%). Hollow steel piles (diameter D = 12.5 mm), L/D = 20, 30 and 40, were driven and CRP test-loaded at rates of 6.58 mm/min and 6.5 × 10^{-3} mm/min. All model piles had head and toe load cells and the latter also measured a mean pile-hoop stress and hence the total radial stress (σ) at the pile toe. The precision involved in the whole cycle of loading operations can best be appreciated by noting that the full working load vertical displacement of a single pile was typically 40 μm. All output was data logged and the total programme comprised 13 sets of tests covering single piles, and 2 × 2 and 3 × 3 groups at spacings (S) of 2.5 D and 5.0 D, with both floating and bedded rigid caps (overhang of 1.25 D is 'standard', and of 4.25 D is 'extended'), all under vertical centric and eccentric loads. (Eccentricity ratios (e/B) were $^1/_{12}$, $^1/_6$ and $^1/_3$ usually, where B is overall cap breadth.)

RESULTS

6. Although a number of intriguing secondary issues were highlighted by the tests (one being that the interaction and consolidation induced by successively driven piles in the closely spaced groups means that a perfectly symmetrical group of even four piles cannot be constructed this way; another being that installed and unloaded piles still carried residual toe loads and residual total radial toe stresses which were about 46% and 76% respectively of their maximum values, the latter figure being 5.5c_u (average) by comparison with about 6c_u measured more precisely by Butterfield and Johnston[11] and theoretical predictions[12] of 7.2c_u (for β = 150); a third issue being that the maximum toe loads correlated well with $N_c \approx 11.5$), this Paper concentrates on the load–displacement results from the complete test series and the load distribution between the piles in a group and the pile cap.

Linearity of the load–displacement curves

7. The linearity of the load–displacement curves was remarkably good (Fig. 1) up to a nominal working load of 0.5Q_{ult}, where Q_{ult} for an N pile group was arbitrarily assessed at N times the load capacity for a single similar pile. For all load tests the coefficients of correlation (r) from linear regression analyses were usually better than 0.98 and, in fact, the linearity extended to about 0.6Q_{ult} with around 98% displacement recovery on unloading.

Shear modulus values

8. *Single piles.* The basic G value[10] determined from some 20 single-pile load–displacement curves, interpreted via PGROUP, was G = 15.4 MN/m^2 ($\bar{\sigma}$ = 1.75 MN/m^2), corresponding to $\beta \approx 154$ ($\bar{\sigma}$ = 17.5).

9. When individual piles within groups were test-loaded singly they became successively stiffer. For example, piles in the 2 × 2 groups were tested individually up to working load at less than 1 h, and at 72 h, after installation. The fourth pile driven was usually some 15% stiffer than the first but when re-tested after completion of all the group load cycles this factor had increased to about 20% for *all* piles. By quick and careful recovery of small, pile-shaft moisture content samples, a reduction in soil moisture content adjacent to the pile was detected averaging 1.5% near the ground surface (diminishing to zero at the pile

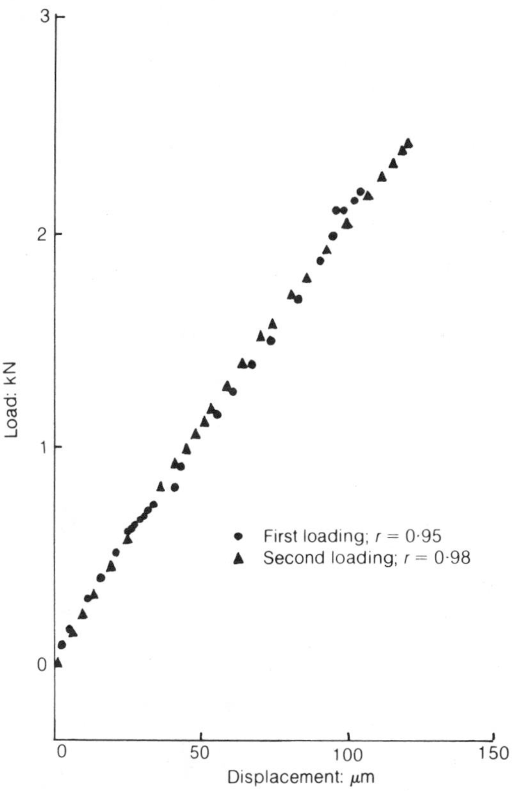

Fig. 1. Load–displacement curve for 3 × 3 floating capped pile group; L/D = 30, S/D = 5.0

toe). The mean value of c_u along the pile was thereby appreciably increased,[10] and although only nine such results were obtained they suggested that, if the 'final' assessed consolidated $\bar{c}_u$ values were used, G would be about 5 + 70c_u MN/m^2. This expression is probably of less practical help than $\beta \approx 150$, but, interestingly, Abdrabbo[9] arrived at a closely similar expression when analysing model pile tests in much softer Kaolin clay. (There is, of course, no *a priori* reason why an empirical, locally fitted expression for G should not be of the form $G = a + \beta c_u$.)

10. *Centrally loaded groups.* Figure 1 shows the typically linear response of a 3 × 3 floating cap group loaded up to 2.5 kN at which the displacement was only 125 μm. From all such tests (Table 1) the mean, best fit PGROUP shear modulus was 17.8 MN/m^2, although if the often-repeated set are weighted only as single tests this drops to 16.4 MN/m^2 with a standard deviation of 1.7 MN/m^2. Figs 2 and 3 show some of these results in a different form which indicates not only the extent of the scatter in the experimental measurements but also how their general pattern conforms to the analytical predictions for groups of different sizes and spacings.

11. It is also of interest to compare the load sharing between different piles in the 3 × 3 groups and between the piles and the cap for the bedded cap tests. For the 3 × 3 floating capped groups the corner piles carried more load than the others, each typically 13–14% of the load, with some 9–10% on the mid-side piles—all essentially as predicted by the elastic analyses, although the centre pile itself carried an appreciably higher proportion of the working load than calculated. When the pile cap was contacting the ground (bedded in plaster) the vertical stiffness was only minimally increased, as predicted,[4] although the

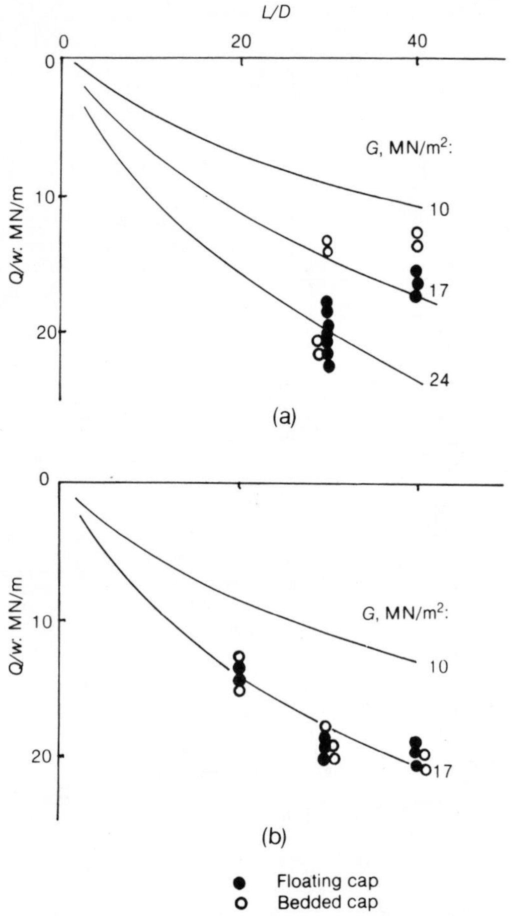

Fig. 2. Stiffness versus L/D for 2 × 2 groups: (a) S/D 2.5; (b) S/D = 5.0

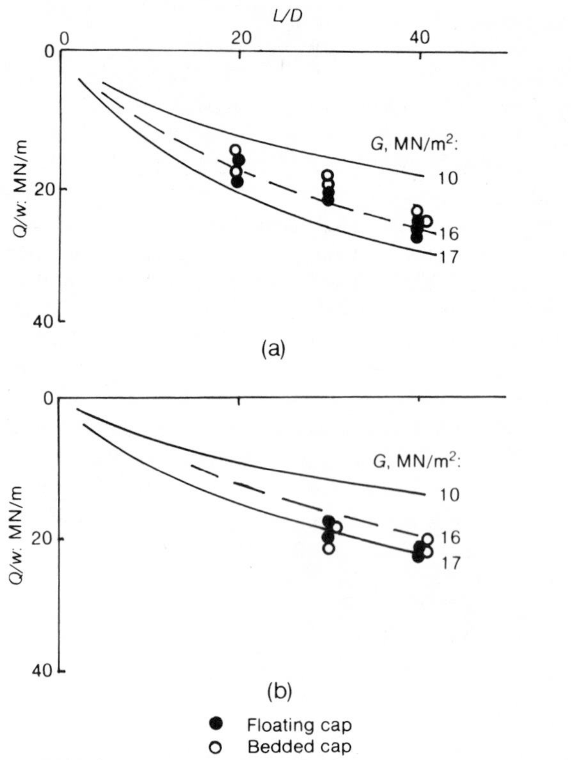

Fig. 3. Stiffness versus L/D for 3 × 3 groups: (a) S/D 5.0; (b) S/D = 2.5

caps did carry a significant proportion of the total load (Table 2). The mean, back-calculated shear moduli were, for standard caps (19 tests) G = 16.3 MN/m^2 ($\overline{\sigma}$ = 1.6 MN/m^2) and for extended caps (20 tests) G = 15.6 MN/m^2 ($\overline{\sigma}$ = 2.0 MN/m^2). Since all cap bedding imperfections must result in a reduction of the experimental group stiffnesses and cap loads, the measured cap loads, summarized in Figs 4 and 5, are not only rather scattered but also fall below the values predicted from PGROUP. Nevertheless, the load support contribution of all the caps was far from negligible, resulting in the piles carrying appreciably reduced loads – distributed between them in essentially the same proportions as in the floating cap tests.

12. *Eccentrically loaded groups.* Pile groups of similar dimensions were also loaded eccentrically in order to investigate whether the moment–rotation curves were reasonably linear and could be interpreted by PGROUP using the

Table 2. Cap loads

	L/D	S/D	Mean cap load, %			
			Standard cap		Extended cap	
			Centric	Eccentric	Centric	Eccentric
2 × 2 groups	20	2.5	—	—	41	—
	30	2.5	30	33	36	—
	40	2.5	—	—	31	—
	20	5.0	40	44	60	—
	30	5.0	35	38	56	—
	40	5.0	30	33	51	—
3 × 3 groups	20	2.5	30	32	35	41
	30	2.5	20	21	25	34
	40	2.5	16	18	22	30
	20	5.0	36	36	57	—
	30	5.0	25	25	47	—
	40	5.0	21	23	45	—

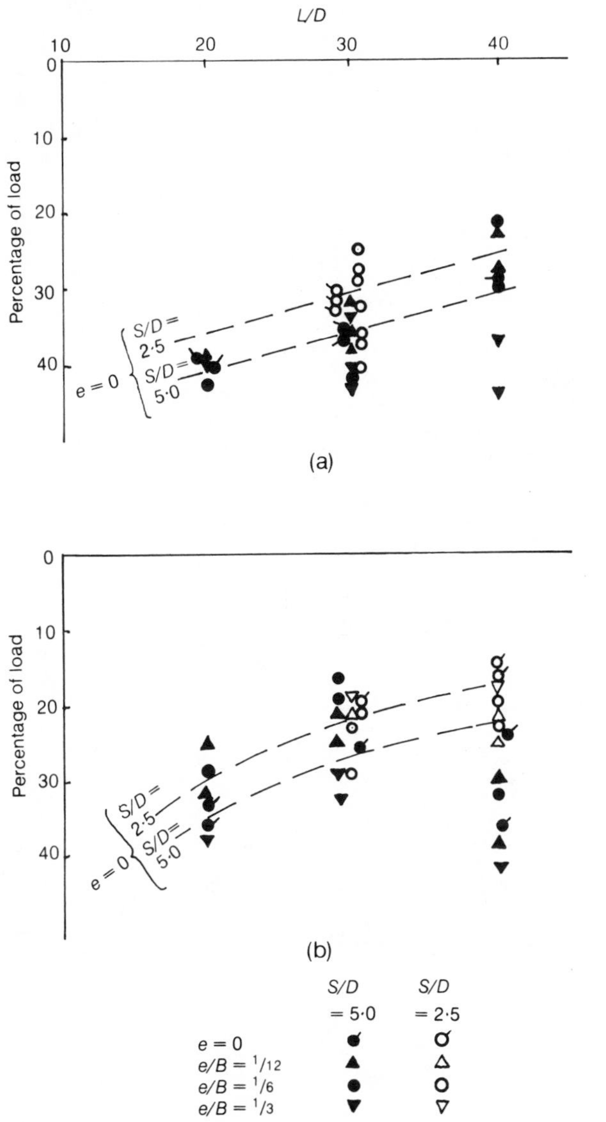

Fig. 4. Percentage of total load carried by standard capped pile groups–centric and eccentric loads: (a) 2 × 2 groups; (b) 3 × 3 groups

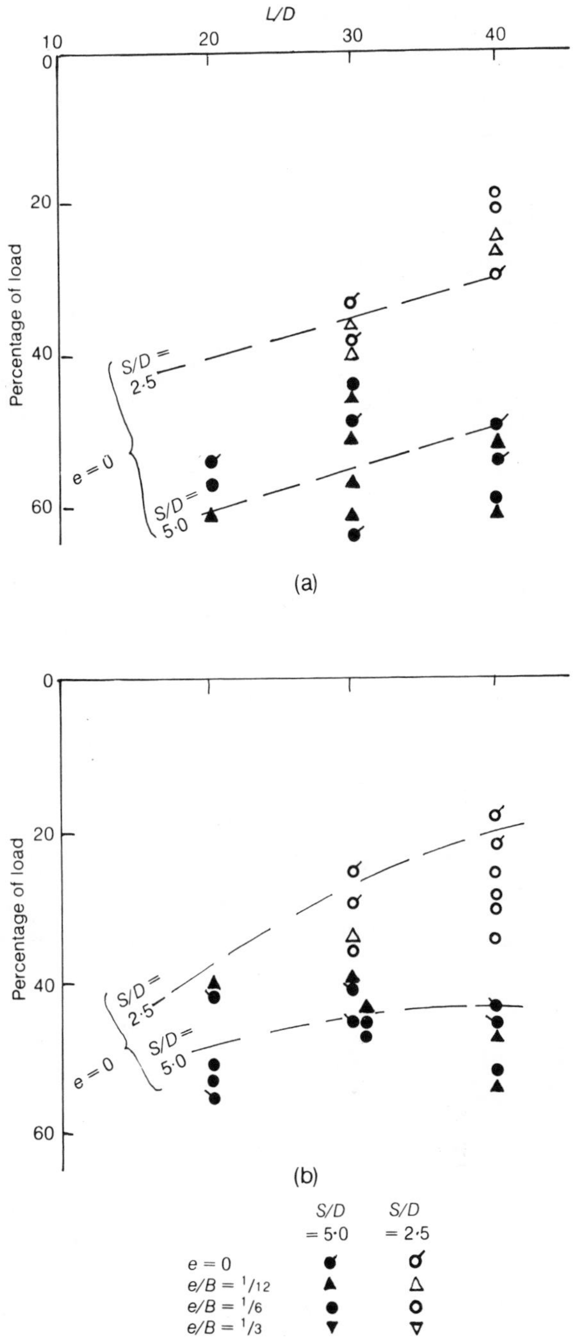

Fig. 5. Percentage of total load carried by extended capped pile groups–centric and eccentric loads: (a) 2 × 2 groups; (b) 3 × 3 groups

single-pile G values, and to investigate whether the principle of superposition could be applied (i.e., whether the vertical response to a particular vertical load was independent of the load eccentricity).

13. The vertical stiffness measurements (floating cap) are summarized in Table 1: taken together with the centrically loaded results, they support a grand mean G of 17.5 MN/m^2 ($\bar{\sigma}$ = 2.2 MN/m^2) for all the group tests, which is to be compared with the single-pile value[10] of 15.4 MN/m^2 ($\bar{\sigma}$ = 1.25 MN/m^2).

14. Figures 6 and 7 demonstrate the good linearity of the measured moment–rotation curves for 2 × 2 and 3 × 3 groups at different vertical load eccentricities. Table 3 summarizes the rotation stiffnesses of the various groups and the back-calculated PGROUP shear moduli. The overall mean G from the moment–rotation curves is 14.5 MN/m^2 ($\bar{\sigma}$ = 2.0 MN/m^2). This is clearly rather less than that assessed from the vertical stiffness values, as might be anticipated from the greater contribution made by the near-surface, unconfined soil to the rotational (and lateral) stiffness of the system, and from the presence of a small pile–soil gap which is known to extend some $2D$–$3D$ below the ground surface around preformed piles in clay.[13,14]

15. The effect of the smaller vertical stiffness of the upper soil is also reflected in the tests with bedded extended caps in which the analysis overestimates the cap estimate of the equivalent shear modulus, G = 12.0 MN/m^2 ($\bar{\sigma}$ = 3.0 MN/m^2).

16. A trend of increasing rotational stiffness with increasing L/D is shown in Fig. 8, as in Figs 2 and 3. This is again consistent with the analytical results.

Table 3

	L/D	S/D	$e = B/12$		$e = B/6$		$e = B/3$	
			M/θ, kN m/rad	G, MN/m²	M/θ, kN m/rad	G, MN/m²	M/θ, kN m/rad	G, MN/m²
2 × 2 groups	30	2.5	(10.9)	(8.2)	20.3	15.4	26.0	19.9
	30	2.5	16.2	12.2	26.0	19.7	29.4	21.9
	20	5.0	33.9	10.5	40.7	13.2	49.7	15.4
	20	5.0	—	—	—	—	—	—
	30	5.0	50.9	12.3	56.5	13.8	57.6	14.1
	30	5.0	52.0	12.6	61.0	15.0	61.0	15.0
	40	5.0	(44.1)	(9.0)	58.8	11.8	59.9	11.0
	40	5.0	62.2	12.5	64.4	13.1	69.3	14.0
3 × 3 groups	30	2.5	73.9	15.1	79.8	16.3	79.1	16.1
	30	2.5	80.2	16.4	84.8	17.3	83.6	17.0
	40	2.5	64.3	11.8	70.3	12.9	72.9	13.4
	40	2.5	64.3	11.8	72.9	13.4	79.1	14.4
	20	5.0	147	12.3	152	12.8	190	15.9
	20	5.0	142	11.9	—	—	—	—
	30	5.0	(346)	(21.1)	267	16.3	251	15.3
	30	5.0	(388)	(23.6)	264	16.1	251	15.3
	40	5.0	232	12.7	266	14.5	331	18.1
	40	5.0	237	13.0	280	15.3	332	18.1
Mean	—	—	—	12.7	—	14.8	—	15.9
Standard deviation	—	—	—	1.4	—	2.0	—	2.6

Grand mean $G = 14.5$ MN/m², standard deviation = 2.0 MN/m².

17. The eccentric vertical load tests therefore established that, within reasonable practical precision, the system response was linear, and superposition of results could be used and interpreted (under working loads) by a G value determined from vertical load tests on single piles.

18. It is also of interest to examine the way in which the eccentric loads were supported by the piles and the cap. The moment generated by the eccentric load (Qe) was carried partly by the cap (when bedded) and partly by a 'push–pull' mechanism in the piles superimposed on the vertical centric load component. The bending moment developed at the top of any pile was always negligibly small (about 0.2% of Qe). Fig. 9 shows the percentage of the applied moment carried by the cap for two sets of extended cap, 3 × 3 group tests for S/D = 2.5 and 5.0. (The cap dimensions are given by $B = [(n-1)\ S/D + A]D$ for an $N = n \times n$ group with values for A of 2.5 and 8.5 for standard and extended caps respectively.) Not only is there general agreement between calculation and measurement but the extended cap contribution to the moment resistance (≈ 60%) is very high. Detailed results from 56 sets of measurements are summarized in Tables 4 and 5 and Fig. 10 for 2 × 2 and 3 × 3 groups with standard-size, bedded caps. The striking features of these results are again not only the vertical load carrying contribution from the cap but also the fact that not until $e \approx B/3$ do any piles in any of the groups go into tension. If the results are analysed in more detail, using the above equation for B, it will be found that onset of tension corresponds closely to the line of action of the load passing outside the centre-line of the outer row of piles in each case.

Pile group settlement ratios

19. Although the settlement ratio (R)—defined as the ratio of the N group displacement to the single, isolated pile displacement under identical loads per pile—is of considerable interest to the designer there is very little published experimental information on R; indeed, until the appearance of the elastic analyses[1,3] there were no worthwhile theoretical predictions either. Since R is a dimensionless ratio one might expect the experiments and calculations to agree here more closely. This is found to be

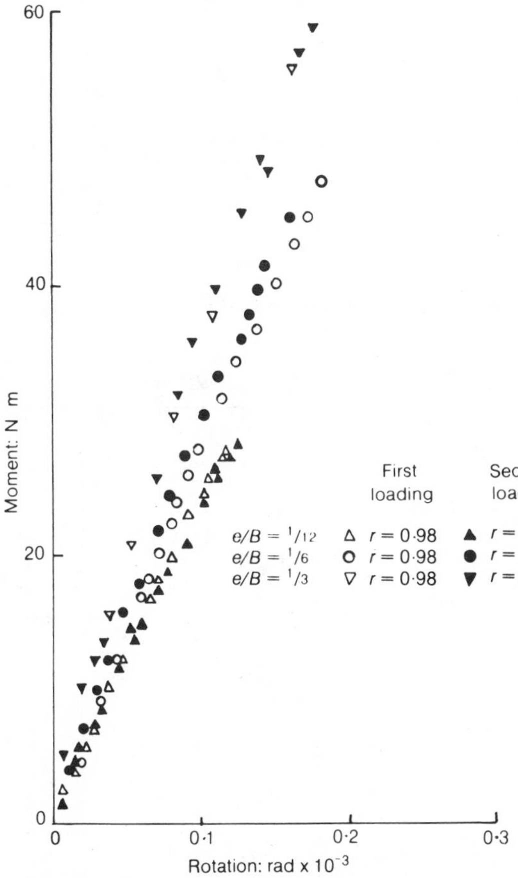

Fig. 6. Moment versus rotation for 3 × 3 long pile groups with standard cap; L/D = 40, S/D = 5.0

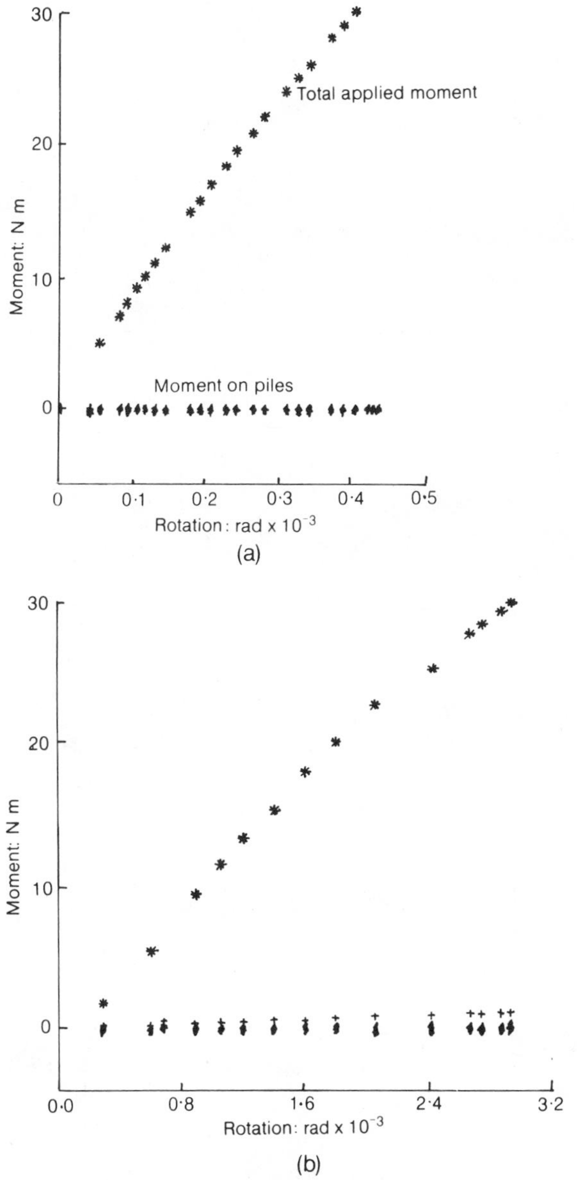

Fig. 7. Typical moment–rotation curves for long pile groups with standard caps; L/D = 40, S/D 2.5; pile head moments also shown: (a) 3 × 3 group; e/B = $^1/_6$; (b) 2 × 2 group; e/B = $^1/_3$, r = 0.988

case and Figs 11–13 summarize the remarkable correspondence established for all the data now availble on 2 x 2 and 3 x 3 bedded and floating cap groups.

CONCLUSIONS

20. Conclusions from the large number of vertical, centric and eccentric, load tests on model pile groups in a stiff clay reported on are as follows.

21. The working load–displacement response, both vertically and rotationally, was very closely linear and elastic.

22. Preformed piles installed subsequently in a group caused an increase in the vertical stiffness of the group due to soil consolidation such that the mean, back-calculated, elastic shear modulus G was 17.5 MN/m^2 ($\bar{\sigma}$ = 2.2 MN/m^2). An identical interpretation (also using PGROUP and $\nu = \frac{1}{2}$) of single, isolated pile tests[10] gave G = 15.4 MN/m^2 ($\bar{\sigma}$ = 1.8 MN/m^2).

23. The vertical component of any eccentric loads could be treated similarly and uncoupled from the applied moment ($M_0 = Qe$). Elastic analysis of the rotational stiffness led to G = 14.5 MN/m^2 ($\bar{\sigma}$ = 2.0 MN/m^2) with the reduction in the psuedo shear modulus attributed to the presence of a pile–soil gap at the pile head and the reduced stiffness of the unconfined, ground surface soil.

24. When the caps were bedded on the gound they carried an appreciable proportion of the applied loading

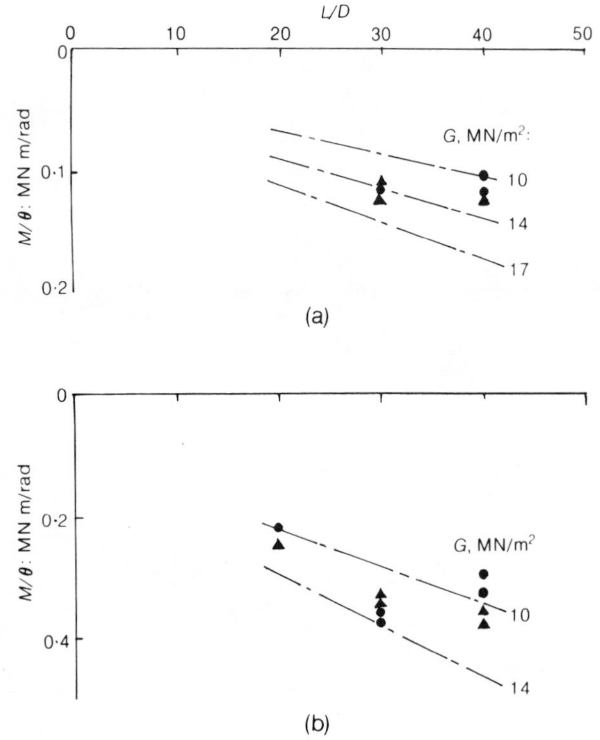

Fig. 8. Rotational stiffness of 3 x 3 extended cap pile groups: (a) S/D = 2.5; (b) S/D = 5.0

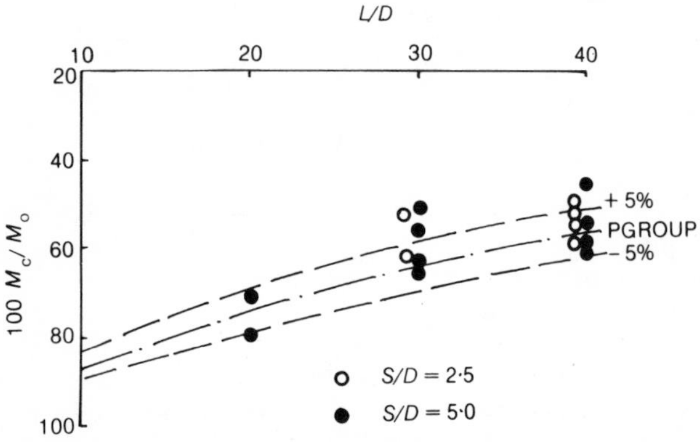

Fig. 9. Percentage moment on cap versus L/D for 3 x 3 extended cap groups

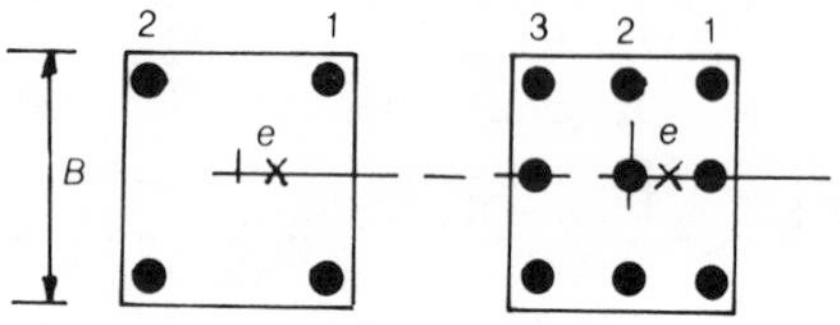

Fig. 10. Definition of load eccentricity e

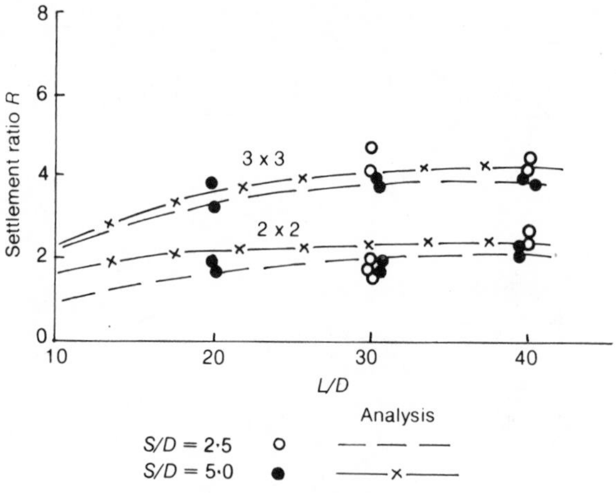

Fig. 11. Settlement ratio versus L/D for floating cap groups

Table 4. Load distribution, 2 x 2 groups

L/D	S/D	e/B	Pile load, %		Cap load, %
			Type 1	Type 2	
20	5.0	1/12 1/6 1/3	21 23 29	10 5 1	38 44 40
30	5.0	1/12 1/6 1/3	21 24 31	11 6 –1	36 40 40
40	5.0	1/12 1/6 1/3	24 23 28	13 10 2	26 34 40
30	2.5	1/10 2/10 3/10	26 28 33	10 5 –2	28 34 38

(typically 25–40% of the vertical load and 50–80% of the applied moment) although in each case this was rather less than predicted analytically. The bedded caps produced, as expected,[4] a negligible increase in the vertical stiffness of the system.

25. The load sharing between the piles in the 3 x 3 groups also fitted the calculated values quite closely, with the exception of the centre pile which, although it supported about twice its predicted load, was still the most lightly loaded pile in the group.

26. The pile-head bending moments developed under eccentric vertical loads were quite negligible, the loads being supported by 'push–pull' action in the piles, and not until the eccentricity was such that the line of action of the load fell outside the centre-line of the outer row of piles ($e \approx B/3$) did any tensile forces develop.

27. The correspondence between the measured and predicted group settlement ratios was exceptionally good, producing strong support for R values of about 2 and 4 for 2 x 2 and 3 x 3 groups of conventional size.

28. In order to assess G empirically from c_u, for single-pile tests, via $G \approx \beta c_u$, the results reported support $\beta \approx 154$.

29. The Authors consider that the foregoing evidence, based on over 100 individual load tests, strongly supports the use of elastic analyses predicting the vertical, working load response of preformed piles in firm-to-stiff clays using elastic moduli determined from precise measurements on single in situ prototype piles. (The linearity of such systems in soft clays and under horizontal loads will be much less good.[9,15]) However, we are not advocating the use of elaborate computer programs as design tools; simple charts for this purpose should be avaliable shortly, developed from parametric studies using PGROUP.[15]

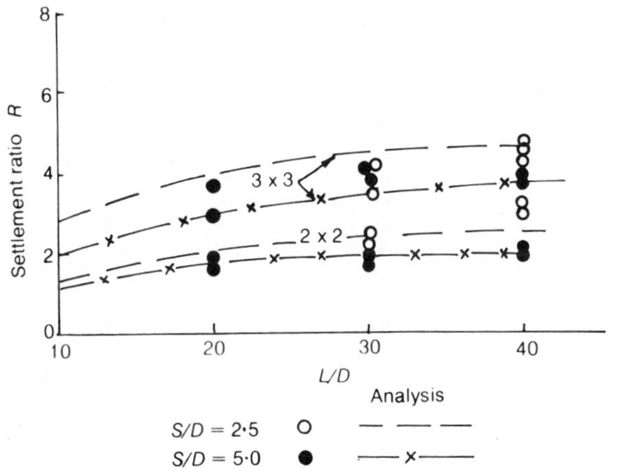

Fig. 12. Settlement ratio versus L/D for bedded cap groups

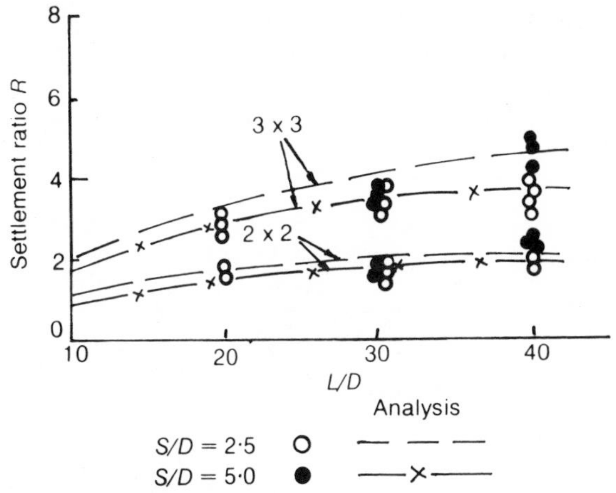

Fig. 13. Settlement ratio versus L/D for extended caps, $e/B \neq 0$

Table 5. Load distribution, 3 x 3 groups

L/D	S/D	e/B	Pile load, %			Cap load, %
			Type 1	Type 2	Type 3	
20	5.0	1/12	10	7.5	6	29.5
		1/6	12	7	5	28
		1/3	15	6.5	–3	54.5
30	5.0	1/12	11	7	7.5	23.5
		1/6	15	7	5.5	17.5
		1/3	17	5.5	–0.5	34
40	5.0	1/12	12	5.5	5	32.5
		1/6	12	5.5	3	38.5
		1/3	14.5	4.5	–0.5	54.5
30	2.5	1/12	13.5	6	7.5	19
		1/6	16	6.5	2	26.5
		1/3	22.5	8	–2.5	16
40	2.5	1/12	12.5	8	6	20.5
		1/6	16	8.5	3	17.5
		1/3	23.5	8.5	–4.5	17.5

REFERENCES

1. POULOS H. G. Analysis of the settlement of pile groups. Géotechnique, 1968, XVIII, 449–471.

2. BANERJEE P.K. and DRISCOLL R.M.C. A programme for the analysis of generally loaded, planar raked pile groups. Department of the Environment, London, 1975, HECB/B/7.

3. BUTTERFIELD R. and BANERJEE P. K. Analysis of axially loaded compressible piles and pile groups. Géotechnique, 1971, XXI, No. 1, 43–60.

4. BUTTERFIELD R. and BANERJEE P. K. The problem of pile cap and pile group interaction. Géotechnique, 1971, XXI, No. 2, 135–142.

5. BANERJEE P. K. and DAVIES T. G. Analysis of pile groups embedded in Gibson soil. Proc. 9th Int. Conf. Soil Mech., Tokyo, 1977.

6. BANERJEE P. K. and DAVIES T. G. The behaviour of axially and laterally loaded piles embedded in non-homogeneous soils. Géotechnique, 1978, XXVIII, No. 3, 309–326.

7. SKEMPTON A. W. and HENKEL D. J. Tests on London Clay from deep borings at Paddington, Victoria, Southbank. Proc. 4th Int. Conf. Soil Mech., 1957, 1, 100.

8. GHOSH N. A model scale investigation of the working load stiffness of single piles and groups of piles in clay under centric and eccentric vertical loads. PhD thesis, Southampton University, 1976.

9. ABDRABBO F. K. A model scale study of single vertical piles and pile groups under general planar loads. PhD thesis, Southampton University, 1976.

10. BUTTERFIELD R. and GHOSH N. The response of single piles in clay to axial load. Proc. 9th Int. Conf. Soil Mech., Tokyo, 1977, 2/18, 451–457.

11. BUTTERFIELD R. and JOHNSTON I. W. The stresses acting on a continuously penetrating pile. Proc. 8th Int. Conf. Soil Mech., Moscow, 1973, 3/7, 39–45.

12. BUTTERFIELD R. and BANERJEE P. K. The effect of pore water pressures on the ultimate bearing capacity of driven piles. Proc. 2nd SE Asian Conf. Soil Eng., Singapore, 1970, 385–394.

13. TOMLINSON M. J. Results of loading tests, adhesion of piles in stiff clays. Construction Industry Research and Information Association, London, 1970, report 26.

14. PEERLESS I. Measurement of the pile head gap in cohesive soils. UG report (unpublished), Southampton University, 1977.

15. DOUGLAS R. The slender pile. PhD thesis, Southampton University, 1979.

15. An approach for the analysis of offshore pile groups

H. G. POULOS, BE, PhD, DScEng, MIEAust, MASCE (Reader in Civil Engineering, University of Sydney, Australia)

The Paper describes a method of analysis of offshore pile groups subjected to axial, lateral and moment loading. The method utilizes the concept of pile interaction factors to develop a system of equations relating the pile group response to the response of single piles to axial and lateral loading. While elastic theory provides the basis for the analysis, allowance can be made for the effects of pile–soil slip and lateral pile–soil yielding, so that non-linear load–deflexion relationships can be developed. Non-homogeneous soil conditions and battered piles can be readily handled, but the present analysis is limited to piles of equal length and uniform diameter, although these limitations could be overcome without excessive difficulty. While the analysis necessarily involves a number of simplifying assumptions, it provides a convenient and relatively economical preliminary means of analysing offshore pile groups, and in particular, of investigating the influence of parameter variations on the behaviour of the group. The selection of input parameters for the analysis is discussed, and solutions are presented for a typical offshore group in which the effect of various factors on the group response is studied. It is found that improvement of group performance is more rapidly achieved by increasing pile spacing rather than increasing pile size or stiffness. The selection of suitable soil parameters, making allowance for cyclic loading effects, has a profound effect on the predicted group behaviour.

INTRODUCTION

The increasing use of fixed offshore platforms supported by pile foundations has encouraged the development of more rational methods of analysis of pile groups subjected to combined axial and lateral loading. Early methods of analysis considered only interaction between the piles through the pile cap and ignored pile interaction through the soil, while the response of an individual pile to load was generally determined from linear subgrade reaction theory.[1–4] Recognition of the importance of pile interaction through the soil has subsequently led to the incorporation of this aspect of behaviour into the analysis, via the use of elastic theory to determine the displacements within a mass due to a loaded pile,[5, 6] although the response of individual piles is still determined by a non-linear subgrade reaction analysis (the 'p–y' approach).

2. Since the use of elastic theory is required to evaluate pile–soil–pile interaction, it appears logical to adopt a unified approach and also use elastic theory to determine the individual pile response to various types of loading. Linear analyses of group behaviour based entirely on elastic theory have been presented by Banerjee and Driscoll,[7] Poulos and Madhav[8] and Poulos.[9] However, because of the non-linearity of pile response in real soils, particularly when subjected to lateral loading, a more realistic analysis requires modification of the purely elastic analysis to allow for local yielding or pile–soil slip along the pile shaft.

3. This Paper describes an elastic-based approach to the analysis of pile groups subjected to axial and lateral loading. Elastic theory is used to evaluate pile–soil–pile interaction and the axial and lateral response of individual piles, but for the latter evaluation, allowance is made for local yield and slip along the pile, so that non-linear load–deflexion relationships can be developed for a group. The implementation of the analysis through a computer program is described, and the determination of appropriate soil parameters for the analysis is discussed. Finally, an example of the application of the analysis to an offshore group is given, and the effects of some soil and pile parameters on group performance are examined.

4. It is implicitly assumed in this Paper that the loads acting on the group are known, but this will not always be the case, as the pile group (or groups) and the supported structure form an interactive system. It is possible to carry out a complete analysis taking account of structure–foundation interaction, in which case the applied loads and moments on the pile group become unknowns and are evaluated by imposing compatibility between the deformations of the structure and the pile group. The analysis presented in this Paper deals with the determination of deformations of the pile group and could be incorporated into a structural analysis, although an iterative solution would be necessary because of the non-linear behaviour of the pile group.

DEVELOPMENT OF ANALYSIS

Concept of interaction factors

5. Considerable simplifications to pile group analyses can be made by utilizing the concept of interaction factors. An interaction factor is the ratio of the increase in deflexion (or rotation) of a pile caused by an identical adjacent loaded

pile, to the deflexion (or rotation) of an isolated pile. Elastic theory can be used to derive interaction factors for axial loading and lateral loading; and by superposition of interaction factors, the deflexions of a group can be expressed in terms of the corresponding deflexions of an isolated single pile. The superposition of interaction factors is approximate only, as the reinforcing effect of intervening piles in a group of piles is ignored. In addition, all piles in the group are assumed to be identical. These approximations can be removed by simultaneous consideration of all elements of all the piles within a group (Banerjee and Driscoll[7]). However, one of the great advantages of the interaction factor approach is that various pile configurations can be considered without the necessity for a total re-analysis of each group. Furthermore, the use of interaction factors allows the analysis of considerably larger groups than is usually feasible with the Banerjee and Driscoll analysis.

Treatment of battered piles

6. The following assumptions are made in the analysis regarding battered piles.

(a) The piles are battered in the same plane and the horizontal load acts in this plane.
(b) Interaction between two battered piles is identical to the interaction between two vertical piles at an equivalent spacing s_e, where s_e is taken here as one-third of the pile length (Fig. 1(a)). The value of s_e probably varies with pile stiffness and the above value will tend to underestimate interaction for long flexible piles, and overestimate interaction for relatively stiff piles.
(c) The axial displacement due to axial load on a battered pile is equal to the vertical displacement due to vertical load on a vertical pile (Fig. 1(b)); similarly, the normal deflexion due to normal load on a battered pile is equal to the horizontal deflexion due to horizontal load on a vertical pile.[8, 10]

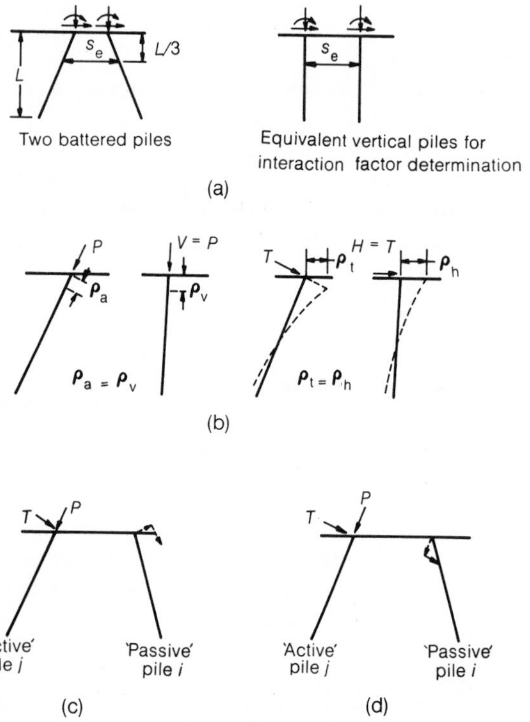

Fig. 1. Assumptions made for analysis of battered piles: (a) interaction between two battered piles; (b) deflexions of battered piles; (c) direction of movements at passive pile due to active pile, using Poulos and Madhav[8] assumption; (d) direction of movements at passive pile due to active pile, using Poulos[9] assumption

7. An assumption must also be made regarding the deflexions of a battered pile caused by another pile. If the reciprocal theorem is to be satisfied, it is necessary to assume that an axial load in a pile j will cause axial deflexion in pile i (in the direction of the axis of pile i) and that normal load in pile j will then be assumed to cause a normal deflexion at pile i (Fig. 1(c)). The above assumptions were made by Poulos and Madhav.[8] On the other hand, it might appear more reasonable to assume that an axial load in pile j will cause a deflexion at pile i which is in the direction of the axis of pile j, while a normal load on pile j would cause a deflexion at pile i in a direction normal to the axis of pile j (Fig. 1(d)). This latter assumption was made by Poulos.[9] For a group composed entirely of vertical piles, or piles which are battered in the same direction, both the above assumptions give identical results, but in the general case slightly different answers will result. The analysis here is described in terms of the first assumption, and the modifications required for the second assumption are outlined.

Derivation of group equations

8. Consideration is given first to two piles, i and j, in a group, as shown in Fig. 2. Taking positive batter as being in the direction of the applied horizontal load, the axial load P_j and normal load T_j are given by

$$P_j = V_j \cos\psi_j + H_j \sin\psi_j \tag{1}$$

$$T_j = -V_j \sin\psi_j + H_j \cos\psi_j \tag{2}$$

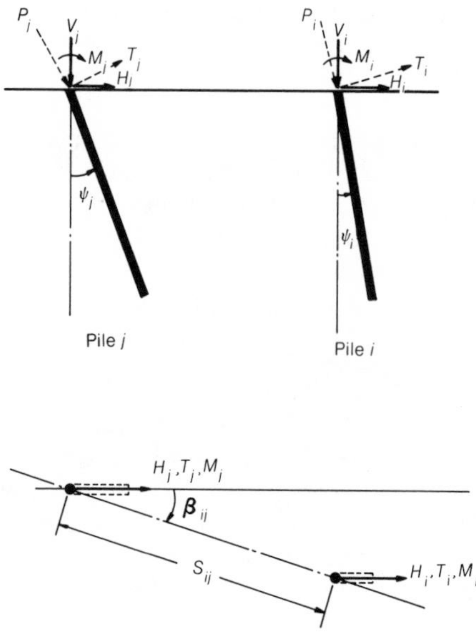

Fig. 2. Two typical piles in a group (note – pile batter is positive in direction of applied horizontal load on group)

9. Following the assumption of Poulos and Madhav,[8] the axial deflexion at i due to axial load on pile j, ρ_{aij}, can be expressed as

$$\rho_{aij} = \alpha_{ij} P_j \rho_{v1} \tag{3}$$

where

α_{ij} is the axial displacement interaction factor corresponding to the spacing s_{ij} between piles i and j
P_j is the axial load in pile j
ρ_{v1} is the axial deflexion of pile i due to unit axial load.

Substituting for P_j from equation (1):

$$\rho_{aij} = \alpha_{ij} (V_j \cos\psi_j + H_j \sin\psi_j)\rho_{v1} \tag{4}$$

10. The normal deflexion at pile i due to pile j, ρ_{tij}, can be expressed as

$$\rho_{tij} = \alpha_{\rho Hij} T_j \rho_{hH1} + \alpha_{\rho Mij} M_j \rho_{hM1} \tag{5}$$

where

$\alpha_{\rho Hij}$ is the interaction factor for normal deflexion due to normal load, corresponding to spacing s_{ij} and angle β_{ij} between piles i and j
$\alpha_{\rho Mij}$ is the interaction factor for normal deflexion due to moment, corresponding to spacing s_{ij} and angle β_{ij} between piles i and j
T_j, M_j are the normal load and moment on pile j
ρ_{hH1} is the normal deflexion due to unit normal load
ρ_{hM1} is the normal deflexion due to unit moment.

Substituting for T_j from equation (2):

$$\rho_{tij} = \alpha_{\rho Hij} (-V_j \sin\psi_j + H_j \cos\psi_j)\, \rho_{hH1} + \alpha_{\rho Mij} M_j \rho_{hM1} \tag{6}$$

11. Now the vertical deflexion at the head of pile i due to pile j, ρ_{vij}, is

$$\rho_{vij} = \rho_{aij} \cos\psi_i - \rho_{tij} \sin\psi_i \tag{7(a)}$$

and the horizontal deflexion, ρ_{hij}, is

$$\rho_{hij} = \rho_{aij} \sin\psi_i + \rho_{tij} \cos\psi_i \tag{7(b)}$$

The values of ρ_{aij} and ρ_{tij} may be substituted from equations (4) and (5). In a similar manner, an expression can be derived for the rotation θ_{ij} of pile i due to pile j. By writing the equations for all piles in the group, one obtains the following:

$$\begin{bmatrix} \mathbf{A_v} & \mathbf{B_v} & \mathbf{C_v} \\ \mathbf{A_h} & \mathbf{B_h} & \mathbf{C_h} \\ \mathbf{A_\theta} & \mathbf{B_\theta} & \mathbf{C_\theta} \end{bmatrix} \begin{Bmatrix} V \\ H \\ M \end{Bmatrix} = \begin{Bmatrix} \underset{\sim}{\rho_v} \\ \underset{\sim}{\rho_h} \\ \underset{\sim}{\theta} \end{Bmatrix} \tag{8}$$

where

V, H, M are the vertical load, horizontal load and moment on the pile heads
$\underset{\sim}{\rho_v}, \underset{\sim}{\rho_h}, \underset{\sim}{\theta}$ are the vertical deflexion, horizontal deflexion and rotation at the pile heads

and the coefficients of the flexibility sub-matrices are given by

$$A_{vij} = \alpha_{ij}\, \rho_{v1} \cos\psi_i \cos\psi_j + \alpha_{\rho Hij}\, \rho_{hH1} \sin\psi_i \sin\psi_j$$

$$B_{vij} = \alpha_{ij}\, \rho_{v1} \cos\psi_i \sin\psi_j - \alpha_{\rho Hij}\, \rho_{hH1} \sin\psi_i \cos\psi_j$$

$$C_{vij} = \alpha_{\rho Mij}\, \rho_{hM1} \sin\psi_i$$

$$A_{hij} = \alpha_{ij}\, \rho_{v1} \sin\psi_i \cos\psi_j - \alpha_{\rho Hij}\, \rho_{hH1} \cos\psi_i \sin\psi_j$$

$$B_{hij} = \alpha_{ij}\, \rho_{v1} \sin\psi_i \sin\psi_j + \alpha_{\rho Hij}\, \rho_{hH1} \cos\psi_i \cos\psi_j$$

$$C_{hij} = \alpha_{\rho Mij}\, \rho_{hM1} \cos\psi_i$$

$$A_{\theta ij} = -\alpha_{\theta Hij}\, \theta_{H1} \sin\psi_j$$

$$B_{\theta ij} = \alpha_{\theta Hij}\, \theta_{H1} \cos\psi_j$$

$$C_{\theta ij} = \alpha_{\theta Mij}\, \theta_{M1}$$

and $\alpha_{\theta Hij}$, $\alpha_{\theta Mij}$ are rotation interaction factors for normal loading and moment, respectively.

12. The three equilibrium equations and the appropriate pile head boundary conditions can be incorporated into equation (8), and the resulting equations solved for the $3n + 3$ unknowns. For example, for a rigid cap, the unknowns are n vertical loads, n horizontal loads, n pile head moments, a reference vertical pile head deflexion, the horizontal deflexion of the group and the group rotation. If known pile head loads and moments are specified, the deflexions and rotation of each pile can be directly evaluated.

13. If the alternative assumption of Poulos[9] is made, the vertical and horizontal deflexions at pile i depend only on the batter angle of pile j. Consequently, in the flexibility submatrices in equation (8), $\sin\psi_i$ and $\cos\psi_i$ are merely replaced by $\sin\psi_j$ and $\cos\psi_j$. Otherwise, the equations are identical.

IMPLEMENTATION OF ANALYSIS

14. The preceding analysis involves no assumptions with respect to the method of calculating the interaction factors or the single pile responses. Several approaches are possible; for example, that used by Focht and Koch[5] which combines the p–y approach for single pile response with elastic theory for calculation of inter-pile interaction. However, an analysis based entirely on elastic theory has the advantage that a consistent set of soil parameters can be used for both single pile response and interaction, and the effects of variations in soil stiffness on group behaviour can readily be evaluated.

15. In order to avoid undue complications, a number of simplifications have been made in implementing the group analysis, as discussed below. A computer program, DEFPIG, has been developed to carry out the analysis.

Calculation of interaction factors

16. The soil has been taken to be purely elastic for the calculation of the interaction factors, but provision has been made to input different values of soil modulus for axial and lateral responses of the piles. The basis of the calculations of axial interaction factors is described else-

where,[11, 12] and involves discretization of the piles into a number of cylindrical shaft elements and annular base elements. The presence of a pile cap in contact with the soil is simulated by circular annular elements at the soil surface. The lateral interaction factor calculation follows that described by Poulos,[13] in which the pile is represented by a thin strip divided into a series of rectangular elements. The present program does not allow for consideration of the effect on lateral response of a pile cap in contact with the soil surface, although such a modification would be possible. To further simplify the analysis, it is assumed that the effects of axial and lateral loading are uncoupled. Axial and lateral interaction factors are evaluated at a number of pile spacings and interpolation is used to obtain values at intermediate spacings.

Single pile responses for group analysis

17. The single pile axial and lateral responses are obtained using elastic-based theories, but non-linear response is allowed for by incorporation of the effects of pile–soil slip in evaluation of axial response[14] and the effects of lateral pile–soil yield on lateral response.[15] Provision is made for group effects on the ultimate skin friction and pile–soil yield pressures by specification of efficiency factors which are applied to the values for a single isolated pile. It is also assumed that the single pile deflexions under unit load are the same for all piles and correspond to the average axial and lateral load and moment levels in the group. An iterative procedure is required to determine these single pile deflexions. Another assumption is that the single pile unit deflexion and rotation due to horizontal load are independent of the moment, and vice versa. While this is true for elastic conditions, it will not generally be true when some pile–soil yielding has occurred. However, the assumption will generally be conservative, as positive horizontal loads at the pile heads are frequently accompanied by negative moments due to pile cap restraint. Consequently, there is less possibility of local yield than if either horizontal load only, or moment only, were applied.

Limiting axial loads in piles

18. The analysis contains a check on the magnitude of the axial load within each pile, so that when the computed axial load in any pile exceeds the ultimate value, the axial displacement compatibility equation for that pile is replaced by one setting the axial load on that pile equal to its ultimate value. The solution is then recycled until all axial loads are less than or equal to the ultimate values.

Soil non-homogeneity

19. Non-homogeneity of the soil is taken into account in the manner described by Poulos.[16] In the evaluation of the influence of one loaded element on another, the value of soil modulus in the Mindlin elastic equations for soil response is taken as the mean of the values at the two elements.

Partially embedded piles

20. The analysis considers only fully embedded piles, but partially embedded piles with free-standing portions above the soil surface can readily be analysed by ascribing very small values of 'soil' modulus, adhesion and lateral yield pressure to the free-standing portions. The piles are then treated as being in a non-homogeneous medium, with the top (air) layer being extremely weak. The results for pile head deflexions from this approach are almost identical with those obtained by adding the deflexions of the free-standing portions to the ground line deflexions.

Load and moment distributions in individual piles

21. It is often of interest to know the distribution of loads, moment and deflexion in individual piles within the group. In analyses in which individual elements of each pile in the group are considered,[7, 17] these distributions are computed as a matter of course, but in the present analysis, only the pile head loads and moments are determined, so further assumptions are necessary in order to calculate the load and moment distributions along each pile. The simplest approach would be merely to evaluate the loads and moment along the pile from the applied loads and moment at the pile head, using the analyses for a single pile, but this ignores the interaction which occurs between the piles through the soil. A more satisfactory (though still approximate) approach is to use elastic theory to sum the soil displacements at the pile in question due to the stresses acting on all elements of all piles in the group. The form of the stress distribution along each pile is assumed to be the same as that along the nominated pile, with the actual magnitude of the stress being scaled by the magnitude of the appropriate load or moment at the pile head.

22. With both the axial and lateral response determined in the above manner, the calculated head displacements generally differ from those calculated from the group analysis because of the different approximations made in each calculation. However, this difference is generally less than 10% unless the piles are close to failure.

Comparison with alternative analysis

23. In an attempt to make some assessment of the accuracy of the present analysis, comparisons were made with elastic solutions obtained by Banerjee and Driscoll.[7] For three different laterally loaded pile groups reasonable agreement between the two solutions was obtained, particularly for the lateral deflexions and axial pile loads. Some differences were found in relation to the distribution of head moment and horizontal load, but overall the agreement was sufficiently close to suggest that the present analysis should provide a reasonable basis for examining pile behaviour. For the group in which the piles were battered at different angles, the Poulos and Madhav formulation gave a slightly larger lateral deflexion than the Poulos[9] formulation. The former was in better agreement with Banerjee and Driscoll's solution. Much scope remains for further detailed comparisons with the results of tests on model and full-scale pile groups.

ESTIMATION OF REQUIRED INPUT PARAMETERS

24. The main parameters required in the analysis are

(a) Young's modulus of the soil along the piles; different values may be relevant for axial and lateral loading;
(b) the limiting pile–soil stresses for axial loading; i.e., the skin friction along the pile shaft, the ultimate pressure on the pile base and, if necessary, the ultimate vertical pressure beneath the pile cap;
(c) the limiting lateral pile–soil yield pressures along the pile.

For all parameters, recognition must be given to the effects of cyclic loading, and group effects must also be considered in the estimation of the limiting axial and lateral stresses.

25. The current state of knowledge of the above parameters is, at best, incomplete, particularly in relation to the effects of cyclic loading. While considerable research has been done on the behaviour of soils under cyclic loading, much work remains to be done to relate this behaviour to the response of piles to cyclic loading. At present, the most feasible design approach appears to be to seek a quasi-static approximation to the behaviour of a pile under an infinitely large number of loading cycles.[18] Most of the current design procedures for this approach have been developed by Reese, Matlock and their co-workers on the basis of field and model tests and are presented in the form of p–y curves for lateral loading, or the equivalent curves for axial loading. However, it is also possible to interpret some of their design recommendations in terms of the elastic-based approach, and Table 1 shows this interpretation for piles in clay. Design parameters for static loading conditions are summarized and modifications for cyclic loading conditions are described. These parameters are expressed in terms of total stress, and consequently severe reduction factors occur, especially for stiff clays. It must be emphasized that many of the parameters are derived from very limited data, particularly in the case of those for cyclic loading, and should be used with due caution. In addition, the response calculated using these parameters will represent an envelope to the long term behaviour. If the cyclic response of the group, or its behaviour under transient loading, is required, the soil parameters used should reflect the effect of a relatively rapidly applied loading as well as the previous cyclic loading history of the soil.

26. The influence of group effects on the limiting axial pile–soil stresses may be estimated by the usual methods of pile group bearing capacity analysis. In clays, use may be made of the Terzaghi–Peck approach of taking the group capacity as the lesser of the values for single pile failure and of failure of the block; alternatively, a variety of empirical efficiency formulae are available.

27. Group effects on the ultimate lateral pressure p_y are less easily estimated although the lateral analogue of the Terzaghi–Peck approach has been found to give reasonable results for model pile groups. Alternative approaches can be used, such as that described by Yegian and Wright,[19] who analysed a two-dimensional plan model of a group using a non-linear finite element analysis. This approach has some shortcomings, but should provide an estimate of the lateral efficiency factor.

BEHAVIOUR OF A TYPICAL OFFSHORE GROUP

28. To illustrate the application of the analysis to a practical problem, an eleven-pile group for a North Sea platform, similar to that described by McClelland,[27] has been analysed. Details of the group are summarized in Fig. 3, together with the assumed loads and a typical soil profile and distribution of undrained shear strength distribution. The vertical cyclic load has been taken as P and the lateral load and groundline moment have been related to this value. On the basis of the available information, values of τ_a, p_y and E_s were determined for the analysis using the procedure shown in Table 1, classifying the soil profile as a

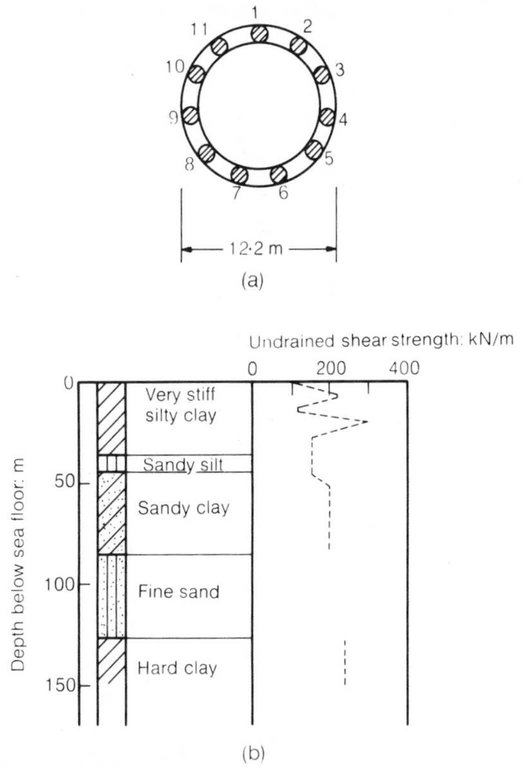

Fig. 3. Pile and soil details: (a) plan of group; (b) simplified soil profile. (Piles are 1.37 m dia., 50 mm wall, 72 m penetration; portion above sea floor (24 m long) grouted into templet sleeves; templet leg and piles battered at 8½° to vertical. From available information and making other assumptions, the following load conditions on the group are assumed: dead load 40 MN, vertical live load ±P MN, lateral live load ±0.4P MN, moment at mud line ±96.6P MN m)

stiff clay. A cyclic load factor of 0.7 was applied to the static τ_a values as the cyclic vertical loading is predominantly one way, although net tensile forces can occur. For the static E_s for lateral loading, a value of 150 c_u was adopted. A constant Poisson's ratio of 0.3 was taken in both cases (to reflect long term drained conditions) and a rigid base was assumed at 250 m below the sea floor. Eight elements were used to divide the pile for axial response and 18 elements for lateral response. The axial group efficiency factors for the shaft and base were taken as unity while the lateral efficiency factor was taken as 0.8. The piles were assumed to be rigidly attached to the template sleeves at the ground line, although the sleeves were assumed to be free to rotate.

29. Because the piles are battered the group response will be dependent on the direction of the applied loading, and a spectrum of load–deflexion curves should be considered. Fig. 4 shows load–deflexion relationships for three loading directions: with the batter (i.e., a pile batter angle of + 8.5°), against the batter (a pile batter angle of –8.5°) and across the batter (equivalent to a zero batter angle). It is clear that loading against the batter produces the largest deflexions and that significant permanent deformations are likely to occur in this loading direction. The load–deflexion behaviour becomes markedly non-linear beyond a load P of about 15 MN, and such a value would represent the maximum desirable design loading.

30. For loading across the batter, the influence of in-

Table 1. Design parameters for offshore piles in clay

Loading	Parameter	Expression or method of determination	Remarks
Static	E_s (axial response)	See E_s versus c_u correlations of Poulos[14]	Values are generally higher than for lateral response
	E_s (lateral response)	(100–150) c_u	Reese and Desai,[20] Banerjee;[21] the higher values are more relevant to stiffer clays
	τ_a	αc_u (total stress basis) or $F\sigma_v'$ (effective stress basis)	See Tomlinson[22] for values of α; for normally consolidated clays, $F = 0.2–0.3$ (Burland[23]); for over-consolidated clays, much larger values of F occur, up to 5 for high OCR values (Vesic[24])
	τ_{bu}	$9\ c_u$	–
	p_y	Lesser of $(3 + \gamma z/c_u + 0.5z/dc_u)\ c_u$ (shallow) and $9c_u$ (deep)	Reese and Desai;[20] can determine depth of transition ($z = z_r$) from shallow to deep case by equating the two expressions; γ = effective unit weight
Cyclic	E_s (axial and lateral response)	For soft clays, take equal to values for static loading; for stiff clays, apply a factor of 0.4 to values for static loading	Holmquist and Matlock,[25] Parry[26]
	τ_a and τ_{bu}	For soft clays apply a factor of 0.7 to static values for one-way cyclic or transient loading (at load levels approaching static value) and a factor of 0.4 for two-way cyclic loading	Based on Holmquist and Matlock;[25] may be conservative for cyclic load levels less than half ultimate static values; data not available for stiff clays, but use of residual values may be appropriate
	p_y	For soft clays: (a) if $z < z_r$, multiply static value by $0.72z/z_r$; (b) if $z > z_r$, multiply static value by 0.72	Matlock;[18] z_r is transition depth for static case (see above), z is depth below surface; strain softening behaviour suggested by Matlock cannot be easily modelled, ultimate values used
		For stiff clays, multiply static value by 0.24	Based on mean of peak and ultimate values suggested by Reese and Desai;[20] very marked strain softening of p–y curves indicated; values of p_y based on residual strength may be appropriate

creasing pile wall thickness from 50 mm to 75 mm and diameter from 1.37 m to 1.68 m was investigated. As would be expected, the deflexions decreased as compared with the original piles, but the effect was generally only marginal. Neither measure could be regarded as an effective or economical means of reducing deflexions, although the use of a larger diameter did increase the ultimate group capacity slightly.

31. The influence of the pile head condition was investigated for the group of vertical piles by carrying out additional analyses in which the pile heads were assumed first to be fixed, and then pinned. As compared with the previous case (Fig. 4), the fixed head pile group suffers a smaller horizontal deflexion (approximately 14% less at P = 15 MN) and also smaller vertical deflexion of the front piles because of the absence of head rotation. On the other hand, a pinned head pile group has much the same vertical deflexion, a 10% smaller rotation, but a significantly larger horizontal deflexion (83% greater at P = 15 MN).

32. An analysis was made of a group of vertical piles in which the piles were located on a circle of 15 m (rather than the original 10.8 m), and it was found that the rotations were substantially reduced (by about 50%), while the lateral and vertical deflexions were decreased by about 10–15%. Increasing the pile spacing would also tend to increase the group efficiency, although in the analysis performed, the lateral efficiency was kept at 0.8.

33. The effect of leaving out one of the piles was also investigated, and found to be relatively small. For example, if pile 1 were left out, the vertical deflexion of pile 4 at P = 15 MN increased from 26 mm to 27 mm, the horizontal deflexion from 83 mm to 91 mm, and the rotation from 0.00230 rad to 0.00232 rad. Removal of pile 9 instead of pile 1 had an almost identical effect.

34. The importance of considering pile–soil–pile interaction was investigated by carrying out an analysis in which all interaction factors were input as zero. For the group having all vertical piles, the results of this analysis are compared with the original analysis (including interaction) in Table 2. Ignoring interaction has little effect on the distribution of vertical and horizontal loads within the group, but increases the pile head moment by about

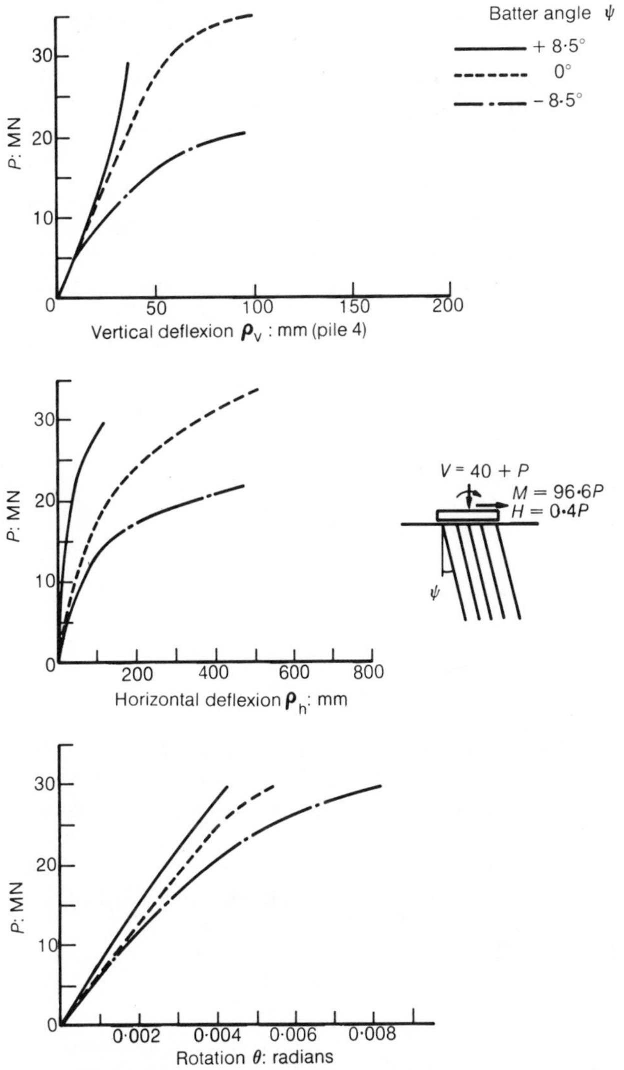

Fig. 4. Effect of batter angle and loading direction on pile group response

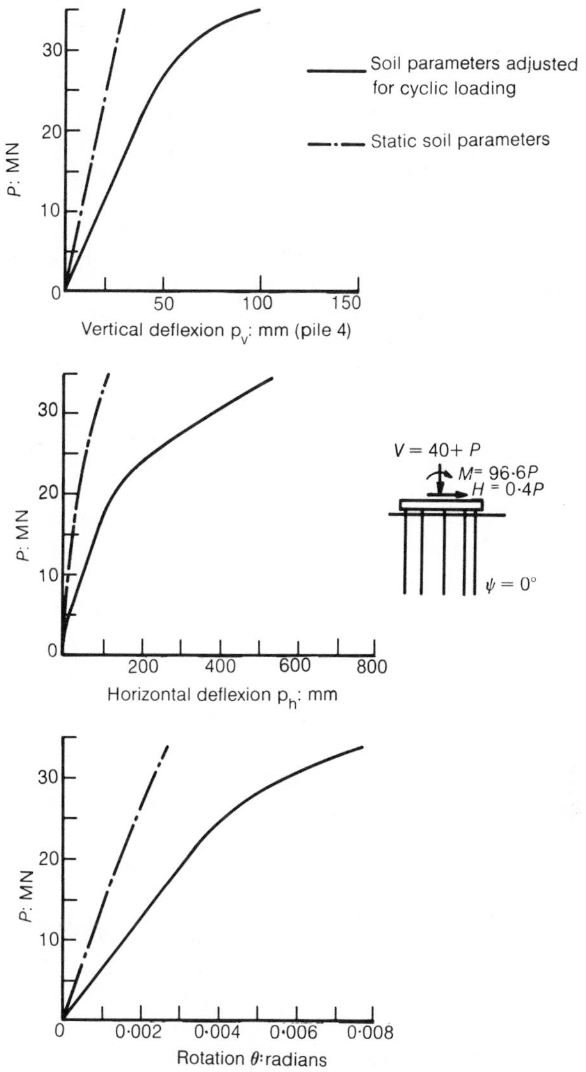

Fig. 5. Effect of soil parameters on group response

20% and the pile head rotation by about 10%, and almost doubles the horizontal and vertical deflexions.

35. Finally, the importance of the input soil parameters was examined by the performance of an analysis using the estimated pile–soil strength and modulus parameters for static loading. Thus, the soil modulus values for both lateral and axial response were 2.5 times the values previously adopted for cyclic loading, the τ values were 1.43 times greater, and the p_y values were 2.5 times greater. Fig. 5 compares the response computed for the two different sets of soil parameters for a group with vertical piles. The group response using the static parameters is almost linear up to $P = 30$ MN, and the large difference between this solution and that using the parameters adjusted for cyclic loading highlights the importance of selecting appropriate parameters. On the basis of the analysis using static parameters, an allowable design value of P of at least 25 MN would be possible, as compared with the value of about 15 MN determined from the analysis using the parameters adjusted for cyclic loading. The calculations therefore suggest that, of all the parameters examined, the soil moduli and pile–soil strength parameters are the most critical.

Table 2. Effect of considering pile–soil–pile interaction: all piles vertical; $P = 15$ MN

Value	Analysis without interaction	Analysis with interaction
Vertical load on pile 4, MN	10.4	10.6
Horizontal load on pile 4, MN	0.55	0.54
Head moment at pile 4, MN m	−1.65	−2.16
Vertical deflexion of pile 4, mm	14	26
Horizontal deflexion, mm	40	82
Head rotation, rad	0.0021	0.0023

CONCLUSION

36. The analysis described in this Paper provides a consistent, if approximate, basis for the analysis of pile group behaviour, incorporating pile–soil–pile interaction and non-linearity of pile response. The method is based on elastic theory and utilizes the concept of interaction factors.

While this analysis does impose certain restrictions (e.g., all piles are identical and have uniform diameter and stiffness), it compensates for this restriction in its ability to provide an economical means of examining various group configurations, pile batters and load combinations, without the necessity for a complete re-analysis of each different group. The entire load–deflexion behaviour to failure can be computed, and the load and moment distribution along individual piles within the group may also be determined approximately.

37. The example described in the Paper illustrates the importance of including pile–soil–pile interaction in the analysis, and shows that improvement of group performance can be achieved more readily by increasing pile spacing rather than increasing the pile size or stiffness; however, structural constraints may often dictate the spacing to be used. It also indicates that, with groups containing piles which are not symmetrically battered, significant permanent deflexions are likely to develop under randomly orientated cyclic loading.

38. As with all theoretical analyses, the success of the present method depends critically on the selection of the input parameters, particularly those of the soil. For offshore pile groups, the problem of selection is accentuated by the existence of cyclic loading and its effects on the soil parameters. The example calculations highlight the vital importance of choosing appropriate strength and deformation parameters if reliable predictions of offshore group response are to be made.

ACKNOWLEDGMENTS

39. The work described in this Paper forms part of a research programme into the deformation behaviour of all types of foundations being carried out at the University of Sydney under the general direction of Professor E. H. Davis. This research is supported by a grant from the Australian Research Grants Committee. The Author gratefully acknowledges the advice and suggestions of Professor Davis, the assistance of Miss D. R. Perkins in developing the program DEFPIG, and the many useful comments and revelations of shortcomings of the original program made by Dr L. M. Kraft of McClelland Engineers, Inc., Houston.

REFERENCES

1. FRANCIS A. J. Analysis of pile groups with flexural resistance. J. Soil Mech. Fdns Div. Am. Soc. Civ. Engrs, 1964, 90, No. SM3, 1–32.

2. PRIDDLE R. A. Load distribution in piled bents. Trans. Instn Engrs Aust., 1963, CE5, No. 2, 1.

3. SAUL W. E. Static and dynamic analysis of pile foundations. J. Struct. Div. Am. Soc. Civ. Engrs, 1968, 94, No. ST5, 1077–1100.

4. REESE L. C. et al. Generalized analysis of pile foundations. J. Soil Mech. Fdns Div. Am. Soc. Civ. Engrs. 1970, 96, No. SM1, 235.

5. FOCHT J. A. and KOCH K. J. Rational analysis of the lateral performance of offshore pile groups. Proc. 5th Offshore Technology Conf., Houston, 1973, 2, 701–708.

6. O'NEILL M. W. et al. Analysis of three-dimensional pile groups with nonlinear soil response and pile–soil–pile interaction. Proc. 9th Offshore Technology Conf., Houston, 1977, 2, 245–256.

7. BANERJEE P. K. and DRISCOLL P. M. Three-dimensional analysis of raked pile groups. Proc. Instn Civ. Engrs, Part 2, 1976, 61, Dec., 653–671.

8. POULOS H. G. and MADHAV M. R. Analysis of the movement of battered piles. Proc. 1st Australia–New Zealand Conference on Geomechanics, Melbourne, 1971, 1, 268–275. Inst. Engrs Aust.

9. POULOS H. G. Analysis of pile groups subjected to vertical and horizontal loads. Australian Geomech. J. 1974, G4, No. 1, 26–32.

10. EVANGELISTA A. and VIGGIANI C. Accuracy of numerical solutions for laterally loaded piles in elastic half space. Conf. Numerical Methods in Geomechanics, Blacksburg, 1976, 3, 1367–1370.

11. POULOS H. G. Analysis of the settlement of pile groups. Géotechnique, 1968, 18, No. 4, 449–471.

12. POULOS H. G. and MATTES N. S. Settlement and load distribution analysis of pile groups. Australian Geomech. J., 1971, G1, No. 1, 18–28.

13. POULOS H. G. Behaviour of laterally loaded piles—II: pile groups. J. Soil Mech. Fdns Div. Am. Soc. Civ. Engrs, 1971, 97, No. SM5, 733–751.

14. POULOS H. G. (DESAI C. S. and CHRISTIAN J. T. (eds)). Settlement of pile foundations. In: Numerical methods in geotechnical engineering. McGraw Hill, New York, 1977, chapter 10.

15. POULOS H. G. Load deflection prediction for laterally loaded piles. Australian Geomech. J., 1973, G3, No. 1, 1–8.

16. POULOS H. G. Settlement of piles in non homogeneous soil. University of Sydney, 1978, research report R333.

17. OTTAVIANI M. Three-dimensional finite element analysis of vertically loaded pile groups. Géotechnique, 1975, 25, No. 2, 159–174.

18. MATLOCK H. Correlations for design of laterally loaded piles in soft clay. Proc. 2nd Offshore Technology Conf., Houston, 1970, 1, 577–594.

19. YEGIAN M. and WRIGHT S. G. Lateral soil resistance–displacement relationships for pile foundations in soft clays. Proc. 5th Offshore Technology Conf., Houston, 1973, 2, 663–676, paper OTC1893.

20. REESE L. C. and DESAI C. S. (DESAI C. S. and CHRISTIAN J. T. (eds)). Laterally loaded piles. In: Numerical methods in geotechnical engineering. McGraw Hill, New York, 1977, chapter 9.

21. BANERJEE P. K. (SCOTT C. R. (ed.)). Analysis of axially and laterally loaded pile groups. In: Developments in soil mechanics. Applied Science Publishers, London, 1978, chapter 9.

22. TOMLINSON M. J. Some effects of pile driving on skin friction. Behaviour of piles. Institution of Civil Engineers, London, 1970, 59–66.

23. BURLAND J. B. Shaft friction of piles in clay, a simple fundamental approach. Ground Engng, 1973, 6, No. 3, 35–40.

24. VESIC A. S. Design of pile foundations. Transportation Research Board, Washington, 1977, NCHRP synthesis 42.

25. HOLMQUIST D. V. and MATLOCK H. Resistance–displacement relationships for axially-loaded piles in soft clay. Proc. 8th Offshore Technology Conf., Houston, 1976, 553–569, paper OTC 2474.

26. PARRY R. H. G. Pile design. Offshore soil mechanics. University of Cambridge, 1976, 177–223.

27. McCLELLAND B. Design of deep penetration piles for ocean structures. J. Geotech. Engng Div. Am. Soc. Civ. Engrs, 1974, 100, No. GT7, 705–747.

16. Axial analysis of piles using a hysteretic and degrading soil model

H. MATLOCK (Director of Research and Development, Fugro, Inc.) and S. H. C. FOO (Staff Engineer, Fugro Gulf, Inc.)

A method has been developed to analyse the driving of foundation piles by impact or vibration, plus a variety of problems dealing with static or dynamic axial loading of bars. The method has been coded in computer program DRIVE 7. A discrete-element mechanical analogue represents the pile member. A hysteretic, degrading support model is used to describe the non-linear inelastic behaviour of the soil. Strength degradation is provided as a function of deflexion and of the number of reversals of deflexion in the range beyond an initially elastic condition. Any variation with depth can be described. Hammer blows can be applied at any point along the pile length. The driving system has the capability to include a mandrel or follower in the analysis. Furthermore, measured force–time relationships can be described as an alternate method of simulating the action of the hammer on the pile. Stability and accuracy are maintained in the analysis by employing an implicit (Crank–Nicolson) type of numerical solution. To illustrate some of the capabilities of the method, three different aspects of the installation and loading of an anchor pile are considered: three consecutive hammer blows during driving of the anchor pile; a static tension pull-out test of the anchor pile immediately after driving; and subsequent cyclic tension loading, under two different load levels, leading to both the progressive stabilization and sudden failure of the anchor pile.

INTRODUCTION

A recently developed computer program, DRIVE 7,[1–3] provides a flexible tool for simulating both pile driving and a variety of other problems dealing with axial loading of bars. Residual stresses after static or dynamic loading can be examined.

2. Currently, most pile driving analyses[4, 5] have been based on work by E. A. L. Smith.[6] The soil model employed in these analyses still lacks generality. Recently, Holloway[7] included a hyperbolic load–deformation curve in addition to the bilinear elasto-plastic soil model used in the conventional wave equation analysis. However, the unloading path, reloading path, and subsequent hysteresis are not fully explained. It is also pointed out by Holloway that in predicting the load–settlement behaviour, a non-linear soil model performs better than the bilinear soil model. This further confirms the need for a more plausible, fully non-linear and inelastic soil model.

3. Also, in the conventional wave equation analysis, the total soil resistance during driving is assumed to be constant. This may be a principal cause of many drivability prediction failures. Some recently published field data[8] clearly demonstrates the progressive reduction in the soil resistance as the pile is being driven. Due to volume change, temporary pore pressure build-ups, near-field liquefaction, and subsequent reconsolidation, the total resistance must be variable with time and with the history of pile displacement. In DRIVE 7, a degrading soil model is included in which soil resistance is progressively lost as reversals of displacement occur during impact and rebound of the pile. The soil model may be easily modified or refined to accommodate other information on degradation which may be developed.

4. In most cases, drivability analyses are based on a single-blow analysis. Multiple-blow solutions have not been found in the published literature. Due to residual stress effects and changes in soil resistance under multiple hammer blows, the difference in the pile response under various blows may be significant. This is especially true when analysing hard-driving cases, as demonstrated by the example problem presented.

METHOD OF ANALYSIS

Elastic pile model

5. A mechanical analogue, very similar to the one used by Smith,[6] represents the elastic pile member. The pile is divided into a series of discrete elements of equal length. The mass of each element in the pile model is lumped at a node. The elastic stiffness of each element is represented by a spring situated between the nodes. A dashpot may be used, parallel to each spring, to represent internal viscous damping of the pile material. The discrete-element representation is illustrated in Fig. 1. This method of modelling the behaviour of a pile under axial loading has been used in relation to static axial loading[9–11] and dynamic axial loading.[6]

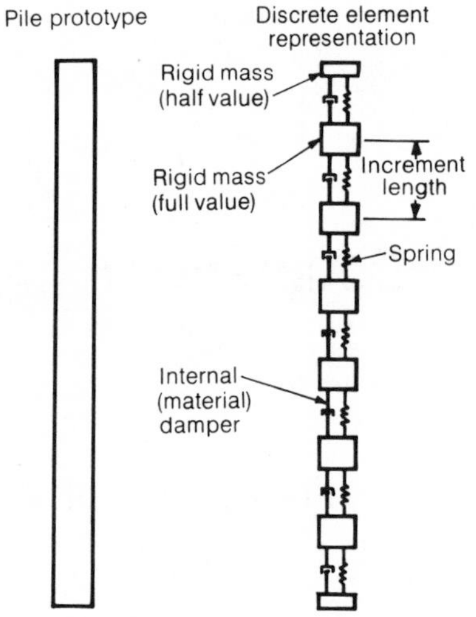

Fig. 1. Physical model of pile used in DRIVE 7

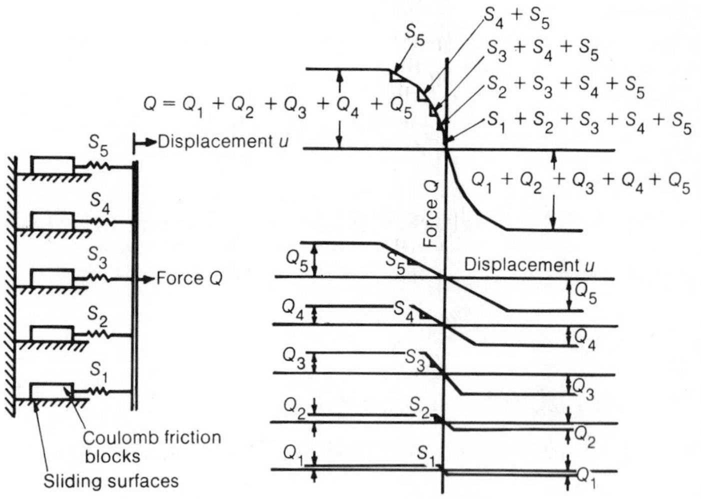

Fig. 2. Sub-element model used in DRIVE 7 (after Chan and Matlock)

Driving system

6. At present, DRIVE 7 does not have a sophisticated hammer routine such as the one used by WEAP[5] to account for the full thermodynamic cycle of a diesel hammer. When modelling a drop type steam hammer, the same discrete-element model used for the pile is applied to the hammer. In modelling the hammer, the springs between the hammer elements can be optionally separated. The DRIVE 7 hammer model has the versatility of being placed at any location along the pile, either with or without a mandrel. When a mandrel is being used, the axial motions in both the pile and the mandrel are computed. Such an example has been presented previously.[2] Force–time relationships produced during impact, either measured or computed, can be used as an alternate method of describing the driving system. The force–time curve can be described at any location along the pile. An example case with two different forcing functions, representing the conventional drop hammer and a hydro-block hammer, has been presented.[1]

Rheological soil model

7. The rate-independent part of the axial support is modelled by an assemblage of elasto-plastic sub-elements at each node to represent any desired non-linear inelastic behaviour of the axial support. This representation is shown in Fig. 2, which demonstrates that the total resistance at any deflexion is equal to the sum of the sub-element forces. Each sub-element at a pile model station is considered individually and the resulting effect at that station is the contribution of all sub-elements. Both stiffness and force, which correspond to a tangent modulus and intercept of the total curve, are transferred to the pile solution at each iteration.

8. The hysteretic response of a symmetric non-linear curve under an arbitrarily imposed motion is illustrated in Fig. 3. As demonstrated in Fig. 3, the forces developed for the complete curve result from a summation of tracing at each individual sub-element.

9. A linear damping element is provided at each node to represent rate-dependent external resistance and also to simulate energy radiated from the pile.

10. The ultimate resistance of many materials, including soil, will decrease somewhat after a full reversal of yielding in both directions. In the present program, a degradation procedure is applied to each elasto-plastic sub-element separately. A degradation factor λ is applied to the ultimate plastic resistance of each sub-element only on the occurrence of a full reversal of direction of slip of that sub-element. In addition to the initial ultimate resistance Q_u and the factor λ, the user specifies a lower-bound resistance Q_{min} which is asymptotically approached as degradation proceeds. Whenever the reduction is applied, the existing ultimate resistance Q_1 is degraded to a new ultimate resistance Q_2 according to the following relationship:

$$Q_2 = (1-\lambda)(Q_1 - Q_{min}) + Q_{min} \quad (1)$$

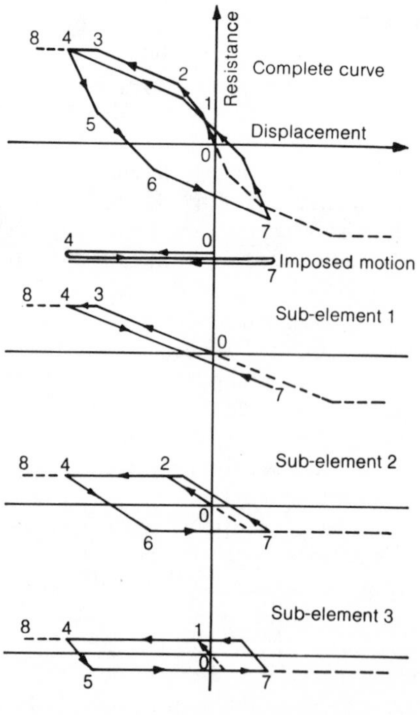

Fig. 3. Illustration of non-linear inelastic support curve during loading, unloading and subsequent reloading using the sub-element concept

Hardening may be simulated by setting Q_{min} greater than Q_u. At present, there is limited knowledge about such degradation. Some experimental research has been done[8, 12, 13] but more is needed in this area. It is emphasized that the hysteretic, degrading soil support model is intended to represent only the soil very near the pile where stresses and strains are concentrated and are very much larger than free-field values.

11. A more versatile but similar support routine has been developed for piles under lateral loading.[14, 15] In addition to all the features discussed here, the lateral routine incorporates the use of gap elements in simulating the lateral moulding-away phenomenon, generally occurring near the soil surface. The bases of soil supports can also be moved in a space-varying and time-varying manner to simulate lateral ground motions in an earthquake analysis. For some problems, it may be desirable to incorporate these capabilities into the axial analysis.

Numerical scheme

12. The dynamic axial model used in DRIVE 7 was developed as the axial counterpart of the dynamic bending model of Salani, Chan and Matlock,[16, 17] and is illustrated in Fig. 4. The governing equation of the solution is based on the summation of forces about pile station j of Fig. 4, and is detailed elsewhere.[1, 2] An implicit formulation of the Crank–Nicolson type[18] is employed for the numerical solution. The implicit operator is shown in Fig. 5. All three displacement values at each new time step $k + 1$ are unknown; therefore, simultaneous solution is required. With the implicit method of formulation, compatibility of displacement is enforced along the column at each time step. Also, the time increment length is independent of the pile increment length used, therefore there is greater freedom in selecting these variables. Mathematical stability is discussed elsewhere.[16, 17]

Iterative solution of non-linear foundation

13. For a non-linear inelastic foundation, a tangent modulus method is used to adjust both the stiffness and the load terms from one iteration to the next. For the iteration under consideration at any time, the pile displacements estimated from the previous iteration are utilized to obtain the tangent modulus stiffness and associated resistance intercept from the resistance–displacement curves. The stiffness and load values are then incorporated into the linear pile stiffness and load matrices to solve for the pile displacements for this iteration. Successively computed pile displacements from each iteration are compared until a user-specified closure tolerance is satisfied. The closure check is applied to all pile stations to achieve full compatibility at each time step.

14. The above iterative improvement procedure has been used in many current pile analyses,[9, 11, 14, 16, 17] and has been repeatedly shown to converge rapidly for most problems.

Notes on output interpretation

15. A recent review of several conventional wave equation programs reveals that a criterion in interpreting the permanent set and corresponding blow count is built into most of the programs. In most cases, the quake value input for the end bearing resistance is subtracted from the maximum pile tip displacement to obtain the permanent set. This criterion in effect assumes that the pile tip rebounds to a point of zero residual stress. Since residual compression loads after driving at the pile tip may not necessarily be zero,[19] such an assumption may cause significant error in the drivability prediction. In DRIVE 7, the final penetration is selected on the basis of the complete time history of the pile tip movement.

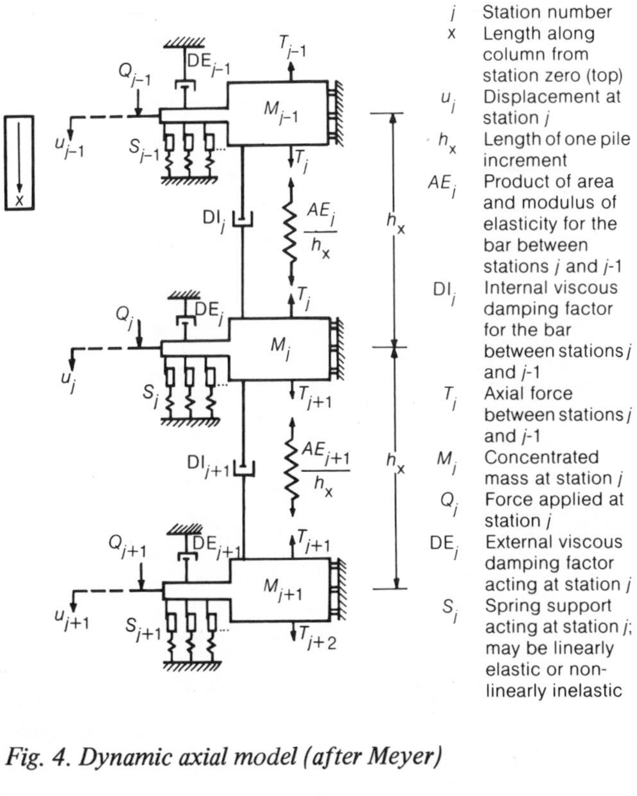

Fig. 4. Dynamic axial model (after Meyer)

EXAMPLE PROBLEMS

16. Three example problems are presented to simulate a sequence of responses of an axially loaded anchor pile for a tension-leg structure. The installation of the anchor pile

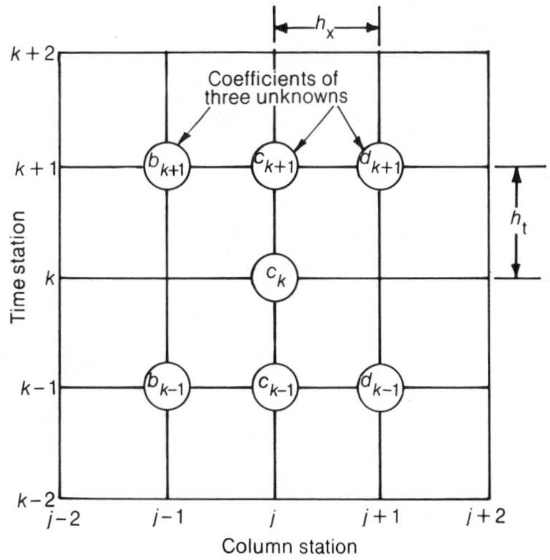

Fig. 5. Implicit operator of Crank–Nicolson type used in DRIVE 7 (after Meyer)

is studied in the first example problem, a multiple blow analysis being considered. Residual stresses along the pile are retained after each hammer blow, for consideration in analyses of subsequent hammer blows. In the second example problem, a tensile, pull-out test of the anchor pile is simulated. Residual stresses in the pile at three different stages of the test are compared. In example problem 3, the behaviour of the anchor pile under cyclic tensile load, with a bias tension, is studied. The performance of the pile–soil system under two different levels of applied load shows both the stabilization and the abrupt failure of the anchor pile. (English units were used in the original set-up and computation of the example problems. All units were subsequently converted to SI units.)

Example problem 1

17. A 0.61 m o.d. anchor pile with a wall thickness of 0.025 m is considered in this example. A 91.44 m follower is employed in the installation of the anchor pile. Dimensions for the follower are taken to be the same as those used for the pile. For the present analysis, the design penetration for the anchor pile is assumed to be 106.68 m below sea floor. Pile driving analysis was performed with the pile tip being located approximately at the design penetration. The general description of the anchor pile and the follower is illustrated in Fig 6(a). A Vulcan 340 offshore steam hammer is used to drive the pile. Mechanical efficiency of the hammer is assumed to be 70%. Nine alternate layers of 0.025 m thick wire rope and 0.0095 m thick steel plates are used as cushion material. The elastic modulus of this combination is taken as 2275 N/m^2. With this elastic modulus and the appropriate dimensions of the pile cap as illustrated in Fig. 7, the spring constant is computed to be 2.385 × 10^9 N/m. Coefficient of restitution for the cushion is assumed to be 0.80.

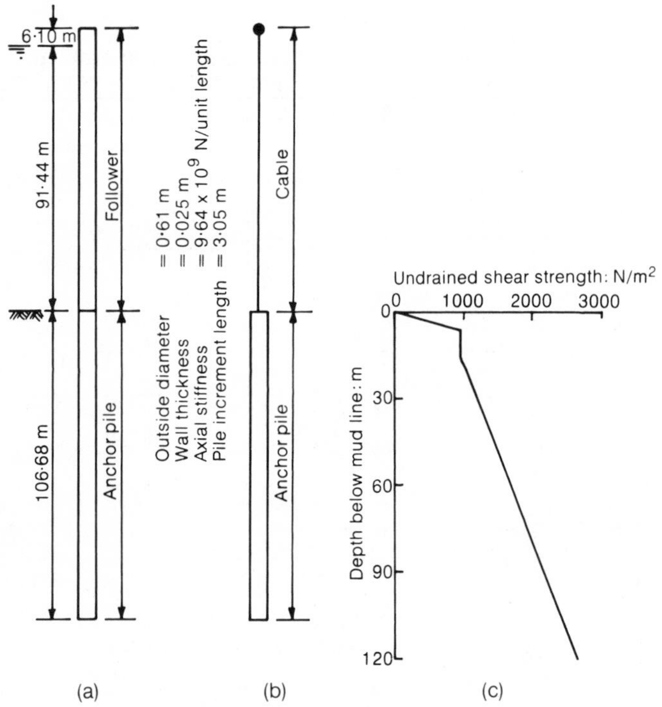

Fig. 6. General description of anchor pile and soil shear strength profile used in example problems

18. The soil considered in this series of problems is primarily a medium strength clay. A profile of undrained shear strength is shown in Fig 6(c). This shear strength profile is used to compute both the side resistance and the end bearing resistance. The load–movement relationships used to represent the end bearing resistance and side resistance are illustrated in Fig. 8. For the present problem, the ultimate bearing capacity q_f is taken as $9cd$ where c is the undrained shear strength at the pile tip location and d is the pile diameter. In conventional wave equation analysis, both the side resistance and the end bearing resistance are

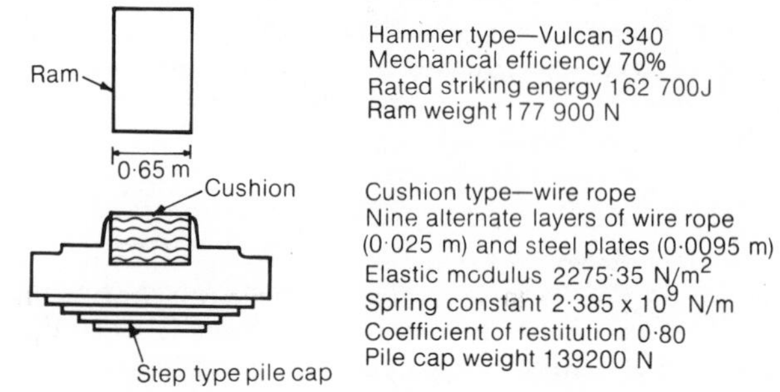

Fig. 7. General descriptions of ram, cushion material and pile cap used in example problem 1

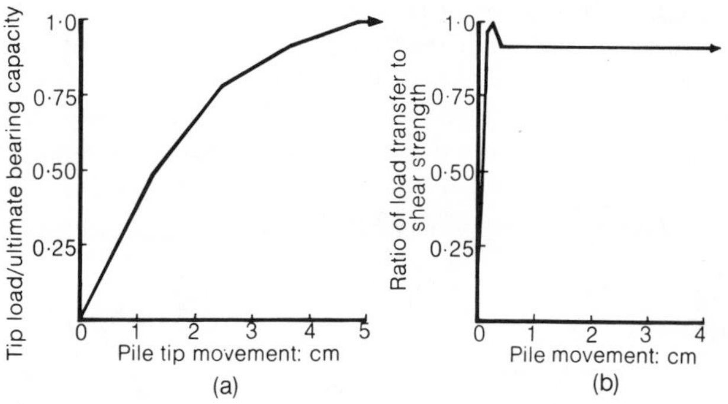

Fig. 8. Load–movement relationship for end bearing and side resistance

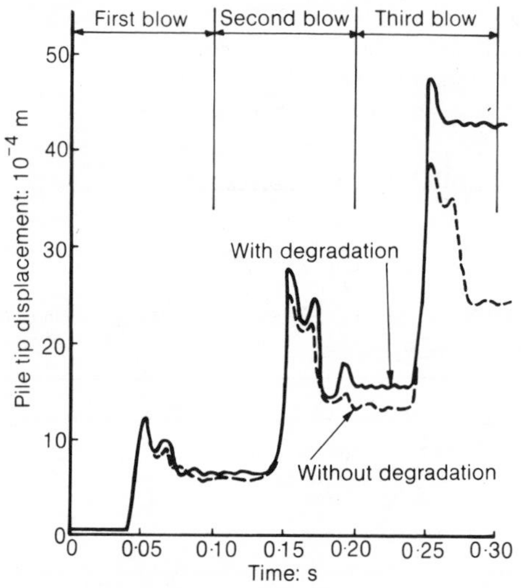

Fig. 9. Pile tip displacement–time history

approximated by an elasto-plastic load–movement curve. The yield point is generally taken to be 0.0025 m for both the side resistance and the end bearing resistance. Previous experimental work on axially loaded piles[10, 11] and deep foundations in general[20] indicate that the behaviour of the soil response is highly non-linear. Furthermore, it is shown that the amount of movement required to mobilize the full resistance of the soil may be significantly less for the side resistance than for the end bearing resistance.

19. A pile increment length of 3.05 m and a time increment length of 0.0005 s are used in this problem. The lower-bound resistance is taken to be 40% of the initial ultimate resistance, and a degradation factor of 0.50 is used. Since this example is intended to simulate redriving, the soil resistance in the analysis for the first hammer blow is taken to be the total available ultimate soil resistance. As reversals of direction of slip continue to occur and subsequent penalties in resistance continue to be applied under each blow, the available resistance to driving will approach asymptotically the lower-bound resistance. This continuous decrease in the available soil resistance during driving is reflected in the increased pile penetration for subsequent hammer blows. To illustrate, the results from the first three hammer blows are presented in Fig. 9 in the form of pile tip displacement versus time. Both degradation in soil resistance during driving and the residual stresses from the previous blow will affect the response of the pile under the present blow. The pile tip displacement–time histories are presented both with and without degradation. A review of Fig. 9 indicates that the presence of both residual stress and strength degradation results in higher penetrations, thus lower blow counts, for the two later blows. In this problem the effect of residual stress alone caused only a small increase in the pile tip penetration in the last blow. However, solutions for other problems indicate that the effect of residual stress may be significantly greater and it appears that the results are quite problem-dependent and cannot be generalized on the basis of one example.

Example problem 2

20. A tensile pull-out test of the anchor pile is simulated in this example. The test is assumed to be performed immediately after the installation of the anchor pile. Residual stresses incurred in the pile during driving are taken into account. In performing the tension test, the load is applied to the anchor pile through a cable. This arrangement is illustrated in Fig. 6(b).

21. The results of the loading are shown in Fig. 10 where the pile head movement is plotted against the applied force. Point A in Fig. 10 corresponds to the after-driving stage. A residual compression of 0.027 m at the pile head is noted. The anchor pile is then loaded to point B, rebounded to point C, reloaded to point D, rebounded again to point E, and finally loaded to failure at point F. Residual stress distributions were computed for points A, C and E, and are illustrated in Fig. 11. At point A (corresponding to the after-driving condition) only compressive forces were present in the pile.

22. The failure load during this simulated pull-out test was computed to be 1.3×10^7 N (Fig. 10). This is lower than the theoretical static ultimate pile capacity which was computed to be 1.46×10^7 N. The observation that the failure load is higher than the lower-bound soil resistance (40% of the ultimate capacity) indicates that the degradation of soil resistance was not complete after three hammer blows and the subsequent cycles of test loading. However, significant effects on drivability have been noted.

23. The presence of residual stress in the pile after driving indicates that, for instrumented field tests, it is important to have a good set of zero readings before pile installation, in order to obtain good interpretation of the test results.

Example problem 3

24. The forces which are applied to the foundation piles for a tension leg platform arise from bouyant forces on the floating platform. Cables of sufficient stiffness are generally employed to tie the floating structure to the anchor piles. Cables are insured against going slack by the

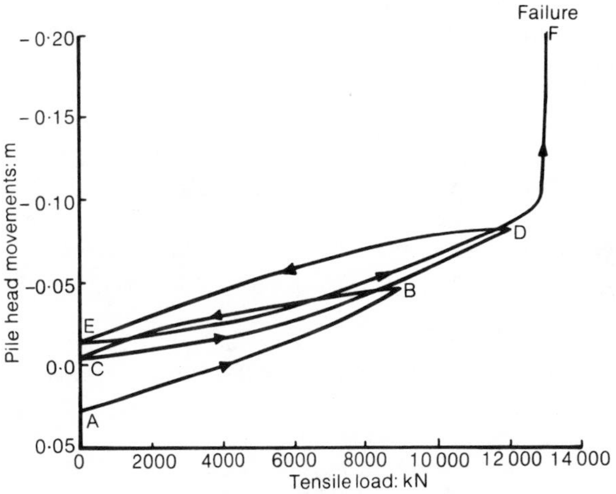

Fig. 10. Computed load–settlement curve of anchor pile under tension load

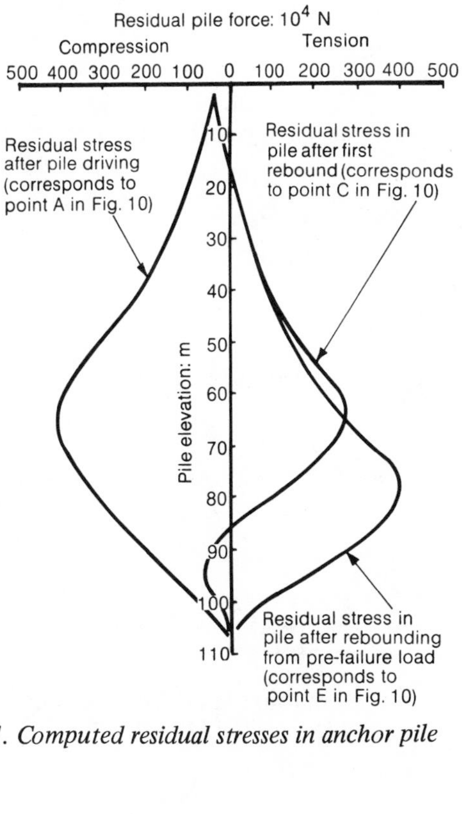

Fig. 11. Computed residual stresses in anchor pile

application of a bias tension. Seasonal storm waves cause large cyclic variations in the forces transmitted to the piles.

25. The interaction between the pile and the soil near the interface is such that, due to up and down movements of the upper portions of the pile, reversals in shear slip may occur in the clay adjacent to the pile even if the cyclic loads on the pile head are unidirectional. The reversals of shear in the soil can cause significant progressive losses of the soil resistance, with the possibility of a sudden and catastrophic failure by pull-out if adequate reserve capacity is not available at greater depth.

26. Because of the complex behaviour of the cyclic pile–soil interaction and the severity of strength losses which have been observed to occur in laboratory studies of cyclic loading,[12] conventional static axial pile capacity predictive techniques and traditional concepts of safety factors are of limited value. The behaviour of the pile–soil system and the cyclic degradation are roughly comparable to the fatigue behaviour of metals; the system will fail (i.e., fracture or pull out) or it will not fail. The wider the excursion of tensile load, the sooner failure will be reached.

27. DRIVE 7 provides a tool for rationally investigating the behaviour of piles under cyclic axial loads. Three soil properties are required at each node point along the length: these are the initial non-linear inelastic resistance–displacement characteristics of the pile–soil coupling, the maximum friction force which will be developed, and the minimum friction force which will be maintained after many reversals of displacement.

28. In the present example problem, it is assumed that sufficient set up has taken place in the soil elements around the pile before the application of the load. Thus, the soil is degraded progressively from its ultimate capacity. Also, for the static loadings the full soil degradation was assumed to occur after only one reversal of slip for each sub-element ($\lambda = 1.0$). Although this leads to artificially increased rates of degradation and progressive pile movements, the amount of degradation would be the same as for an analysis extended over many cycles of loading, with only partial degradation in each cycle. Computer time is conserved by this procedure.

29. The predicted behaviour of the anchor pile is shown in Fig. 12. The static bias tension was taken as 2.224×10^6 N, and the cyclic component was $\pm 4.448 \times 10^6$ N. The small amount of progressive upward movement due to the cyclic loading indicates that some degradation has occurred, but the pile movement stabilizes rapidly, showing that the degradation process has been halted, and that an equilibrium response has been reached.

30. The cyclic component was increased to $\pm 4.670 \times 10^6$ N and a similar solution was performed. The results are shown in Fig. 13. Progressive upward movement occurred on each load cycle, and the soil strength was sufficiently reduced by the degradation process that an abrupt failure occurred on the fifth application of the load. With a smaller value of λ and correspondingly less strength loss in each cycle, many more cycles would have been required to produce the same behaviour, but the same failure would have been predicted.

31. The effects of cyclic loading on the capacity of the pile may be noted by comparing the conventional static ultimate axial capacity of 1.46×10^7 N with the cyclic load required to fail the pile, which is probably about 1.156×10^7 N, or about 79% of the static capacity. However, this value is still substantially greater than a lower bound of 5.84×10^6 N obtained by assuming arbitrarily that the up and down movement of the entire pile would be great enough to fully degrade the clay to 40% of the ultimate static capacity.

CONCLUSION

32. A general, versatile numerical procedure is presented to facilitate the examination of an axially loaded pile, under both static and dynamic loadings. The method of analysis includes a fully non-linear, hysteretic, degrading soil model. By solution of three example conditions of loading of an axially loaded anchor pile, the following three principal conclusions may be reached.

33. For a hard driving case, a multiple-blow analysis indicates that the driving behaviour may be significantly affected by residual stress effects and degradation in soil resistance under continuous driving.

34. To properly interpret the results from instrumented field tests, a good set of zero readings must be taken prior to pile installation to take into account the residual stresses incurred in the pile during pile driving.

35. Progressive loss in resistance along the upper

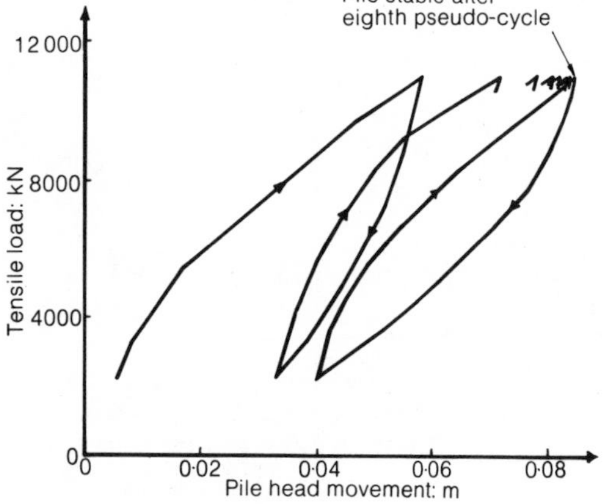

Fig. 12. Computed cyclic axial pile behaviour under tensile loading: maximum load level 1.112×10^7 N

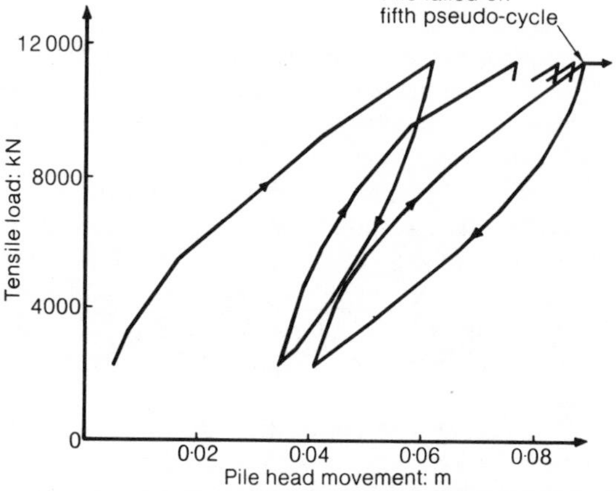

Fig. 13. Computed cyclic axial pile behaviour under tensile loading: maximum load level 1.156×10^7 N

portion of a pile may occur due to reversals of motion even under unidirectional cyclic loading. The behaviour of the complete pile–soil system must be examined to determine whether the cyclic response stabilizes or whether it leads progressively to failure. Stabilization of response may occur at loads considerably greater than that given by the simple integration of cyclic minimum resistance over the whole pile length.

REFERENCES

1. FOO S. H. C. Analysis of driving of foundation piles. MS thesis, The University of Texas at Austin, 1978, 146.

2. FOO S. H. C. et al. Analysis of driving of foundation piles. Proc. 1977 Offshore Technology Conf., Houston, Texas, 1977, OTC 2842.

3. MEYER P. L. A discrete-element method of pile driving analysis. MS thesis, The University of Texas at Austin, 1976.

4. HIRSCH T. J. et al. Pile driving analysis by one-dimensional wave theory: state of the art. Highway Research Record, 1970, No. 333, 33–54. (Highway Research Board, National Research Council, Washington.)

5. GOBLE G. G. and RAUSCHE F. Wave equation analysis of pile driving, WEAP program: Vol. 1—Background. National Technical Information Service, Springfield, Virginia, 1976, Report FHWA–IP–76–14.1.

6. SMITH E. A. L. Pile driving analysis by the wave equation. J. Soil Mech. Fdns Div. Am. Soc. Civ. Engrs, 1960, 16, Aug., SM4, 35–61.

7. HOLLOWAY D. M. and DOVER A. R. Recent advances in predicting pile driveability. Proc. 1978 Offshore Technology Conf., Houston, Texas, 1978, OTC 3273.

8. HEEREMA E. P. Predicting pile driveability: Heather as an illustration of the "friction fatigue" theory. Proc. European Offshore Petroleum Conf. London, 1978, Paper EUR 50.

9. BOGARD D. and MATLOCK H. A computer program for the analysis of beam-columns under static axial and lateral loads. Proc. 1977 Offshore Technology Conf., Houston, Texas, 1977, OTC 2186.

10. COYLE H. M. and REESE L. C. Load transfer for axially loaded piles in clay. J. Soil Mech. Fdns Div. Am. Soc. Civ. Engrs, 1966, 2, Mar. SM2, 1–25.

11. MEYER P. L. et al. Computer predictions for axially-loaded piles with nonlinear supports. Proc. 1975 Offshore Technology Conf., Houston, Texas, 1975, OTC 2186.

12. HOLMQUIST D. V. and MATLOCK H. Resistance–displacement relationships for axially-loaded piles in soft clay. Proc. 1976 Offshore Technology Conf., Houston, Texas, 1976, OTC 2474.

13. GUYTON W. L. Soil resistance to vibratory pile driving. PhD dissertation, The University of Texas at Austin, 1968, 167.

14. MATLOCK H. et al. Simulation of lateral pile behavior under earthquake motion. Proc. American Society of Civil Engineers Specialty Conf. Earthquake Engineering and Soil Dynamics, Pasadena, California, 1978.

15. MATLOCK H. et al. Example of soil–pile coupling under seismic loading. Proc. 1978 Offshore Technology Conf., Houston, Texas, 1978, OTC 3310.

16. CHAN J. H. C. and MATLOCK H. A discrete-element method for transverse vibrations of beam-columns resting on linearly elastic or inelastic supports. Proc. 1973 Offshore Technology Conf., Houston, Texas, 1973, OTC 1841.

17. MATLOCK H. and SALANI H. J. Finite difference methods for plate vibration problems. J. Struct. Div. Am. Soc. Civ. Engrs, 1969, 95, Mar., ST3, 441–456, Paper 6477.

18. CRANK J. and NICOLSON P. A practical method for numerical evaluation of solutions of partial differential equations of heat conduction type. Proc. Camb. Phil. Soc. 1947, 43, 50–67.

19. HUNTER A. H. and DAVISSON M. T. Measurements of pile load transfer. In: Performance of deep foundations. American Society for Testing and Materials, 1969, 106–117, ASTM STP 44.

20. SKEMPTON A. W. The bearing capacity of clays. Proc. Building Research Congress, London, 1951, Division I, Part III, 180–189.

17. Unified method for analysis of laterally loaded piles in clay

W. R. SULLIVAN, MSCE (recently Staff Engineer, ARMAC Engineers, Inc., Tampa, Florida, now Geotechnical Engineer, Golder Associates, Atlanta, Georgia), L. C. REESE, PE, PhD, (T. U. Taylor Professor of Civil Engineering and Associate Dean, College of Engineering, The University of Texas at Austin) and C. W. FENSKE, PE, MSCE (Manager of Engineering, McClelland Engineers, Inc., Houston, Texas)

The behaviour of a pile under lateral loading is obtained by solving an equation by digital computer. The equation includes a term at each depth for the soil modulus. The soil response is given by a family of curves giving soil resistance as a function of pile deflexion (p–y curves). The soil moduli are secants to the p–y curves and can vary in any arbitrary manner with depth. Two major experimental programmes have been performed in the field on full-sized, instrumented piles in clay that is submerged. One series of experiments was performed in soft to medium clay and the other in stiff clay. Results from these two studies have been reanalysed and a unified method has been proposed for predicting p–y curves for clay. The properties of the clay are given by the undrained shear strength, the submerged unit weight, and the shape of the stress–strain curve. The method can treat the cases of short term and repeated loading and can be applied to the analysis of piles of any geometry, stiffness, and pile head fixity. Predictions of pile behaviour were made using the unified method and were compared with results from experiments. The unified method has been shown to be an acceptable approach. A number of practical problems can be attacked with considerable success. Agreement between predictions with the analytical method and experimental results are good or show the analytical method to be somewhat conservative.

INTRODUCTION

While axially loaded piles frequently may be designed satisfactorily by simple static methods, the design procedure for laterally loaded piles is more complex, involving the solution of a fourth-order differential equation. The solution must ensure that conditions of equilibrium and compatibility are satisfied. The problem of laterally loaded piles is further complicated because of the non-linear soil response.

2. The differential equation to be solved, as derived from conventional beam theory,[1] is

$$EI\frac{d^4y}{dx^4} + P_x\frac{d^2y}{dx^2} - p = 0 \qquad (1)$$

where EI is flexural rigidity of pile, y is deflexion of pile, x is length along pile, P_x is axial load, and p is soil reaction per unit length. Equation (1) may be solved conveniently by a digital computer;[2] however, non-dimensional methods may sometimes be employed to yield an acceptable solution.[3] Both methods of solution give all the necessary design information, including the moment, deflexion and shear at desired lengths along the pile.

SOIL RESPONSE

3. The approach to solving the problem of laterally loaded piles is based on the assumption attributed to Winkler.[4] The soil surrounding a pile is depicted as a set of non-linear elastic springs, and the Winkler assumption states that each spring acts independently. The depression of one spring has no effect on an adjacent spring. It is obvious that the Winkler assumption is not valid for soils; however, the errors involved in the use of the assumption are relatively small.

4. It is convenient to think of the soil response in terms of a p–y curve. Fig. 1 defines the concept. Part (a) of the figure shows the depth at which the soil behaviour is considered. Part (b) depicts the probable earth pressure distribution prior to any lateral loading. Part (c) shows a possible earth pressure distribution after the section under consideration has deflected a distance y_1. As the deflexion increases, the soil resistance on the section of the shaft changes. A possible family of p–y curves is shown in Fig. 2; each curve represents the soil behaviour at a different depth.

5. For convenience in solving equation (1), a secant modulus of soil reaction, E_s, is often used (Fig. 3):

$$E_s = -p/y \qquad (2)$$

As may be understood from Figs 2 and 3, E_s can vary in an arbitrary manner with depth and with deflexion.

6. Two major experimental programmes have been carried out on full-sized, instrumented piles that were installed in clays below a water surface. The two programmes have led to recommendations for the prediction of families of p–y curves. Brief descriptions of the programmes and their results are presented in the following sections.

EXPERIMENTAL PROGRAMME WITH PILES IN SOFT CLAY

7. Matlock[4] performed lateral load tests employing a steel pipe pile 325 mm in diameter and 12.8 m in length. It was driven into clays near Lake Austin that had a shear strength of about 38 kPa. The pile was recovered, taken to Sabine Pass, Texas, and driven into clay with a shear strength that averaged about 14 kPa in the significant upper zone.

8. Families of bending moment curves were obtained from electrical resistance strain gauges that were attached to the interior of the pile. Sets of curves were obtained for short term static loading and other sets were obtained for cyclic loading. The effects of soil creep were minimized in the short term testing. In the cyclic tests, a load of a given magnitude was repeated until the deflexions and bending moment reached an equilibrium condition. The data were analysed and experimental p–y curves were obtained.[5] Principles of mechanics were employed to the extent possible and equations were developed for prediction of p–y curves on the basis of soil properties, pile geometry, and nature of loading.

9. Matlock[4] recommended the following procedure for

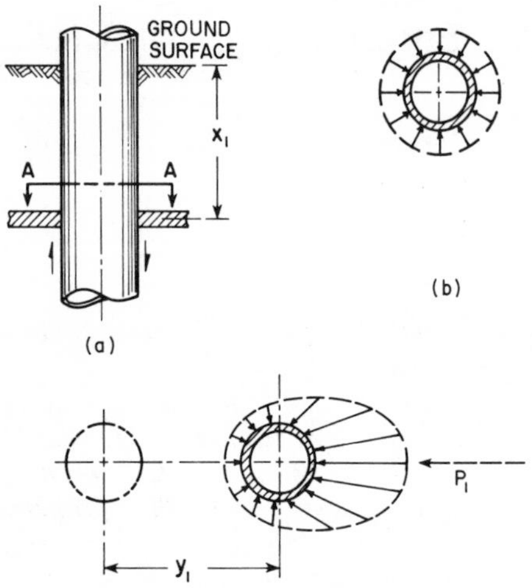

Fig. 1. Graphical definition of p and y: (a) pile elevation; (b) view AA – earth pressure distribution prior to lateral loading; (c) view AA – earth pressure distribution after lateral loading

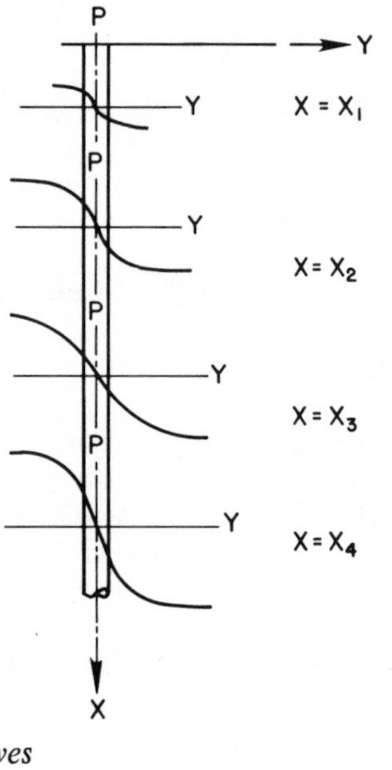

Fig. 2. p–y curves

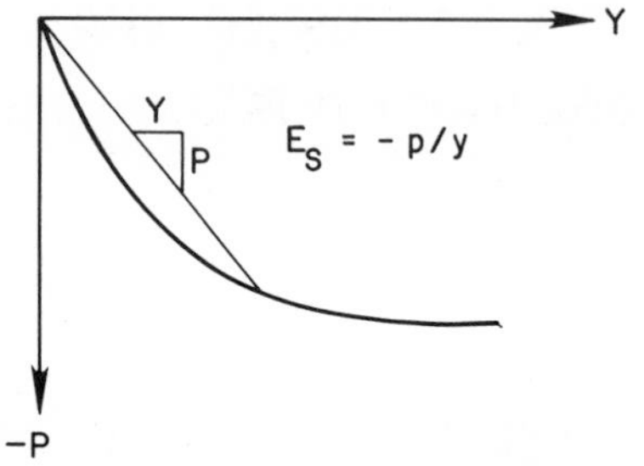

Fig. 3. Illustration of the secant modulus

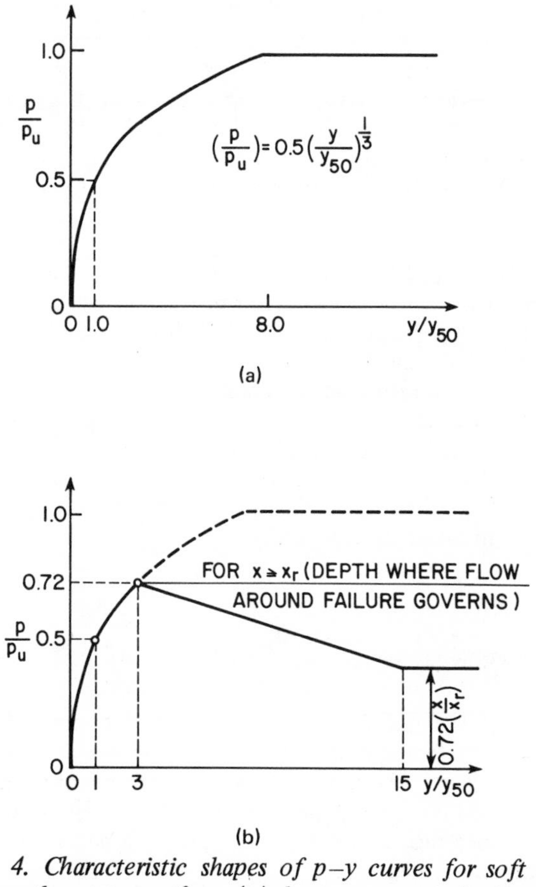

Fig. 4. Characteristic shapes of p–y curves for soft clay below the water surface: (a) short term static loading; (b) cyclic loading

p–y curves for soft clay for short term static loading (Fig. 4 (a)).

(a) Obtain the best possible estimate of the undrained shear strength and effective unit weight with depth. Also obtain the value of ϵ_{50}, the strain corresponding to one half the maximum principal stress difference. If no values of ϵ_{50} are available, typical values suggested by Skempton[6] are given in Table 1.

(b) Compute the ultimate soil resistance per unit length of the pile, p_u; use the smaller of the values given by the equations below:

$$p_u = \left\{3 + \frac{\bar{\gamma}}{s_u}x + \frac{0.5}{b}x\right\} s_u b \qquad (3)$$

$$p_u = 9\,s_u b \qquad (4)$$

where $\bar{\gamma}$ is average effective unit weight from ground surface to depth of p–y curve, x is distance from ground surface to depth of p–y curve, s_u is undrained shear strength at depth x, and b is width of pile. The value of p_u is computed at each depth where a p–y curve is desired, based on shear strength at that depth.

(c) Compute the deflexion, y_{50}, at one half the ultimate soil resistance from the following equation:

$$y_{50} = 2.5\,\epsilon_{50} b \qquad (5)$$

(d) Points describing the p–y curve are now computed from the following relationship:

$$p/p_u = 0.5\,(y/y_{50})^{1/3} \qquad (6)$$

The value of p remains constant at p_u beyond a y value of $8y_{50}$.

10. The following procedure is for cyclic loading (Fig. 4(b)).

(a) Construct the p–y curve in the same manner as for short term static loading for values of p less than $0.72p_u$.

(b) Solve equations (3) and (4) simultaneously to find the depth, x_r, where the transition occurs. If the unit weight and shear strength are constant in the upper zone, then

$$x_r = \frac{6s_u b}{\gamma b + 0.5 s_u} \qquad (7)$$

If the unit weight and shear stregth vary with depth, the value of x_r should be computed with the soil properties at the depth where the p–y curve is desired.

(c) If the depth of the p–y curve is greater than or equal to x_r, then $p = 0.72p_u$ for all values of y greater than $3y_{50}$.

(d) If the depth to the p–y curve is less than x_r, then the value of p decreases from $0.72p_u$ at $y = 3y_{50}$ to the value given by the following expression at $y = 15y_{50}$:

Table 1. Suggested values of ϵ_{50} for soft clay

Consistency of clay	ϵ_{50}
Soft	0.020
Medium	0.010
Stiff	0.005

NOTATION

A	coefficient used to define the shape of the p–y curve for the unified clay criteria
A_c, A_s	empirical parameters for developing p–y curves for stiff clay below water surface
b	width of pile
EI	flexural rigidity of pile
E_s	secant modulus of soil reaction
E_{sc}	slope of portion of p–y curve for stiff clay below water surface
E_{si}	slope of initial portion of p–y curve for stiff clay below water surface
E_{ss}	slope of portion of p–y curve for stiff clay below water surface
$(E_s)_{max}$	limiting maximum value of soil modulus on p–y curve for unified clay criteria
F	coefficient used to define deterioration of soil resistance at large deformations for short term static loading for the unified clay criteria
k	coefficient of horizontal subgrade modulus
k_c	coefficient of initial soil modulus for cyclic loading for stiff clay below water surface
k_s	coefficient of initial soil modulus for short term static loading for stiff clay below water surface
LI	liquidity index
N_p	soil resistance factor employed in development of unified clay criteria for stiff clay below water surface
PI	plasticity index
P_x	axial load on pile
p	soil reaction per unit length of pile
p_R	peak cyclic resistance on p–y curve
p_u	ultimate soil resistance per unit length of pile
p_{CR}	residual soil resistance on cyclic p–y curve for unified clay criteria
p_{offset}	quantity used in developing p–y curve for stiff clay below water surface
s_u	undrained shear strength at depth x
$(s_u)_{avg}$	average undrained soil shear strength above depth x
w_L	liquid limit
x	length along pile or distance from ground surface to depth at which p–y curve applies
x_r	depth below ground surface to transition in ultimate soil resistance equations
y	deflexion of pile
y_p	value of y at transition point on p–y curve for cyclic loading for stiff clay below water surface
y_{50}	deflexion at one half the ultimate soil resistance
γ	effective unit weight of soil
$\bar{\gamma}$	average effective unit weight from ground surface to depth at which p–y curve applies
ϵ_{50}	strain corresponding to one half the maximum principal stress difference

$$p = 0.72p_u (x/x_r) \quad (8)$$

The value of p remains constant at the value given by equation (8) beyond a y value of $15y_{50}$.

11. For determining the shear strength of the soil required in the p–y construction, Matlock[4] recommended the following tests in order of preference:

(a) in-situ vane shear tests with parallel sampling for soil identification;
(b) unconsolidated undrained triaxial compression tests having a confining stress equal to the overburden pressure with the shear strength being defined as half the total maximum principal stress difference;
(c) miniature vane tests of samples in tubes;
(d) unconfined compression tests.

12. Tests must also be performed to determine the unit weight of the soil.

EXPERIMENTAL PROGRAMME WITH PILES IN STIFF CLAY

13. Reese et al.[7] performed lateral load tests employing steel pipe piles 610 mm in outside diameter and 15.2 m in length. The piles were driven into stiff clay at a site near Manor, Texas. The clay had an undrained shear strength ranging from about 96 kPa at the ground surface to about 290 kPa at a depth of 3.66 m.

14. Experimental and analytical procedures were followed that were similar to those employed by Matlock.[4] The following procedure was recommended for p–y curves for stiff clay for short term static loading (Fig. 5).

(a) Obtain the best possible estimate of the undrained shear strength and effective unit weight with depth. If no values of ϵ_{50} are available, use values from Table 2.

Table 2. Recommended value of ϵ_{50} for stiff clays

	Average undrained shear strength, kPa		
	50–100	100–200	200–400
ϵ_{50}	0.007	0.005	0.004

(b) Compute the average undrained soil shear strength, $(s_u)_{avg}$, over the depth x.
(c) Compute the ultimate soil resistance; use the smaller of the values given by the equations below:

$$p_u = \left\{2 + \frac{\gamma}{(s_u)_{avg}} x + \frac{2.83}{b} x\right\} (s_u)_{avg} b \quad (9)$$

$$p_u = 11 s_u b \quad (10)$$

(d) Establish the initial straight-line portion of the p–y curve,

$$p = kxy = E_{si} y \quad (11)$$

Use the appropriate value of k_s or k_c from Table 3.

(e) Compute the following:

$$y_{50} = \epsilon_{50} b \quad (12)$$

(f) Choose an appropriate value of A_s from Fig. 6 for the particular non-dimensional depth.
(g) Establish the first parabolic portion of the p–y curve,

$$p = 0.5p_u (y/y_{50})^{0.5} \quad (13)$$

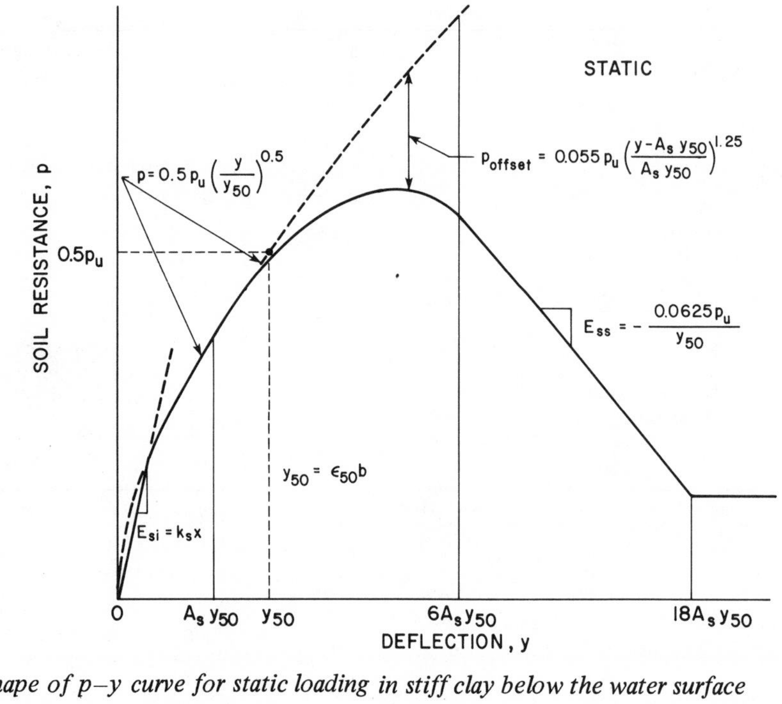

Fig. 5. Characteristic shape of p–y curve for static loading in stiff clay below the water surface

Equation (13) should define the portion of the p–y curve from the point of the intersection with equation (11) to a point where $y = A_s y_{50}$ (see note in paragraph 15).

(h) Establish the second parabolic portion of the p-y curve,

$$p = 0.5p_u(y/y_{50})^{0.5} - 0.055p_u\left\{\frac{y}{A_s y_{50}} - 1\right\}^{1.25} \quad (14)$$

Equation (14) should define the portion of the p–y curve from the point where $y = A_s y_{50}$ to a point where $y = 6A_s y_{50}$ (see note in paragraph 15).

(i) Establish the next straight-line portion of the p–y curve,

$$p = 0.5p_u(6A_s)^{0.5} - 0.411p_u - 0.0625\,(p_u/y_{50})(y - 6A_s y_{50}) \quad (15)$$

Equation (15) should define the portion of the p–y curve from the point where $y = 6A_s y_{50}$ to a point where $y = 18A_s y_{50}$ (see note in paragraph 15).

(j) Establish the final straight-line portion of the p–y curve,

$$p = 0.5p_u(6A_s)^{0.5} - 0.411p_u - 0.75p_u A_s \quad (16)$$

Equation (16) should define the portion of the p–y curve from the point where $y = 18A_s y_{50}$ and for all larger values of y (see following note).

15. Note: The step-by-step procedure is outlined, and Fig. 5 is drawn, as if there is an intersection between equations (11) and (13). However, there may be no intersection of equation (11) with any of the other equations defining the p–y curve. Equation (11) defines the p–y curve until it intersects with one of the other equations or, if no intersection occurs, equation (11) defines the complete p–y curve.

16. The following procedure is for cyclic loading (Fig. 7).

(a) Use step (a) as for the static case.
(b) Use step (b) as for the static case.
(c) Use step (c) as for the static case.
(d) Choose the appropriate value of A_c from Fig. 6 for the particular non-dimensional depth.

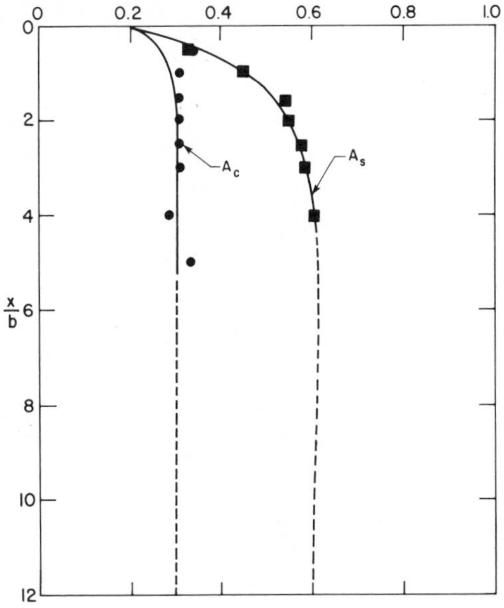

Fig. 6. Values of parameters A_s and A_c

(e) Use step (e) as for the static case.
(f) Compute the following:

$$y_{50} = \epsilon_{50} b \quad (17)$$

$$y_p = 4.14A_s y_{50} \quad (18)$$

(g) Establish the parabolic portion of the p–y curve,

$$p = A_c p_u\left(1 - \left\{\frac{y - 0.45y_p}{0.45y_p}\right\}^{2.5}\right) \quad (19)$$

Equation (19) should define the portion of the p–y curve from the point of the intersection with equation (11) to where $y = 0.6y_p$ (see note in paragraph 17).

(h) Establish the next straight-line portion of the p–y curve,

$$p = 0.936A_c p_u - 0.085\,(p_u/y_{50})(y - 0.6y_p) \quad (20)$$

Equation (20) should define the portion of the p–y curve from the point where $y = 0.6y_p$ to the point where $y = 1.8y_p$ (see note in paragraph 17).

Table 3. Recommended values of k for stiff clays

	Average undrained shear strength,* kPa		
	50–100	100–200	200–400
k_s (static), MN/m^3	135	270	540
k_c (cyclic), MN/m^3	55	110	220

*The average shear strength should be computed from the shear strength of the soil to a depth of 5 pile diameters. It should be defined as half the total maximum principal stress difference in an unconsolidated undrained triaxial test.

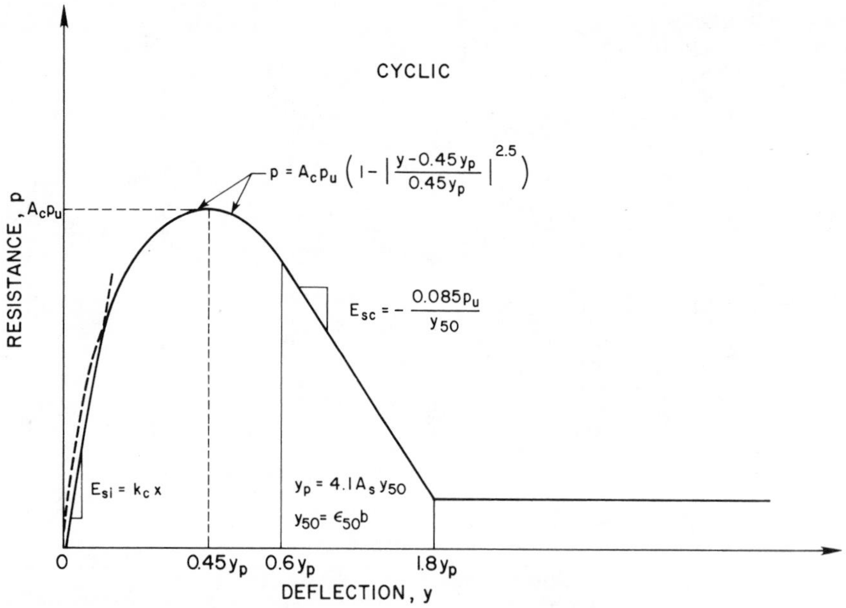

Fig. 7. Characteristic shape of p–y curve for cyclic loading in stiff clay below the water surface

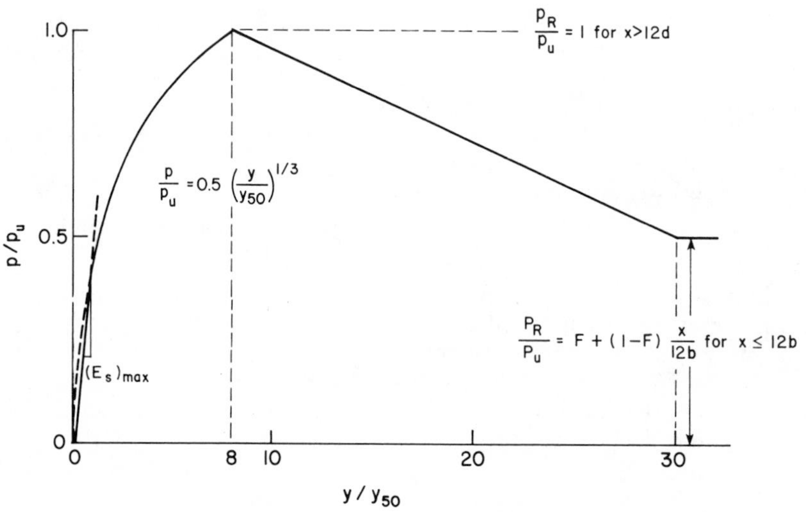

Fig. 8. Characteristic shape of p–y curve for unified clay criteria for short term static loading

Table 4. Values of N_p at the ground surface

Reference	N_p
Reese[8]	2
Hansen[9]	2.57
Thompson[10] (finite elements)	3.12
Matlock[4]	3
Reese et al.[7]	0.4

Table 5. Values of N_p at a depth of three pile diameters

Reference	N_p
Reese[8]	10.5
Hansen[9]	6.26
Thompson[10] (finite elements)	5.70 – 10.93
Matlock[4]	4.5
Reese et al.[7]	6.2

Table 6. Values of N_p at great depth

Reference	N_p at $x = 12b$	Limiting value of N_p
Reese[8]	–	12
Thompson[10] (block flow)	–	7 – 12
Hansen[9]	7.51	8.14
Thompson[10] (finite elements)	5.70 – 10.93	10.93
Matlock[4]	9	9
Reese et al.[7]	6.6	6.6

(i) Establish the final straight-line portion of the p–y curve,

$$p = 0.936 A_c p_u - 0.102 (p_u/y_{50}) y_p \qquad (21)$$

Equation (21) should define the portion of the p–y curve from the point where $y = 1.8 y_p$ and for all larger values of y (see following note).

17. Note: The step-by-step procedure is outlined, and Fig. 7 is drawn, as if there is an intersection between equations (11) and (19). However, there may be no intersection of those two equations and there may be no intersection of equation (11) with any of the other equations defining the p–y curve. If there is no intersection, the equation should be employed that gives the smallest value of p for any value of y.

18. The undrained shear strength of the soil and other soil properties should be obtained from appropriate laboratory or in situ tests.

DEVELOPMENT OF UNIFIED METHOD

19. The two sets of recommendations that have been presented for clays were used to compute the behaviour of the test piles, and the computed behaviour agreed well with that from the experiments. Furthermore, the recommendations were generalized as much as possible by using the principles of mechanics with the view of providing guidance for designers. However, designers have little or no guidance in solving problems where the properties of the clay are different than those at the experimental sites. In view of the need to have recommendations, at least in preliminary form, that can be employed to design piles under lateral loading in any clay soils, studies were made of the results of the two experimental programmes with a view to developing a unified approach. Those studies and their results are presented in the following sections.

20. A study was made of the ultimate soil resistance p_u that can be expected to develop against a pile that is subjected to lateral load. Two types of behaviour of the soil are postulated. Near the ground surface a wedge of soil is assumed to move up and out at the failure condition, and at depth the soil is assumed to flow around the pile at failure and a plane strain analysis can be employed. The value of p_u can be expressed in general form for failure near the ground surface as follows:

$$p_u = N_p (s_u)_{avg} b \qquad (22)$$

The average value of the undrained shear strength is employed because of the assumption of the wedge type failure. At the depth where plane strain behaviour is assumed, the shear strength at the particular depth should be employed:

$$p_u = N_p s_u b \qquad (23)$$

Table 4 shows values of N_p at the ground surface suggested by various investigators and Table 5 shows values suggested by the same authors at a depth of three pile diameters. The works cited in Tables 4 and 5 were studied and the major experimental studies[4, 7] were carefully considered. The following equations are proposed for the ultimate soil resistance near the ground surface:

$$p_u = \left\{ 2 + \frac{\bar{\gamma}}{(s_u)_{avg}} x + \frac{0.833}{b} x \right\} (s_u)_{avg} b \qquad (24)$$

$$p_u = \left\{ 3 + \frac{0.5}{b} x \right\} s_u b \qquad (25)$$

Equation (25) is similar to the one proposed by Matlock[4] except that the term involving submerged unit weight is omitted.

21. Equations (24) and (25) have some theoretical basis as indicated previously.[8] Table 6 gives values of N_p at great depth as recommended in the sources used for Tables 4 and 5. A value of N_p of 9 was adopted for the unified method, leading to the following equation for the ultimate resistance at great depth:

$$p_u = 9 s_u b \qquad (26)$$

The smallest of the values of ultimate resistance computed by equations (24)–(26) should be employed.

22. The characteristic shape of the p–y curve for short term static loading is shown in Fig. 8. The initial portion of the curve is linear and is described by equation (27):

$$p = (E_s)_{max} y \qquad (27)$$

The parameter $(E_s)_{max}$ is the limiting maximum value of soil modulus. When no other method is available, $(E_s)_{max}$ can be estimated using equation (28):

$$(E_s)_{max} = kx \qquad (28)$$

Representative values for k are given in Table 7.

23. The curved portion of the p–y curve is described by equation (29):

$$p = 0.5 p_u (y/y_{50})^{1/3} \qquad (29)$$

Equation (29) is identical to the equation recommended by

Table 7. Representative values for k for unified criteria

s_u, kPa	k, MN/m^3
12 – 25	8
25 – 50	27
50 – 100	80
100 – 200	270
200 – 400	800

Table 8. Representative values for ϵ_{50} for unified criteria

s_u, kPa	ϵ_{50}
12 – 25	0.02
25 – 50	0.01
50 – 100	0.007
100 – 200	0.005
200 – 400	0.004

Matlock[4] to define the soil resistance for the initial portion of the p–y curve.

24. The equation for computing y_{50} is

$$y_{50} = A\epsilon_{50}b \tag{30}$$

Values of ϵ_{50} may be taken from Table 8. The method for estimating A is discussed later.

25. The soil resistance after large deformation is given by equations (31) and (32). The smaller of the values computed by the two equations should be employed:

$$p_R = p_u \left\{ F + (1-F)\frac{x}{12b} \right\} \text{ for } x < 12b \tag{31}$$

$$p_R = p_u \qquad \text{for } x > 12b \tag{32}$$

The parameter F is a function of the stress–strain characteristics of the soil and is discussed later. The selection of $30y_{50}$, the deflexion at which the residual resistance is reached, and the increase of p_R/p_u with depth are arbitrary. However, the use of the shape shown in Fig. 8 for the residual portion of the curve gives good agreement between the measured and computed values for experiments, as shown later.

26. The characteristic shape of the p–y curve for cyclic loading is shown in Fig. 9. The decrease in the soil resistance with depth due to cyclic loading is consistent with the recommendation made by Matlock.[4] The recommended shape of the cyclic p–y curve is completely empirical. However, its use does give satisfactory agreement between measured and computed values for full-scale experiments, as shown below.

27. Equations (30) and (31) for defining the p–y curves contain the coefficients A and F. These were determined empirically from results of load tests at Sabine and Manor and are given in Table 9 (where O_R is overconsolidation ratio, S_t is sensitivity, w_L is liquid limit, PI is plasticity index, and LI is liquidity index). The recommended procedure for estimating A and F for other clays is given below.

(a) Determine as many of the following properties of the clay as possible: s_u, ϵ_{50}, w_L, PI, LI, failure strain from stress–strain curve, overconsolidation ratio, degree of saturation, degree of fissuring, ratio of residual to peak shear strength.

(b) Compare the properties of the soil in question with the properties of the Sabine and Manor clays listed in Table 9.

(c) If the properties are similar to either the Sabine or Manor clay properties, use A and F for the similar clay. If the properties are not similar to either, the engineer should estimate A and F using his judgement with Table 9 as a guide.

28. The shapes of the p–y curves for the unified method (Figs 8 and 9) are based on the assumption that there is an intersection between equations (27) and (29). If that intersection does not occur, the p–y curve is defined by equation (27) until there is an intersection between equation (27) and the curves defining the p–y curves at greater pile deflexions.

Table 9. Curve parameters for unified criteria

Site	Sabine River	Manor
Clay description	Inorganic, intact $(s_u)_{avg} = 15$ kPa $\epsilon_{50} = 0.007$ $O_R \approx 1$ $S_t \approx 2$ $w_L = 92$ PI = 68 LI = 1	Inorganic, very fissured $(s_u)_{avg} \approx 115$ kPa $\epsilon_{50} = 0.005$ $O_R > 10$ $S_t \approx 1$ $w_L = 77$ PI = 60 LI ≈ 0.2
A	2.5	0.35
F	1.0	0.5

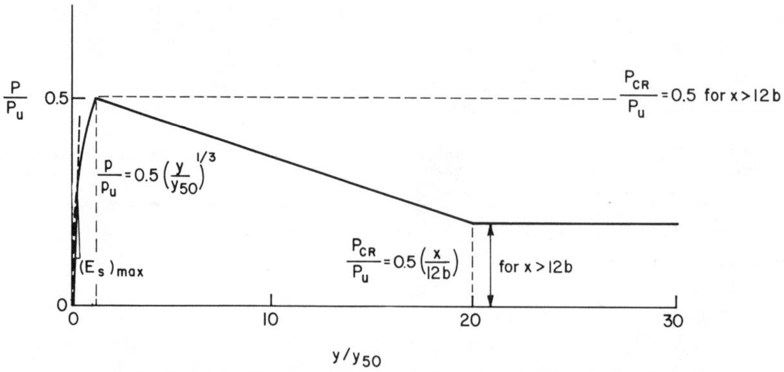

Fig. 9. Characteristic shape of p–y curve for unified clay criteria for cyclic loading

COMPARISON OF RESULTS FROM ANALYSES WITH THOSE FROM EXPERIMENTS

Lake Austin

29. The Lake Austin tests and others in this section were analysed using both the unified criteria and the criteria previously proposed.[4, 7] The parameters A and F for the unified criteria were chosen as 2.5 and 1.0 respectively, based on an average shear strength of 38 kPa as measured by the vane test. The Lake Austin clay is inorganic and slightly fissured and is classified as a CH. It has a liquid limit of 64, a plasticity index of 41, a liquidity index of 0.5, an overconsolidation ratio of about 4, and a sensitivity of about 3. The water table was above the ground surface and the submerged unit weight of the soil was 800 kg/m^3. The value of ϵ_{50} was 0.012.

30. The pile head was free to rotate. The loadings were short term static and cyclic. The maximum moments computed by the unified criteria are in good agreement with the measured values, as shown in Figs 10 and 11. In general, the results using the unified criteria agree slightly better with experimental results than do the results using the Matlock criteria.

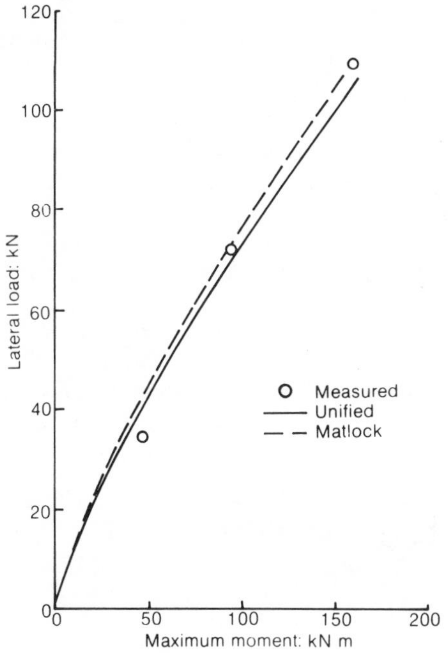

Fig. 10. Comparison of measured and computed maximum moments for Lake Austin for short term static loading

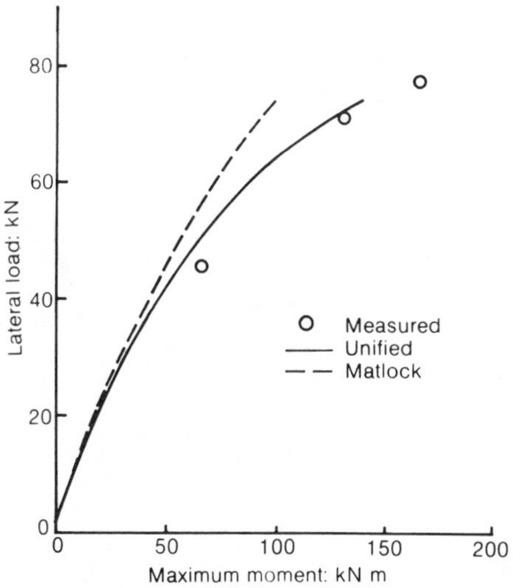

Fig. 11. Comparison of measured and computed maximum moments for Lake Austin for cyclic loading

Sabine

31. The Sabine tests were performed at a site where the clays were described as a slightly overconsolidated marine deposit. The properties of the clay are shown in Table 9. As indicated, the factors A and F were selected as 2.5 and 1.0, respectively.

32. The water table was above the ground surface, the pile head was free to rotate, and the loadings that are analysed in this Paper were short term static and cyclic. The

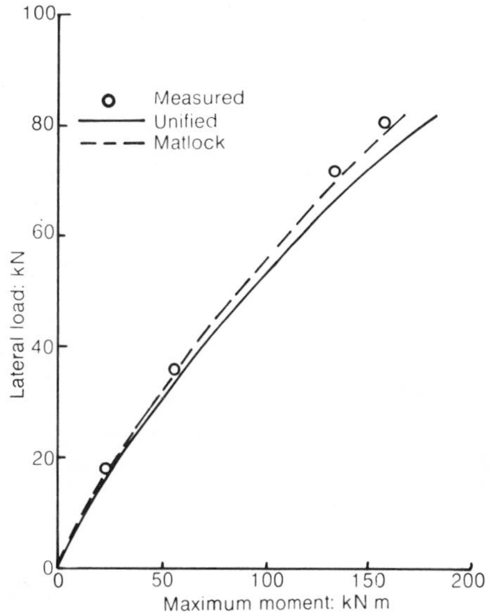

Fig. 12. Comparison of measured and computed maximum moments for Sabine for short term static loading

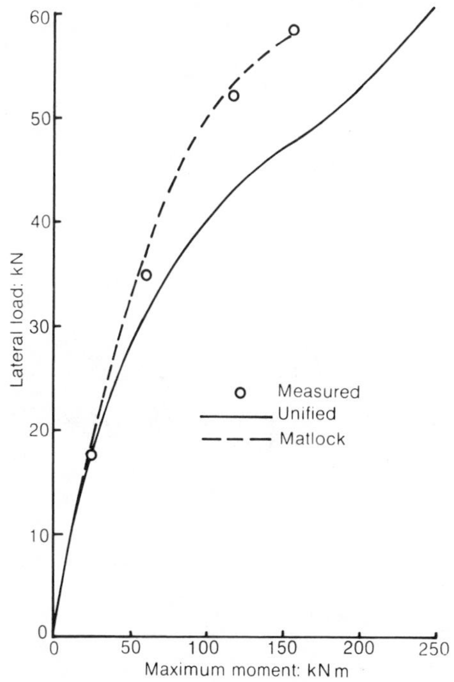

Fig. 13. Comparison of measured and computed maximum moments for Sabine for cyclic loading

results of computations using the unified criteria and the Matlock criteria are compared with experimental results in Figs 12 and 13. Agreements are good in all cases for loads of relatively low magnitude. At the higher loads for the short term static loading, the Matlock criteria give somewhat better agreement with the experiment than does the unified method; and considerably better agreement at the higher loads for the cyclic loading case.

Manor

33. The clay at Manor is highly overconsolidated, inorganic and very fissured. The properties of the clay are given in Table 9. As indicated, the factors A and F were selected as 0.35 and 0.5, respectively.

34. The water table was above the ground surface, the pile head was free to rotate, and the loadings were short term static and cyclic. The results of computations using the unified criteria and the Reese et al. criteria are compared with experimental results in Figs 14 and 15. As can be seen, excellent agreement was obtained in all instances. It is not shown here, but excellent agreement was obtained between computed groundline deflexions using both methods and the deflexions from the experiment.

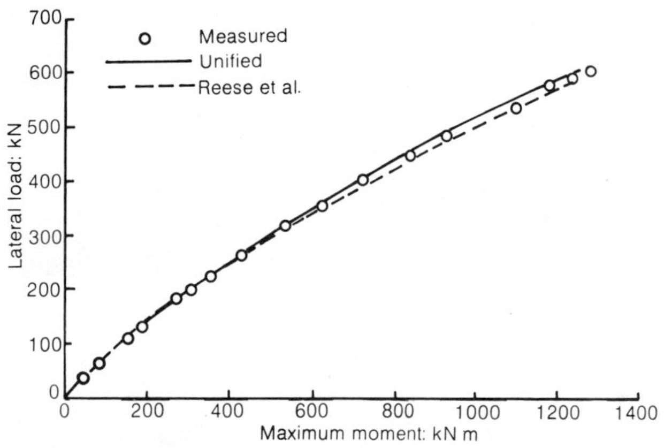

Fig. 14. Comparison of measured and computed maximum moments for Manor for short term static loading

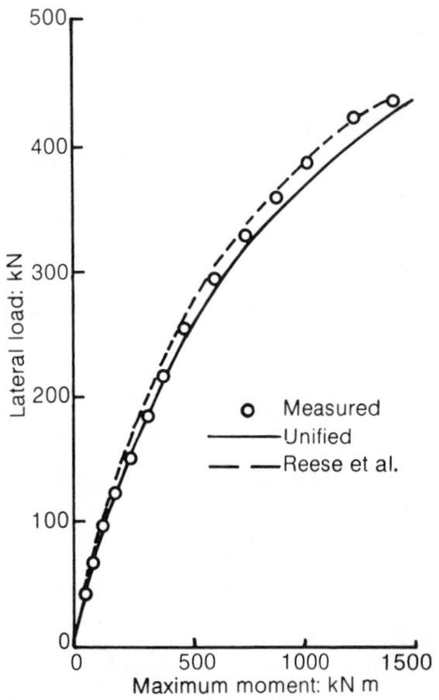

Fig. 15. Comparison of measured and computed maximum moments for Manor for cyclic loading

San Francisco Bay mud

35. Short term static tests, reported by Gill,[11] were performed on steel pipe piles of four different diameters whose properties are given in Table 10. The pile head was free to rotate. The load was applied 800 mm above the ground surface. Groundline deflexions and slopes were measured, but no data were given on slopes. The distribution of shear strength used in the analysis was measured with an in situ vane, and is given in Fig. 16. The soil at the site is described as an insensitive, gray, slightly organic, silty clay and is classified as a CH in the unified classification system. The liquid limit is 71, the plasticity index is 29. For purposes of computation ϵ_{50} was estimated as having a value of 0.01. The submerged unit weight was estimated as 800 kg/m^3.

36. The test site was flooded several days before the test, the water level being raised to approximately 30 mm above the ground surface. The shear strength in Fig. 16 was measured after the strength had stabilized after flooding.

37. The tests in bay mud were analysed using the unified criteria. The parameters A and F were chosen as 2.5 and 1.0, respectively, because the bay mud was felt to be

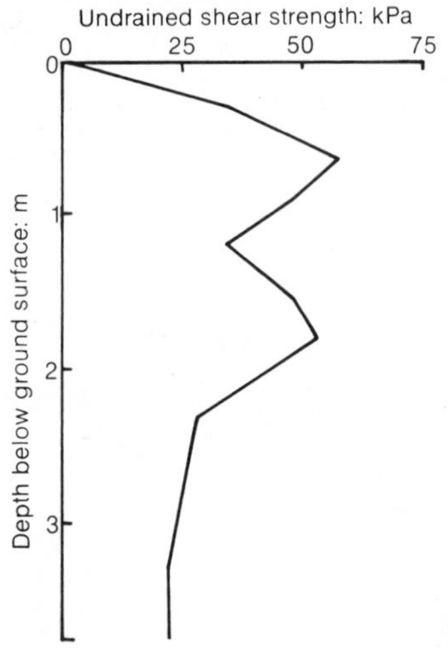

Fig. 16. Distribution of shear strength used for bay mud

Table 10. Pile properties for tests in bay mud

Pile	Diameter, b, mm	Bending stiffness, EI, kN m^2	Penetration, m
P1	114	6.23×10^5	5.55
P2	219	5.26×10^6	6.22
P3	324	3.12×10^7	5.09
P4	406	4.84×10^7	8.14

similar to the Sabine clay, based on the limited soil data. The computed groundline deflexions are less than the measured values in all cases, as shown in Figs 17 and 18. The most probable source of error is in the value of the coefficient A. A larger value of A would lead to a better agreement, and is probably justified.

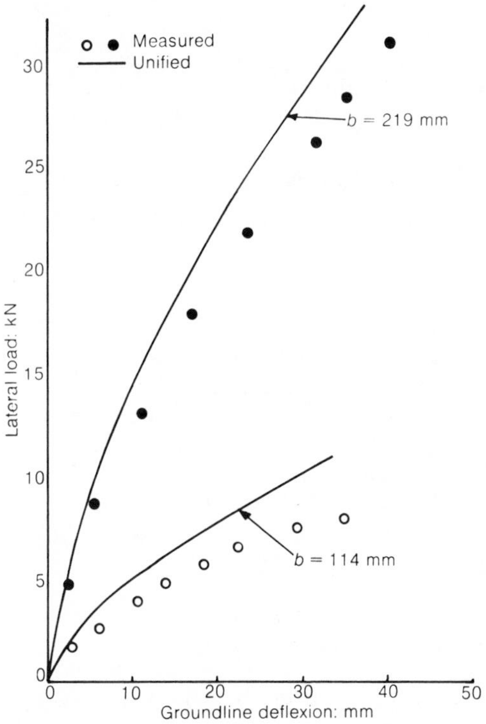

Fig. 17. Comparison of measured and computed groundline deflexion for piles P1 and P2 in bay mud

El Centro clay

38. Short term static tests, reported by Gill and Demars,[12] were performed on steel pipe piles of four different diameters whose properties are given in Table 11. The pile head was free to rotate. The load was applied at 810 mm above the ground surface. Groundline deflexions and slopes were measured. The distribution of shear strength used in the analysis was measured with an in situ vane, and is given in Fig. 19. The soil at the site is described as a silty clay and a clayey silt and is classified as a ML–CL in the unified classification system. The liquid limit is 32 and the

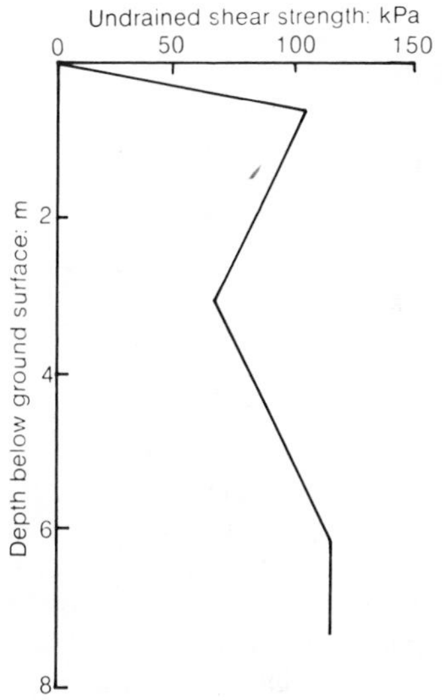

Fig. 19. Distribution of shear strength used for El Centro clay

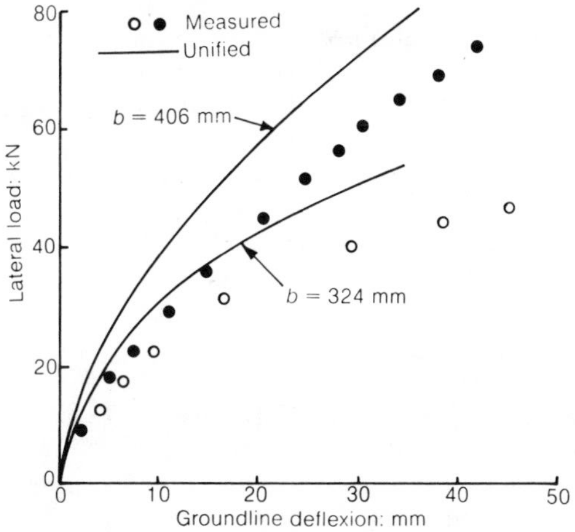

Fig. 18. Comparison of measured and computed groundline deflexions for piles P3 and P4 in bay mud

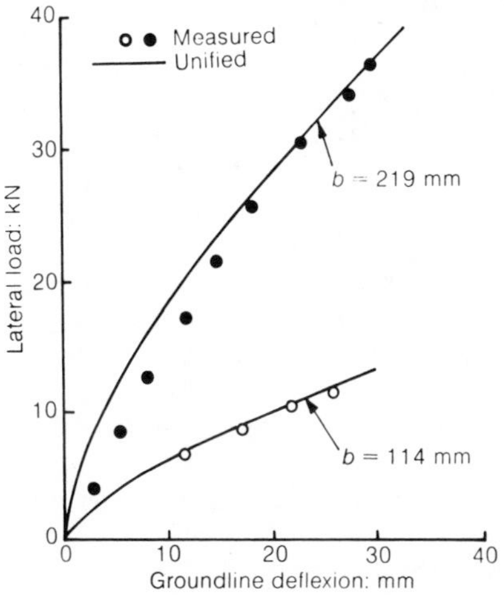

Fig. 20. Comparison of measured and computed groundline deflexions for piles P1 and P2 in El Centro clay

Table 11. Pile properties for tests at El Centro

Pile	Diameter, b, mm	Bending stiffness, EI, kN m^2	Penetration, m
P1	114	6.49×10^5	3.66
P2	219	4.97×10^6	5.18
P3	324	2.41×10^7	6.71
P4	406	4.84×10^7	8.23

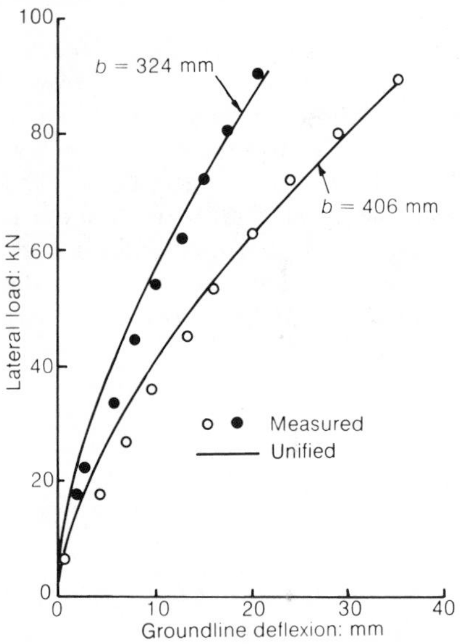

Fig. 21. Comparison of measured and computed groundline deflexions for piles P3 and P4 in El Centro clay

plasticity index is 10. For purposes of computation, ϵ_{50} was estimated as having a value of 0.01. The submerged unit weight was estimated as 800 kg/m^3.

39. The test site was flooded, the same as for the bay mud site, and the strength given in Fig. 19 was measured after the strength had stabilized after flooding.

40. The tests at El Centro were analysed using the unified criteria. The parameters A and F were chosen as 2.5 and 1.0, respectively, because the El Centro clay was felt to be similar to the Sabine clay, based on the limited soil data. The computed groundline deflexions are in good agreement with the measured values, for all four pile sizes, as shown in Figs 20 and 21.

CONCLUDING COMMENTS

41. The ability to predict the behaviour of piles under lateral loading is dependent on being able to derive curves giving the soil response (p–y curves). Two methods of predicting soil response curves for submerged clays have been published: Matlock[4] and Reese et al.[7] Results from the experimental programmes on which the methods of Matlock and Reese et al. were based were reanalysed and a single method, the unified method, was derived. The computed results employing the unified method agreed well with the experimental results reported by Matlock and Reese et al., and also agreed reasonably well with results from other experiments.

42. While the unified method presents a single and a somewhat simpler approach for predicting p–y curves for submerged clays, both under short term static and cyclic loadings, the method involves the selection of two empirical parameters with only a minor amount of guidance from experiments and none from theory. It is plain that additional experiments employing full-sized, instrumented piles in a variety of submerged clays are sorely needed.

43. In spite of the lack of full validation of the methods of predicting p–y curves, the approach described herein is believed to be the best currently available. It is hoped that the unified method presented here can be the vehicle to correlate the results from additional experiments into a single set of p–y criteria for submerged clays.

ACKNOWLEDGEMENTS

44. Most of the work described in this Paper was supported by a grant from the National Science Foundation, ENG. 74-19444. McClelland Engineers also provided financial assistance. The support of the sponsors is gratefully acknowledged.

REFERENCES

1. HETENYI M. Beams on elastic foundation. University of Michigan Press, Ann Arbor, Michigan, 1946.
2. REESE L. C. and MANOLIU I. Analysis of laterally loaded piles by computer. Buletinul Stiintific Al Institutului De Constructii Bucuresti, 1973, XVI, No. 1, 35–70.
3. MATLOCK H. and REESE L. C. Foundations analysis of offshore pile supported structures. Proc. 5th Int. Conf. International Society of Soil Mechanics and Foundation Engineering, Paris, 1961, 2, 91–97.
4. MATLOCK H. Correlations for design of laterally loaded piles in soft clay. Proc. 2nd Annual Offshore Technology Conf. Houston, Texas, 1970, 1, 577–594. Paper OTC 1204.
5. MATLOCK H. and RIPPERGER E. A. Measurements of soil pressure on a laterally loaded pile. Proc. Am. Soc. Test. Mater. 1958, 58, 1245–1259.
6. SKEMPTON A. W. The bearing capacity of clays. Proc. Building Research Congress, Division I, London, 1951, Part III, 180–189.
7. REESE L. C. et al. Field testing and analysis of laterally loaded piles in stiff clay. Proc. 7th Offshore Technology Conf. Houston, Texas, 1975, 2, 473–483, Paper OTC 2312.
8. REESE L. C. Discussion of soil modulus for laterally loaded piles by Bramlette McClelland and John A. Focht, Jr. Trans. Am. Soc. Civ. Engrs, 1958, 123, 1071–1074.
9. HANSEN J. B. The ultimate resistance of rigid piles against transversal forces. The Danish Geotechnical Institute, Copenhagen, 1961, bulletin 12, 5–9.
10. THOMPSON G. R. Application of the finite element method to the development of p–y curves for saturated clays. Master's thesis, The University of Texas, Austin, 1977.
11. GILL H. L. Soil behavior around laterally loaded piles. Naval Civil Engineering Laboratory, Port Hueneme, California, 1968, report R–670.
12. GILL H. L. and DEMARS K. R. Displacement of laterally loaded structures in nonlinearly responsive soil. Naval Civil Engineering Laboratory, Port Hueneme, California, 1970, report R–760.

18. Analysis of the load–deflexion behaviour of offshore piles and pile groups

F. E. TOOLAN, MA, MSc, MICE and M. R. HORSNELL, BSc, MICE (Fugro Ltd, Consulting Geotechnical Engineers, Ruislip)

The Paper descibes the analytical techniques that the Authors have used to calculate the load–deflexion behaviour of the piled foundations of North Sea jackets. The first part of the Paper considers axially and laterally loaded single piles. Comparisons are made between solutions based on the discrete element approach, in conjunction with P–Y and T–Z curves, and solutions based on elastic theory. A rational approach for selecting elastic moduli for the soil is described. The shortcomings of both methods are discussed, and a simple but more rigorous method developed by the Authors is described. For the axial case, results from the three methods are compared with published field records. The second part of the Paper considers the load–deflexion behaviour of pile groups subjected to axial, lateral, moment and torque loading. An analytical technique developed by the Authors is described, which uses as input the single pile load–deflexion curve, the elastic properties of the soil, the pile dimensions and the group layout. The technique can cater for different pile sizes within a group, together with layered soil conditions.

INTRODUCTION

The platform type most widely used by oil companies in offshore oil and gas fields is the steel jacket. This is supported on a piled foundation. The geotechnical consultant is involved in many aspects of the planning, design, installation and acceptance of jackets and in particular their foundations. However, this Paper concentrates on one design topic: the determination of the load–deflexion behaviour of the piles and pile groups under transient loading conditions.

2. The transient load–deflexion analyses of the foundations are important because

(a) they provide some of the data from which the dynamic amplification factors for the jacket loadings may be calculated;

(b) the results are required for computing the deflexions, velocities and accelerations of the structure at deck and other levels;

(c) they allow the determination of the distribution of loads among piles, and hence the stresses in the pile material and in the connections between the piles and jacket;

(d) they indicate the factor of safety of pile groups under combined loading conditions (i.e. axial and moment or shear and torque).

The above are vital and in some cases limiting considerations for the overall design of a jacket.

3. The Authors have been responsible for the foundation design of numerous offshore structures including some very large jackets located in the North Sea. During the course of this work they have developed design guidelines and numerical techniques which, in their opinion, are directly relevant to the requirements of the structural designer of the jacket. This Paper is intended as a practical guide for all those involved in offshore pile design, and presents guidelines and methods for the elasto-plastic analysis of single piles, and the application of the resulting response characteristics to pile group analysis.

Foundation response characteristics for structural design

4. In designing a jacket, one of the major problems faced by the structural designer is the realistic modelling of foundation–structure interaction at the interface between the piles and the legs of the structure.

5. At the mudline interface between pile and leg there are six degrees of freedom governing forces and deflexions which may or may not be independent (Fig. 1). One method of representing the restraint offered by the foundation to the superstructure would be to use a set of three linear translational springs and three linear rotational springs. This restraint would be expressed in matrix form as

$$\begin{Bmatrix} F_1 \\ F_2 \\ F_3 \\ F_4 \\ F_5 \\ F_6 \end{Bmatrix} = \begin{bmatrix} K_{11} & 0 & 0 & 0 & 0 & 0 \\ 0 & K_{22} & 0 & 0 & 0 & 0 \\ 0 & 0 & K_{33} & 0 & 0 & 0 \\ 0 & 0 & 0 & K_{44} & 0 & 0 \\ 0 & 0 & 0 & 0 & K_{55} & 0 \\ 0 & 0 & 0 & 0 & 0 & K_{66} \end{bmatrix} \begin{Bmatrix} U_1 \\ U_2 \\ U_3 \\ U_4 \\ U_5 \\ U_6 \end{Bmatrix} \quad (1)$$

INSTITUTION OF CIVIL ENGINEERS. Numerical methods in offshore piling. ICE, London, 1980, 147–155.

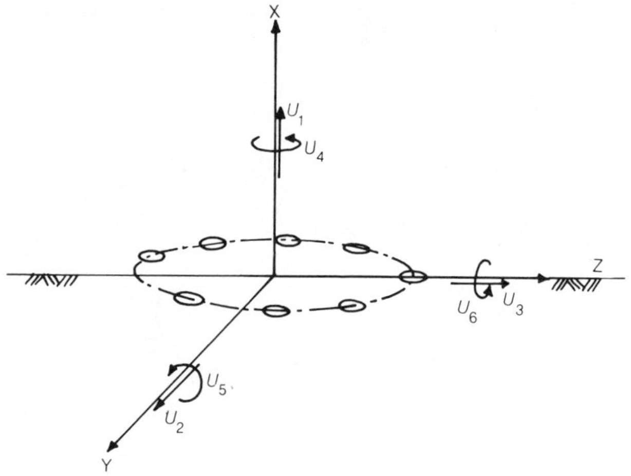

Fig. 1. Degrees of freedom of a typical pile group

6. One disadvantage of using the governing stiffness matrix of equation (1) is that it ignores three important interactions which are known to take place within the pile group itself; these are the interactions between the pile head restraints of the individual piles, between axial load and group moment, and between lateral shear force and group moment. Introduction of terms governing these interactions into equation (1) results in the following stiffness matrix for each pile group:

$$\begin{Bmatrix} F_1 \\ F_2 \\ F_3 \\ F_4 \\ F_5 \\ F_6 \end{Bmatrix} = \begin{bmatrix} K_{11} & 0 & 0 & 0 & K_{15} & K_{16} \\ 0 & K_{22} & 0 & 0 & 0 & K_{26} \\ 0 & 0 & K_{33} & 0 & K_{35} & 0 \\ 0 & 0 & 0 & K_{44} & 0 & 0 \\ K_{51} & 0 & K_{53} & 0 & K_{55} & 0 \\ K_{61} & K_{62} & 0 & 0 & 0 & K_{66} \end{bmatrix} \begin{Bmatrix} U_1 \\ U_2 \\ U_3 \\ U_4 \\ U_5 \\ U_6 \end{Bmatrix} \qquad (2)$$

where

F_1 is axial leg load
F_2 and F_3 are horizontal shear loads at the mudline
F_4 is torque acting about the vertical axis
F_5 and F_6 are leg moments about two orthogonal axes normal to the leg centre line
U_1 is axial deflexion of group
U_2 and U_3 are lateral group deflexions
U_4 is rotation about vertical axis
U_5 and U_6 are rotations about axes normal to leg centre line
K_{11} is axial load per unit group axial deflexion
K_{22} and K_{33} are group lateral loads required to produce unit lateral group deflexions
K_{26} and K_{35} are lateral group loads produced by unit head rotation
K_{44} is torque produced by unit twist about group centre line
K_{51} and K_{61} are moments produced by axial deflexion with no pile cap rotation
K_{53} and K_{62} are moments produced by unit translation of head, head not allowed to rotate
K_{55} and K_{66} are moments produced by unit rotation of head, head not allowed to translate.

7. The method of analysis described here produces coupled non-linear foundation response characteristics (i.e., K_{ij}) in a form which is suitable for use in a structural analysis. A simplified sequence of structural analyses would be as follows.

(a) The structural designer makes a first estimate of magnitude of the loads at foundation level.
(b) Using these loads and the coupled non-linear response characteristics provided by the geotechnical consultant, the structural designer determines the values of K_{ij} in equation (2).
(c) A structural analysis is made with these values of K_{ij} and a revised set of foundation loads is produced.
(d) The new loads are used to produce revised values of K_{ij}. The structural analyses are repeated.
(e) The procedure is continued until compatability is obtained between the loads and deflexions in the superstructure, and the loads and deflexions of the foundations.

8. In the opinion of the Authors, the determination of the foundation response characteristics is not amenable to generalized standard solutions, particularly for structures in the North Sea. This is because the soil conditions are highly stratified and variable, the stress levels in the soil are high, the response characteristics are non-linear, and the design

NOTATION

A	cross-sectional area of pile
E	elastic modulus of pile
E_i	elastic modulus at ith elevation
EI	flexural rigidity of beam column
E_s	elastic soil modulus
F	longitudinal body force
F_i	force
K	a constant peculiar to the sand of a particular site
K_{ij}	pile group interaction stiffness
P_{ax}	axial load on beam column
P_j	load carried by pile j
P–Y	the relationship defining lateral soil–pile interaction
r_i	lever arm of pile i about either a vertical or horizontal axis
T–Z	the relationship defining axial soil–pile interaction
U_i	deflexion
w	deflexion of pile or displacement of soil
x	distance from top of pile
Y_L	unit elastic displacement
a_{ij}	Poulos interaction factor between piles i and j
δ_i	lateral displacement of ith elevation at working load
δ_θ	applied rotation, twist about vertical axis, or rotation about horizontal
ρ_{Gi}	group translation of pile i
ρ_{1i}	single pile movement of pile i (i.e. movement through the soil)
Σ	summation
du/dx	strain
σ_v	effective vertical stress

processes for foundations and superstructure are iterative in themselves and with each other. It should be noted that from the initial feasibility studies through final design stage to post-installation checks, numerous structural and foundation analyses are made for a typical jacket.

9. Based on the considerations set out above, it is apparent that any method for calculating response characteristics must be

(a) able to model all relevant soil conditions;
(b) capable of incorporating all pile properties and geometry;
(c) such that the results are directly applicable to the structural analyses;
(d) capable of providing fast turn-around times;
(e) subject to low computer costs.

The methods described below satisfy these criteria.

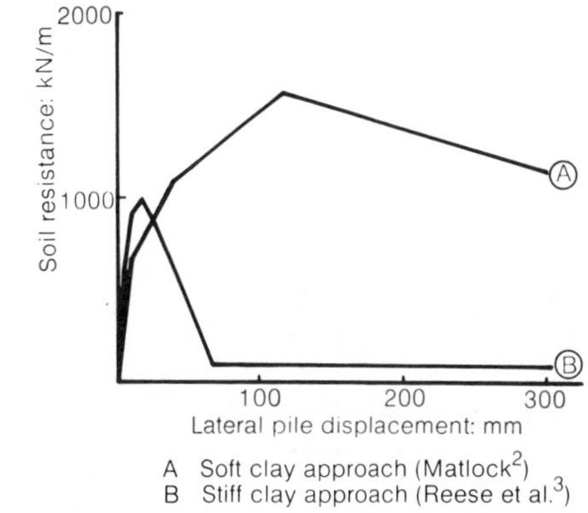

Fig. 2. Comparison of P–Y curves; pile diameter 1.5 m, depth 20 m, shear strength (undrained) 200 kN/m²

SINGLE PILE LOAD–DEFLEXION BEHAVIOUR

10. The response characteristics of single isolated piles are required for the structural analyses of jackets supported on one pile per leg, and as input for pile group analyses for larger jackets. There are four degrees of freedom for a symmetrical pile in symmetrical soil: axial, lateral, moment and twist. The lateral and moment responses are closely coupled via the degree of restraint at the pile head. In this Paper the determination of two of these isolated pile responses, axial and lateral, are considered in detail. The twist response is ignored and the moment response is implicit in the lateral by coupling.

Lateral behaviour

11. The basic differential equation governing the lateral response of an embedded pile modelled as a beam column takes the form

$$EI(x)\frac{d^4 w}{dx^4} + P_{ax}\frac{d^2 w}{dx^2} + E_s(x, w)\, w = 0 \qquad (3)$$

where EI is flexural rigidity of beam column, P_{ax} is axial load on beam column, E_s is soil modulus, w is deflexion of pile or displacement of soil, and x is distance from top of pile.

12. Equation (3) can be solved using either finite difference or finite element methods. It is debatable which approach yields the most practical solution.[1] However, both methods involve division of the pile into a number of discrete elements, with the lateral soil resistance mobilized by movement of the pile being modelled by non-linear 'ground springs' acting at the connection between two elements. These ground springs are referred to as *P–Y* curves.

13. The *P–Y* curves are analytical tools, not fundamental soil properties. Numerous forms of *P–Y* curve have been derived by backfiguring ground springs from lateral pile test results through 'beam–column' programs. For the soil conditions in the North Sea it is common practice to use the *P–Y* curves derived by Matlock[2] for soft clay, Reese et al.[3] for stiff clay, and Reese et al.[4] for sand. The *P–Y* curves should not be used blindly. Those mentioned have been derived for relatively homogeneous soil conditions. They require careful adaptation when used in the highly stratified soil conditions of the North Sea. For example, the 'stiff clay *P–Y* curves' can give a much softer response than the 'soft clay *P–Y* curves' when applied to the same soil parameters for conditions of large strains (Fig. 2). Before finalizing the *P–Y* curves, the geotechnical consultant should examine them in the light of the deflected shape of the pile obtained from the beam–column analysis.

14. Having analysed the response of an isolated pile using *P–Y* data and beam–column analyses, it is desirable to check the results by an independent method. Two methods are available: the first is to install test piles at the proposed location and make lateral load tests; the second is to use an alternative analytical method. Of the two options the first is preferable. Until recently such tests have been prohibitively expensive, but this is now changing. For example, the Authors' parent company has developed a sea bed rig which can perform lateral pile load tests from a soil survey vessel. When relevant pile test data is not available, analytical methods have to be used. The analytical method used by the Authors for checking purposes is the elastic solution. For cyclic loading conditions this is usually, but not always, valid for loads and deflexions within the working range.

15. The most simple form of elastic solution is for a pile in a linearly elastic homogeneous continuum. Generalized solutions for this case have been obtained by Poulos[5] and published in the form of charts of influence factors for pile head displacements for various relative stiffnesses of pile to soil.[6] In applying this method to offshore sites one has to idealize real soil conditions by an equivalent homogeneous elastic soil.

16. Work carried out by Broms[7] indicates that for laterally loaded piles in stiff clay a ratio of elastic soil modulus to undrained shear strength of 200 is appropriate. The relationship can be used to convert the shear strength profile for a clay site into a profile of elastic modulus versus depth. For sites where the soil conditions are fairly uniform the determination of an equivalent elastic modulus causes little problem. However, for more typical North Sea sites where layers of normally and heavily overconsolidated clays may be interspersed with sand layers the equivalent modulus is not as obvious.

17. From the lateral analyses which will have been carried out using *P–Y* curves, the distribution of lateral displacements from the pile head to the point of contraflexure (typically at 10–15 pile diameters beneath the pile head)

will be known for working load conditions. Using these results, an equivalent elastic modulus can be determined from the expression

$$E_s = \frac{\sum_{i=1}^{n} E_i \delta_i}{\sum_{i=1}^{n} \delta_i} \qquad (4)$$

where n is number of elevations considered down to point of contraflexure, E_i is elastic modulus at ith elevation, and δ_i is lateral displacement of ith elevation at working load.

18. In sand strata the Authors have derived elastic modulus values from the stress–strain relationships obtained from consolidated drained triaxial tests. Analysis of several such tests at any one site enables a relationship between elastic modulus and effective overburden pressure to be obtained of the form

$$E_s = K \sigma_v^{1/2} \qquad (5)$$

where K is a constant peculiar to the sand of a particular site. Using equation (5) an elastic modulus for sand can be obtained at any depth for substitution into equation (4).

19. Comparison between the load–deflexion responses calculated using P–Y curves and elastic solutions are shown in Fig. 3 for two typical North Sea sites. As can be seen, over the initial load range the agreement between the two pile head responses is good in both sand and clay for piles of up to 2.13 m dia.

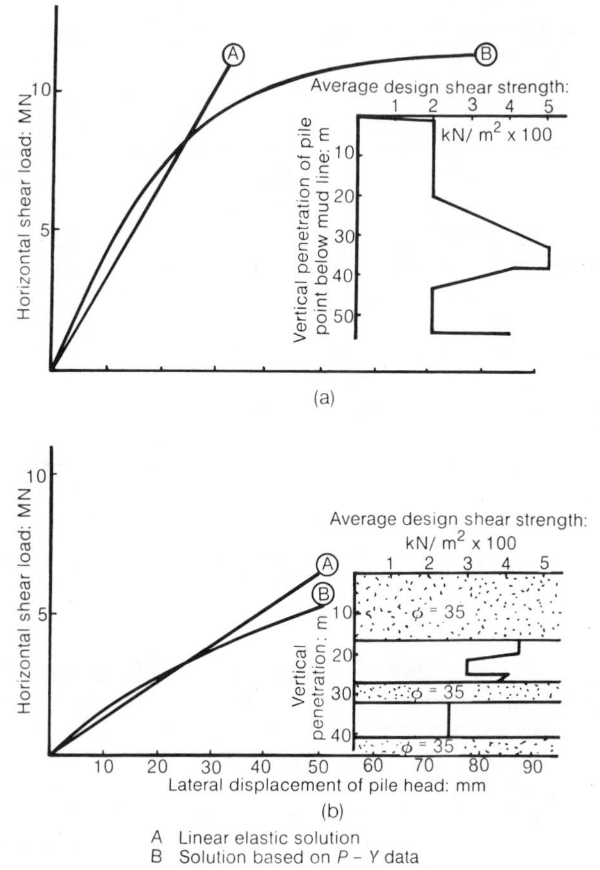

Fig. 3. Comparison of lateral load–deflexion analyses: (a) pile diameter 2.13 m; (b) pile diameter 1.37 m

20. *Effects of scour.* In carrying out the lateral analysis of offshore piles, the effects of scour around the piles must be accounted for. There are two mechanisms by which scour depth can affect the lateral behaviour of a pile. The first is the direct effect of the loss of lateral restraint caused by 'local scour' at the head of the pile. The second is a reduction in overburden pressure on the pile caused by general scour in the vicinity of the platform. This reduction causes a corresponding reduction in the stiffness of the P–Y curves for sand layers at any depth and a reduction in elastic modulus in accordance with equation (5). Both local and general scour lead to higher deflexions and greater bending moments in the pile than if scour had not taken place.

21. From the results of model tests[8,9] and theoretical calculations,[10] predicted scour depths for North Sea environments are expected to be in the region of 1.5 pile diameters, with little or no scour beyond a distance from the centre of the pile of 2.5 pile diameters. Scour will form a cone of depression around the pile, violating the horizontal sea bed conditions on which P–Y curves are based. To overcome the problem of generating P–Y data for non-horizontal sea bed conditions it has become common practice to make some simplifying assumptions. The most appropriate method seems to lie between the following two approaches: applying full local scour and zero general scour; and applying full scour and half this value for general scour. In the Authors' opinion applying full general scour is unduly conservative. Only by careful consideration of prevailing soil, sea bed and marine conditions can the appropriate scour depth be determined.

Axial behaviour

22. The basic differential equation governing the axial response of a pile subjected to axial loading is

$$EA \frac{d^2 u}{dx^2} + F = 0$$

where E is elastic modulus of pile, A is cross-sectional area of pile, du/dx is strain, and F is longitudinal body force. Finite difference or finite element techniques can be employed to solve this equation to yield pile displacements for given loading conditions. Using either technique, the pile has to be divided into a number of discrete elements. The friction between soil and pile is modelled by non-linear ground springs acting at the connection between a pile element and its respective soil element. These ground springs are generally referred to as T–Z curves.

23. Vijayvergiya[11] has developed T–Z curves which are generalized functions relating mobilized side friction and end bearing to axial pile movements. These T–Z curves were obtained by back-analysing numerous reported pile load tests. The majority of these tests were carried out on piles with diameters not exceeding 0.6 m. Obviously the axial response predicted using these curves for offshore piles, where pile diameters may be of the order of 1.5 m or more, should be correlated against typical pile load test data. This data is never available. The only alternative is to repeat the analysis using an alternative method of solution.

As one wants to check the effectiveness of using T–Z curves the alternative method has to be one which does not involve their use. One such method which can be easily carried out is the elastic solution for a pile in a linearly elastic homogeneous continuum. As for laterally loaded piles, Poulos and Mattes[12] have published generalized solutions for this case in the form of charts of influence factors for pile head displacements for various stiffnesses of pile to soil. Application of these solutions to offshore sites requires idealizing real soil conditions by an equivalent homogeneous elastic soil.

24. An appropriate ratio between elastic moduli and undrained shear strength for immediate settlement calculations for piles in overconsolidated clays is indicated[13–15] to be 400. This is for working load levels and is twice the value used for laterally loaded piles. The reason for this difference is stress level. For laterally loaded piles, movements are governed largely by the soil properties along the upper length of the pile shaft. Frequently the soils near sea bed level are in a state of failure. In contrast, displacements of axially loaded piles are governed mainly by the soil properties at the tip and along the lower length of pile shaft. The soils in this zone are relatively lightly stressed. The work of Seed and Idriss,[16] which provides a relationship between stress level and elastic moduli of soils, indicates that the factor of 2 mentioned above is of correct order of magnitude.

25. In sand strata the soil modulus for axial loading is estimated in exactly the same way as for lateral loading; i.e., with equation (5). However, the computed value of the axial modulus is always much higher than that used for lateral loadings. This is because only the upper sand layers are used for lateral analyses whereas the strata along the whole length of the pile shaft are included in the axial computations.

26. The Authors have found that up to working load levels, there is good agreement between axial load–deflexion curves generated by the T–Z approach and displacements calculated by elastic methods for piles with diameters of less than 1.0 m. For pile diameters of more than 1.0 m the T–Z approach predicts a stiffer response than the equivalent elastic solution, the discrepancy increasing with pile diameter. This is illustrated in Fig. 4, where the elastic and T–Z generated single pile axial load response curves are compared for piles of the same length of penetration but with varying diameters at a North Sea location. The soil conditions are those shown on Fig. 3(b).

27. The reasons for the discrepancies between the two solutions and why they increase with diameter lie in some of the assumptions implicit in both methods, and the experimental data on which the T–Z approach is based (i.e., mainly on small diameter piles). The relevant assumptions in the T–Z method are

(a) no transmission of stress form one soil element to another;
(b) slip along pile shaft independent of pile diameter;

whereas in the elastic method the following are assumed:

(a) no slip between soil and pile wall;
(b) shear stress generated between soil and pile dependent only on deflexion and soil modulus;
(c) homogeneous soil or idealized stratification.

28. The ideal solution for analysing large diameter piles lies somewhere between the T–Z and the elastic approach. The main advantages of the two methods are summarized in Table 1.

Table 1

Method of analysis	Advantage
$T-Z$	Realistic modelling of pile compressibility, soil stratification, and shear failure (i.e., 'slip' of pile/soil interface)
Elastic	Realistic modelling of elastic compression set up by stress regime induced into the surrounding soil mass by large diameter piles

29. To combine the advantages of two methods, the Authors use the following sequence of analysis.

(a) The discrete element solution is carried out, but the generalized T–Z curves of Vijayvergiya[11] for shaft friction are replaced by T–Z curves obtained from laboratory and model pile tests which provide only the 'slip' behaviour at the pile/soil interface. These curves can be obtained from the work of Coyle and Reese.[17] The generalized T–Z curve of Vijayvergiya for the end spring is retained. The result yields the pile head displacement due to the pile compressibility, slip behaviour and compressibility of the soil beneath the pile tip. It also provides the load distribution along the pile.
(b) The load distribution along the pile is idealized as a series of point loads, Mindlin's solution[18] being used for a vertical point load within a soil mass to obtain the additional elastic compression at the

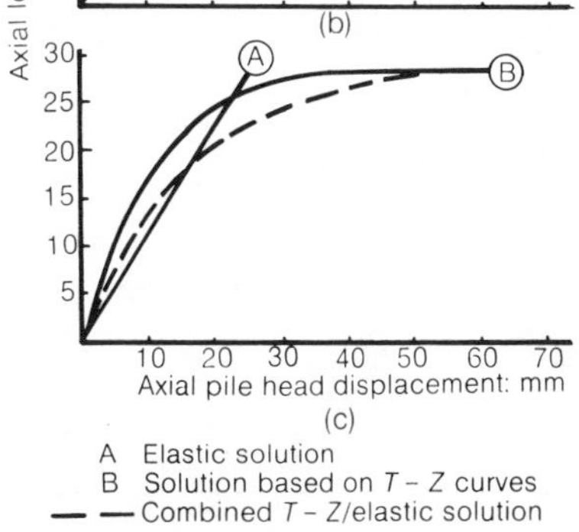

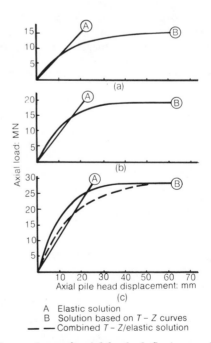

Fig. 4. Comparison of axial load–deflexion analyses; pile wall thickness 0.05 m, penetration 40 m: (a) pile diameter 0.6 m; (b) pile diameter 0.9 m; (c) pile diameter 1.37 m

pile tip. The sum of these additional displacements for each idealized point load yields the additional displacement the pile will experience due to the stress regime set up in the soil by the pile.

(c) The results from (a) and (b) are combined to give the pile's total axial response.

30. This method has been used to analyse the piles whose axial response characteristics have previously been discussed and are included in Fig. 4. As can be seen, at this site the Authors' method results in axial response characteristics for a large diameter pile which are softer than those of the T–Z method but stiffer than those of the elastic solution.

31. The method detailed above has been used to analyse the pile load test carried out by Jelinek et al.[19] on 1.3 m dia. bored piles in an interglacial lacastrine clay deposit, south-east of Munich. Fig. 5 illustrates the variation in axial response predicted by the elastic method, the T–Z method and the combined elastic/T–Z method, and compares the predictions with observed pile response. For the reason already discussed, the T–Z method produces too stiff a response. The elastic method gives reasonable agreement over the initial load range but as shear failure develops along the interface between pile and soil, so the elastic response gets stiffer than observed behaviour. However, the combined elastic/T–Z method gives good correlation between predicted and observed responses over the full load range. It should be noted that the observed curve in Fig. 5 is based on the immediate settlement only. The consolidation settlements which occurred during intervals between load stages have been subtracted from the curve shown in the original reference.[19] Also it is not always the case with offshore piles for the elastic approach to provide a stiffer response than the Authors' method.

GROUP LOAD–DEFLEXION BEHAVIOUR

32. As previously discussed, the structural designer requires the response characteristics of pile groups subjected to combined axial, lateral, moment and torque loadings. The Authors' methods of analysis described below are elasto-plastic in nature, are based on realistic soil properties, and when programmed for a computer are considerably cheaper to run than comparable finite element programs. Unlike pure elastic programs, soil failures both locally and generally are accounted for and non-homogeneous soil conditions and non-uniform piles are easily modelled. The Authors' method is based on the following assumptions:

(a) piles are connected by a rigid pile cap;
(b) for calculating the interaction effects between piles, the soil is linearly elastic and homogeneous to at least the level of the pile tips;
(c) for calculating the interaction effects between piles, the piles may be regarded as fixed-headed;
(d) the pile group is 100% efficient with respect to axial and lateral capacity.

The validity of the last assumption is checked in each case using equivalent pier methods. Also the stress contours around each pile are drawn to ensure that a pile is not shadowed by adjacent piles (Fig. 6). To date all pile groups dealt with by the Authors have had efficiencies of 100%. However, the method can easily be modified to determine the load–deflexion behaviour of less efficient groups.

33. The relationship of axial or lateral group displacement and the axial or lateral loads in the piles can be expressed (after Focht and Koch[20]) as

$$\rho_{Gi} = \rho_{Ii} + Y_L \sum_{\substack{j=1 \\ j \neq i}}^{n} P_j \alpha_{ij} \qquad (6)$$

where ρ_{Gi} is group translation of pile i, ρ_{Ii} is single pile movement of pile i (i.e., movement through the soil), Y_L is unit elastic displacement (after Poulos[5,12]), P_j is load carried by pile j, α_{ij} is Poulos interaction factor between piles i and j, and n is number of piles in the group.

34. The single pile load–deflexion behaviour can be obtained from analyses described earlier in this Paper. Equation (6) can therefore be written as

$$\rho_{Gi} = f\{P_i\} + Y_L \sum_{\substack{j=1 \\ j \neq i}}^{n} P_j \alpha_{ij} \qquad (7)$$

If the initial single pile load and deflexion prior to the application of a small increment of pile group displacement

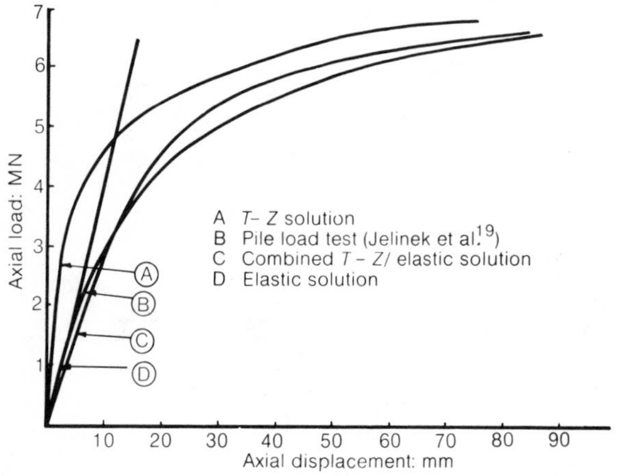

Fig. 5. Comparison of observed and predicted load–displacement behaviour

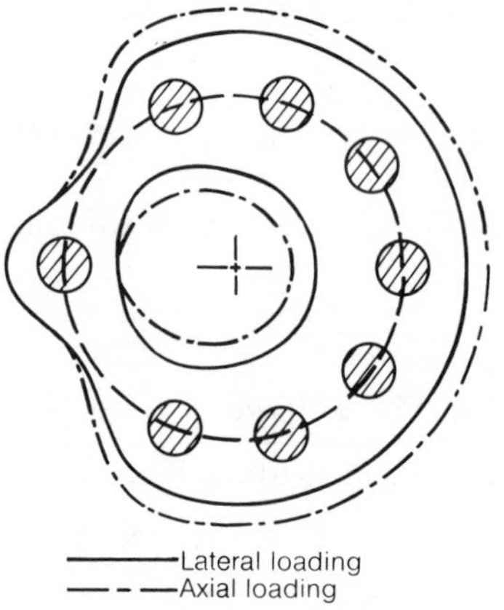

Fig. 6. Idealized zones of working stress around a typical offshore pile group; pile diameter 2.13 m, pitch circle diameter 12.2 m

are known, the relationship between ρ_{Ii} and P_i can be expressed as a linear function; i.e.

$$\rho_{Ii} = m_i P_i + c_i \quad (8)$$

Thus for a group of n piles, n simultaneous equations of the following form can be obtained:

$$\rho_{Gi} = m_i P_i + c_i + Y_L \sum_{\substack{j=1 \\ j \neq i}}^{n} P_j \alpha_{ij} \quad (9)$$

These n simultaneous equations will contain $n+1$ unknowns (i.e., n pile loads and the group displacement). However, if the solution is started at zero conditions (i.e., zero load and zero displacement) and increments in small steps of group displacement, then group displacement becomes a known quantity, and the n simultaneous equations can be solved to obtain the single pile loads and hence, by summation, the group load. Determination of single pile displacement from equation (8), together with the single pile loads, provides information for the determination of the linear functions for the next increment of group displacement.

35. This simplified approach of group translational analysis provides directly the information required by the structural designer for input into his pile group stiffness matrix. It can also be modified to provide information regarding the rotational behaviour of the group by substitution of the following relationship into equation (9):

$$\rho_{Gi} = r_i \, \delta\theta \quad (10)$$

where r_i is lever arm of pile i about either a vertical or horizontal axis, and $\delta\theta$ is applied rotation, twist about vertical axis, or rotation about horizontal.

36. The interaction factors used in the group analysis are obtained from the work of Poulos and Davis.[21] However, these factors are based on elastic solutions and, as was the case for single pile analyses, real soil conditions have to be replaced by an equivalent homogeneous elastic soil. For overconsolidated clays the relationships between elastic soil modulus and undrained shear strength already quoted (200 for lateral loading and 400 for axial loading) are applicable only in zones of working strains. These strains will occur at stresses of 20–50% of failure stress for both axially and laterally loaded piles. Analysis of the stress distribution around axially and laterally loaded piles indicates that the 10% stress contour is located at a distance from the centre of the pile of 2.5 diameters for axial loading and 2 diameters for lateral loading. Therefore the first step in determining interaction factors for any group is to draw on to a plan of the pile groups these zones of working stress. If two zones intersect, then within that zone stresses will be approximately 20% or more of failure stress (i.e., working strain conditions). This is illustrated in Fig. 6, where these zones are shown for a typical offshore pile group. If, in any group, piles exist between which there is no zone of working strain, then interaction will be dictated by a low strain elastic soil modulus of the order of 500–1000 times the undrained shear strength. Only careful consideration of the prevailing soil conditions can determine the appropriate value. For sands the low strain elastic soil modulus is determined from triaxial tests via an equation of identical form to that of equation (5). The K value obtained is higher than that for low strain conditions. These higher values of soil modulus may be found appropriate for calculating foundation responses suitable for use in a fatigue analysis.

37. For the group twist analysis careful consideration should be given to the interaction factors used. For small increments of twist all piles will move at right angles to their pitch circle. Interaction due to neighbouring piles can be resolved into two components, one tangential to the pitch circle, the other radial. The radial component will have no effect on the twist capacity of the group, therefore twist interaction will be obtained from the summation of all tangential components.

38. Coupling of modes using equations (9) and (10) is a simple procedure. For example, the axial group load–deflexion behaviour and the moment–rotation behaviour are coupled by first displacing the pile group vertically and then rotating the group about its neutral axis. The summation of the pile loads at every increment of rotation provides the group axial load. The group moment at any time is computed from the individual pile loads multiplied by their respective lever arms. The results are presented in graphical form. In Fig. 7, curves have been computed for the pile group shown in Fig. 6 and the soil conditions of Fig. 3(a). The higher the group axial load, the softer is the moment–rotation behaviour and the lower is the moment capacity of the pile group. This is because as the vertical load is increased the most heavily loaded piles approach failure. Additional penetration of the piles into the soil, for example by rotation, hardly increases the load in the piles. Thus the moment resistance is low. In the limit a pile group at axial failure has no moment capacity.

39. The moment computed by the above method is not the only moment acting on the pile cap. The lateral displacement of the group induces shear forces and moments at the head of each pile. The shear forces are obtained directly from the analyses based on equation (9). The pile head moments are calculated using Y modifiers, P–Y curves and the beam–column program as described by Focht and

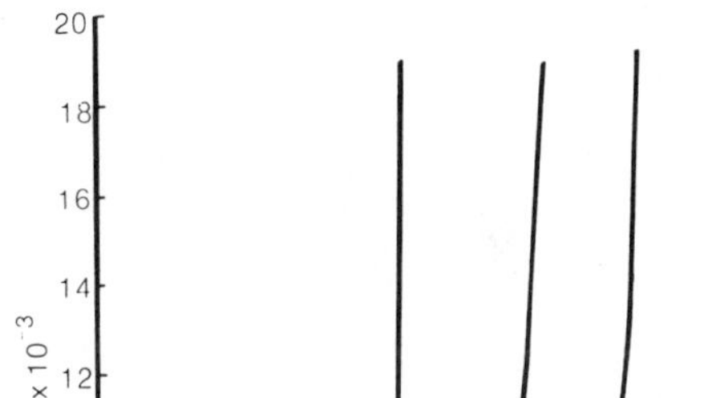

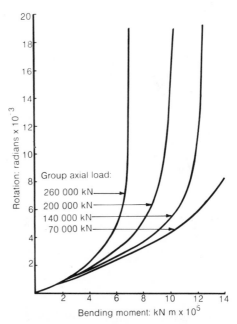

Fig. 7. Moment–rotation behaviour for various vertical loads; pile diameter 2.13 m, penetration 50 m

Koch.[20] Because the shear force distribution among piles is rarely uniform, a different Y modifier has to be applied to each pile, appropriate to its individual loading intensity. Also the rotation at the head of the piles should be equal to that of the pile group, and fixity at the pile heads should be correctly employed. Only by doing this can compatibility between the calculated responses of the structure and its foundation be ensured and the correct pile head moments obtained. The total moment in the pile cap is the sum of the 'pile head moments' and the 'group moment'.

40. Details of the ways in which the structural designers may use the foundation response data provided by the geotechnical consultant are outside the scope of this Paper. Each structural designer has his own approach, which may vary for different types of structure; the approach is tailored to the input requirements of the programs to be used for the structural analyses. Generally the analyses are iterated until compatibility is achieved between the loads, moments, deflexions and rotations at the base of each leg and those calculated for the heads of the piles using foundation response data. After each iteration a new assessment can be made of the loads and moments in the structure, allowing more appropriate foundation behaviour to be selected for the next run. The situation is frequently simplified by the fact that the axial and lateral load–deflexion relationships of piles are relatively uncoupled and each is approximately linear over the working load range. An example which shows this is the group axial load–deflexion curve presented as Fig. 8. The iterations are required principally to ensure compatibility between the moments and rotations which, as explained previously, are coupled to both the axial and lateral modes.

Effects of soil stratification beneath pile tip

41. The method described previously for analysing axial group behaviour is applicable only when there are homogeneous soil conditions below the tips of the piles. For sites where the pile group is underlain by strata which are significantly stiffer or stronger than the founding strata a more rigorous solution is required.

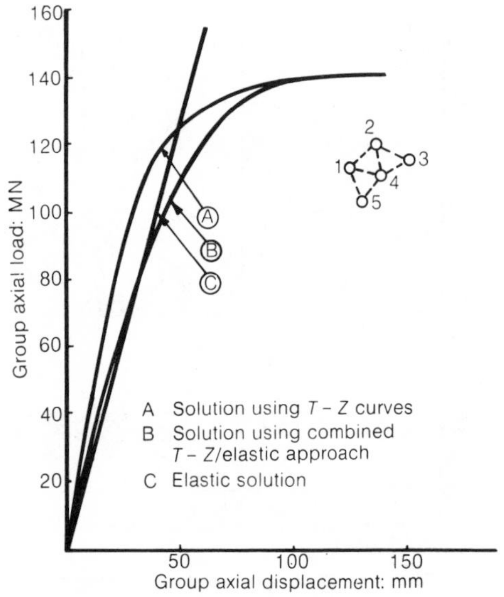

Fig. 8. Comparison between computed axial load–displacement curves; pile diameter 1.37 m, wall thickness 0.05 m, penetration 40.5 m

42. Poulos[22] details a solution for group settlement analysis which accounts for stratification beneath the pile tip. Basically it is as follows:

(a) impose rigid layer conditions at a convenient interface between strata beneath pile group;
(b) carry out single pile analysis accounting for effects of rigid layer;
(c) compute interaction factors for pile group accounting for rigid layer;
(d) carry out group analysis;
(e) determine dimensions of an equivalent pier to replace pile group;
(f) using equivalent pier, compute settlements in soil strata beneath the rigid layer and hence, using the results from (d), compute total group settlement.

43. A variation of this method has been adopted for the analysis of the pile group shown in Fig. 8. Each pile in this group is identical to one of the single piles analysed previously, the results of the analysis being shown in Fig. 4(c). The soil conditions are shown in Fig. 3(b). For computational purposes the rigid layer was imposed at a depth of 48.2 m below sea bed. The effects of the rigid layer on the single pile load–deflexion behaviour have been calculated using the load distribution obtained from the T–Z analysis for the floating single pile. This load distribution has been idealized as a series of vertical point loads acting along the length of the pile at the levels of the ground springs, with load intensities equal to the spring loads. Mindlin's solution[18] for a vertical point load in a semi-infinite elastic half-space can be used to predict the settlement, due to the stress regime set up in the soil, at the pile tip and at the rigid layer itself. The settlement at the pile tip is the difference between these two values plus the compression of the bottom ground spring. For calculating the magnitude of the interaction factors the recommendations of Poulos and Davis[21] are followed. The procedures described previously for utilizing equation (9) can then be applied to determine the load–deflexion curve of a pile group underlain by a rigid layer. To this curve must be added the settlement of the strata beneath the level of the rigid layer due to the equivalent pier.

44. In addition to the load–deflexion curve computed by the above method, the equivalent curve computed by the T–Z approach and Poulos's elastic method are plotted in Fig. 8. For Poulos's method the same elastic soil modulus has been used as in the Authors' method. For the T–Z approach the elastic soil modulus for calculating interaction effects has been backfigured from the single pile load–deflexion curve predicted by the T–Z method (Fig. 4(c)). At the design group axial load of approximately 95 MN the Authors' method predicts a group deflexion of 42 mm, the T–Z method provides a deflexion of 27 mm, and Poulos's method predicts 39 mm. Unfortunately there are, to the Authors' knowledge, no field tests in which the immediate settlement has been measured of pile groups representative of those used offshore. It is therefore impossible to state that the Authors' method is correct and that the T–Z method is providing too stiff a response. However the evidence of the single pile load test presented in this Paper indicates that the T–Z approach underestimates immediate settlements of large diameter piles. It would therefore

appear to be incorrect to backfigure a soil modulus from a load–settlement curve generated by the T–Z method for a large diameter pile.

CONCLUSION

45. The principal conclusions resulting from the work presented in this Paper are as follows.

46. The T–Z method for predicting the axial load–deflexion behaviour of single piles provides realistic results for piles with diameters of less than approximately 1.0 m. For piles with diameters in the order of 1.5 m or more, the T–Z method will seriously underestimate deflexions.

47. The combined T–Z/elastic method presented in this Paper is theoretically valid for all pile diameters. Very good agreement was obtained between the results of a field test on a large diameter pile in overconsolidated clay and the load–deflexion curve predicted by the method.

48. The selection of the values of the elastic moduli must be made with great care with due account being taken of the direction of loading and the stress levels in the surrounding soil. Even for piles in a homogeneous isotropic soil, the appropriate moduli may be different for axial and lateral loading.

49. The way in which the single pile elasto-plastic load–deflexion curves are used in the Authors' pile group programs allows all modes—in particular, the important coupled mode of moment–rotation—to be investigated effectively, quickly and relatively cheaply.

ACKNOWLEDGEMENTS

50. The Authors express their gratitude to Fugro Ltd for permission to publish this Paper. The analysis for determining the load–deflexion behaviour of single piles and pile groups was initiated in response to the requirements of CJB/Earl and Wright.[23] CJB/Earl and Wright were the structural designers of the foundations of a number of jackets for which Fugro were the geotechnical consultants.

REFERENCES

1. SMITH I. M. Installation and performance of piled foundations. Numerical Methods in Geomechanics Conference, Aachen, 1979.
2. MATLOCK H. Correlations for design of laterally loaded piles in soft clay. Proc. 2nd Annual Offshore Technology Conf., Houston, 1970, paper OTC–1204.
3. REESE L. C. et al. Field testing and analysis of laterally loaded piles in stiff clay. Proc. 6th Annual Offshore Technology Conf., Houston, 1975, paper OTC– 2312.
4. REESE L. C. et al. Analysis of laterally loaded piles in sand Proc. 5th Annual Offshore Technology Conf., Houston, 1974, paper OTC–2080.
5. POULOS H. G. The behaviour of laterally loaded piles: I–Single piles. J. Soil Mech. Fdns Div. Am. Soc. Civ. Engrs, 1971, 97, SM5, 711–731.
6. ROWE P. W. The single pile subject to horizontal force. Gēotechnique, 1956, 6, 70–85.
7. BROMS B. B. The lateral resistance of piles in cohesive soils. J. Soil Mech. Fdns Div. Am. Soc. Civ. Engrs, 1964, 90, Mar., SM2, Part I, 27–63.
8. DELFT HYDRAULICS LABORATORY. Scour around circular columns. Delft Hydraulics laboratory, 1965, report M848 (in Dutch).
9. COLEMAN N. L. Analysing laboratory measurements of scour at cylindrical piers in sand beds. International Association for Hydraulics Research XIV Congress, Paris, 1971, Vol. 3.
10. CARSTENS T. and SHARMA H. R. Local scour around large obstructions. International Association for Hydraulic Research XVI Congress, Sao Paulo, 1975.
11. VIJAYVERGIYA V. N. Load–movement characteristics of piles. Ports '77 Conf., Long Beach, California, 1977.
12. POULOS H. G. and MATTES N. S. The behaviour of axially loaded end-bearing piles. Gēotechnique, 1969, 19, No. 2, 285–300.
13. BUTLER F. G. General report and state-of-the-art. Review Session III. Heavily over-consolidated clays. Conference on the settlement of structures, Cambridge, 1974.
14. POULOS H. G. Load–settlement prediction for piles and piers. J. Soil Mech. Fdns Div. Am. Soc. Civ. Engrs, 1972, 98, SM9, 879–897.
15. RANDOLPH M. F. and WROTH C. P. A fundamental approach to predicting the deformation of vertically loaded piles. University of Cambridge Department of Engineering, 1977, report CUED/C-Soils TR 38.
16. SEED H. B. and IDRISS I. M. Soil moduli and damping factors for dynamic response analyses. Earthquake Engineering Research Centre, University of California, Berkeley, 1970, report EERC 70–10.
17. COYLE H. M. and REESE L. C. Load transfer for axially loaded piles in clay. J. Soil Mech. Fdns Div. Am. Soc. Civ. Engineers, 1966, 92, SM2, 1–26.
18. MINDLIN R. D. Force at a point in the interior of a semi-infinite solid. J. Appl. Phys., 1936, 7, No. 5, 195–202.
19. JELINEK R. et al. Load tests on 5 large diameter bored piles in clay. Proc. IX Int. Conf. Soil Mech., Tokyo, 1977.
20. FOCHT J. and KOCH K. Rational analysis of the lateral performance of pile groups. Proc. 4th Annual Offshore Technology Conf., Houston, 1973.
21. POULOS H. G. and DAVIS E. H. Elastic solutions for soil and rock mechanics. John Wiley and Sons, 1974, 279–281 and 291–295.
22. POULOS H. G. Estimation of pile group settlements. Ground Engng, 1977, 10, Mar. No. 2, 40–50.
23. LEWIS G. H. G. Mathematical modelling of pile group behaviour. CJB/Earl and Wright, London, 1978.

19. Effect of radial variation in modulus on stresses after consolidation around a driven pile

S. A. LEIFER, R. C. KIRBY and M. I. ESRIG (Woodward-Clyde Consultants, Clifton, New Jersey)

A closed form elastic solution for radial consolidation around a driven pile has been developed by Randolph and Wroth. In their analysis, constant material properties (Young's modulus and Poisson's ratio) were used throughout the soil mass. As expected, the changes in effective stresses due to reconsolidation (dissipation of excess pore pressures generated by pile driving) were found to be highest at the pile/soil interface and to decrease markedly with radial distance from the pile. However, substantial variations in material properties with radial distance from the pile/soil interface should be expected, because of soil remoulding due to pile driving and because the equivalent elastic modulus of the soil is related to the level of stress applied. An extension of the Randolph–Wroth solution to allow variation in material properties with radial offset is described in this Paper. The stress changes that occur during reconsolidation have been investigated for several assumed distributions of Young's modulus with radial offset and a range in values of Poisson's ratio. An evaluation of the applicability of the results of the parametric studies is presented.

INTRODUCTION

Effective stress approaches to the prediction of the axial capacity of driven piles in clay have received increased attention during the last several years.[1–3] A rational prediction of pile capacity using an effective stress approach requires determination of initial in situ effective stresses, followed by analysis of the changes in effective stress during pile installation, reconsolidation, and pile loading.

2. Determination of in situ effective stresses always involves some uncertainty. It is substantially more difficult for offshore soils than for onshore soils because of problems involved in sampling, measurement of pore water pressures in situ, and the existence of gas pressure in some marine sediments. Methods for estimating in situ stresses are described in considerable detail by Ladd et al.,[4] and the consequence of the existence of gas is discussed by Esrig and Kirby.[5] A brief discussion of the changes in stress due to pile loading is presented by Kirby and Esrig.[6]

3. Pore pressures generated during pile installation are relevant to this Paper in that they serve as the initial conditions for the consolidation problem. Kirby and Esrig[6] and Wroth et al.[7] discuss changes due to pile installation and conclude that expansion of a cylindrical cavity in an ideal elastic–plastic material, under plane strain conditions, provides a reasonable model for pile installation.

4. Expansion of a cavity increases both total stress and pore water pressure. As the pore pressures dissipate, and water flows radially away from the pile, volume decreases occur in the soil adjacent to the cavity (i.e., the pile) and are accompanied by a reduction in total stress. In order to predict the axial capacity of a driven pile, it is necessary to know how much of the total stress increase caused by pile installation remains 'locked in' as effective stress when consolidation is completed. In contrast to the Terzaghi formulation of the consolidation problem, in which the total stress remains constant, analysis of reconsolidation around an expanded cavity requires consideration of the changes in both total stress and pore water pressure during consolidation.

5. A solution to this problem using constant values of Young's modulus, E, and Poisson's ratio, ν, throughout the soil mass is presented by Randolph and Wroth.[8] However, because the soil compressibility is related to the level of stress and to the remoulding of the soil during pile driving, substantial variations in E with radial distance are expected and, in fact, are suggested as occurring by Cooke and Price.[9] Extension of the Randolph–Wroth solution to consider variations in Young's modulus and Poisson's ratio with radial distance is described herein. The analysis presented below considers only the changes in effective stress that occur at the completion of consolidation. Discussion of the variation in stress and pore pressure with time during consolidation is beyond the scope of this Paper.

MATHEMATICAL DEVELOPMENT

6. The installation of a driven pile is modelled as the expansion of a cylindrical cavity under plane strain conditions in a two-phase, elastic–plastic medium which remains at constant volume. Cavity expansion produces two distinct zones within the soil mass: a plastic zone surrounding the cavity in which the soil has experienced, during

cavity expansion, strains large enough to cause yield; and an elastic zone surrounding the plastic zone in which the soil has experienced, during cavity expansion, strains less than that required to cause yield. These two zones are shown schematically in Fig. 1. The radial distance to the elastic/plastic boundary, R, can be estimated from the equation given in the figure caption.

7. Figure 2 shows the predicted variation with radial offset of excess pore pressure caused by pile installation (i.e., cavity expansion). Within the plastic zone the pore pressure distribution is log-linear, starting from zero at the elastic/plastic boundary, and increasing as the pile face is approached. The slope of the pore pressure distribution line is equal to $-2S_u$. Within the elastic zone, the pore pressure is predicted to be zero.

8. For analysing the effective stress changes due to reconsolidation, it is assumed that the initial pore pressures are as predicted by the analysis of the expansion of a cylindrical cavity in an elastic–plastic material but that the soil behaves as a linear elastic material during reconsolidation. The plastic and elastic zones of the cavity expansion analysis are called the inner and outer zones respectively,

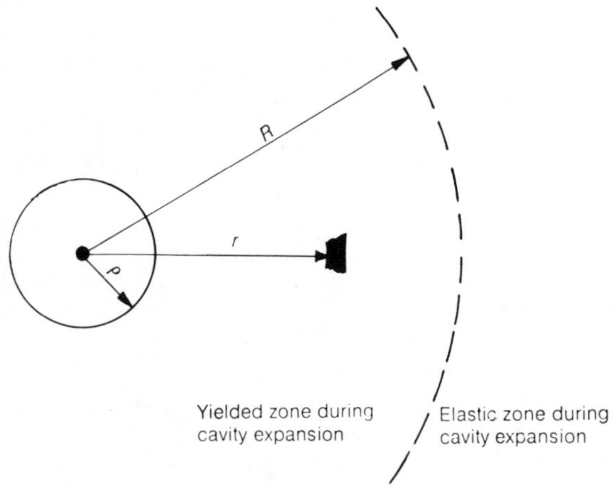

Fig. 1. Definition of problem geometry; $R/\rho = (E_u/3S_u)^{1/2}$, where S_u is undrained shear strength

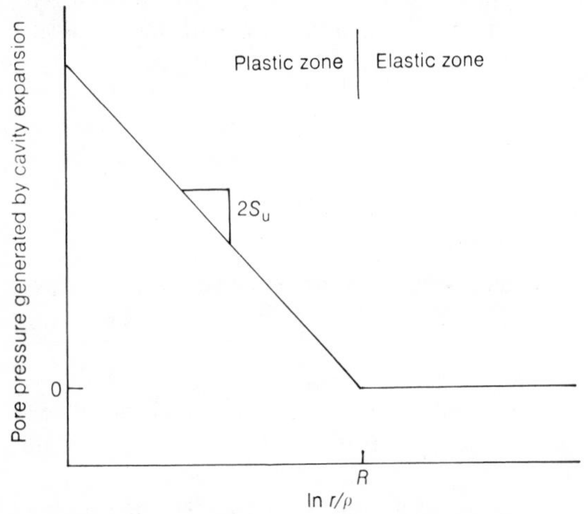

Fig. 2. Initial excess pore pressure distribution based on expanding cavity analysis

for the analysis of stress changes due to reconsolidation, in order to avoid confusion about the assumed constitutive relationships.

9. Assuming the soil behaves as a linear elastic material during consolidation and that zero strain occurs in the vertical direction, the following relationships can be derived for the change in effective stress during consolidation:[8]

$$-\Delta\bar{\sigma}_r = E^* \left\{ (1-\nu)\frac{d\xi}{dr} + \nu\frac{\xi}{r} \right\} \tag{1}$$

$$-\Delta\bar{\sigma}_\theta = E^* \left\{ (1-\nu)\frac{\xi}{r} + \nu\frac{d\xi}{dr} \right\} \tag{2}$$

where

$$E^* = \frac{E}{(1+\nu)(1-2\nu)}$$

E is Young's modulus

ν is Poisson's ratio

ξ is outward radial displacement

$\Delta\bar{\sigma}_r$ and $\Delta\bar{\sigma}_\theta$ are changes in radial and circumferential normal effective stresses.

10. By combining equations (1) and (2) with the equations of equilibrium for the inner and outer zones, the differential equations obtained are, for the inner zone,

$$r\frac{d^2\xi}{dr^2} + \frac{d\xi}{dr} - \frac{\xi}{r} = \frac{2S_u}{E^*(1-\nu)} \tag{3}$$

and for the outer zone

$$r\frac{d^2\xi}{dr^2} + \frac{d\xi}{dr} - \frac{\xi}{r} = 0 \tag{4}$$

The right hand term in equation (3) is associated with the initial excess pore pressure in the inner zone.

11. Integrating equations (3) and (4) gives, for the inner zone,

$$r\xi = \left\{ \frac{r^2}{2}\ln r - \frac{r^2}{4} \right\} A^* + \frac{r^2}{2} B_i + C_i \tag{5}$$

and for the outer zone

$$r\xi = \frac{r^2}{2} - B_o + C_o \tag{6}$$

where

$$A^* = \frac{2S_u}{E^*(1-\nu)}$$

and B_i, C_i, B_o and C_o are constants of integration.

12. The boundary conditions for evaluation of the constants of integration are as follows:

(a) at $r = \rho$, $\xi = 0$ (ρ = pile radius)
(b) at $r = \infty$, $\xi = 0$
(c) at $r = R$, $(\Delta\bar{\sigma}_r)_{\text{inner zone}} = (\Delta\bar{\sigma}_r)_{\text{outer zone}}$
(d) at $r = R$, $\xi_{\text{inner zone}} = \xi_{\text{outer zone}}$.

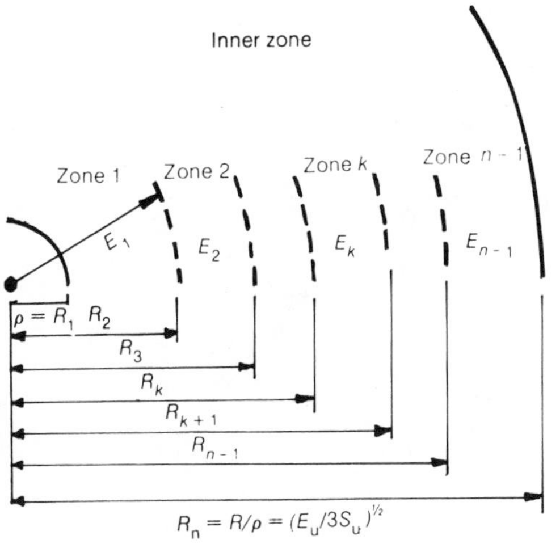

At r = ρ: $\xi = 0$;

$$\frac{\rho^2}{2}B_1 + C_1 = -\left\{\frac{\rho^2}{2}\ln\rho - \frac{\rho^2}{4}\right\}A_1^*$$

Each material boundary within the inner zone

At $r = R$: $\xi_{k-1} = \xi_k$ and $(\Delta\bar{\sigma}_r)_{k-1} = (\Delta\bar{\sigma}_r)_k$;

$$\frac{R_k^2}{2}B_{k-1} + C_{k-1} - \frac{R_k^2}{2}B_k - C_k = \left\{\frac{R_k^2}{2}\ln R_k - \frac{R_k^2}{4}\right\}\left\{A_k^* - A_{k-1}^*\right\}$$

$$\frac{E_{k-1}^*}{2}B_{k-1} + \frac{E_{k-1}^*}{R_k^2}(2\nu - 1)C_{k-1} - \frac{E_k^*}{2}B_k - \frac{E_k^*}{R_k^2}(2\nu - 1)C_k = 0$$

Boundary between the inner and the outer zone

At r = R/ρ: $\xi_{n-1} = \xi_n$ and $(\Delta\sigma_r)_{n-1} = (\Delta\sigma_r)_n$;

$$\frac{1}{2}B_{n-1} - \frac{F}{R_n^2}C_{n-1} = -\left\{\frac{A_{n-1}^*}{4}F + \frac{A_{n-1}^*}{2}\ln R_n\right\}$$

$$\frac{R_n^2}{2}B_{n-1} + C_{n-1} - C_n = -\frac{A_{n-1}^*}{2}\left\{R_n^2\ln R_n - \frac{R_n^2}{2}\right\}$$

where

$$F = \frac{(1 - 2\nu)(E_{n-1}^* - E_n^*)}{E_{n-1}^* + (1 - 2\nu)E_n^*}$$

Fig. 3. Notation and governing equations for multi-material solution

13. The constants of integration in equations (5) and (6) can be determined for these boundary conditions and the displacement function, ξ, evaluated. Randolph and Wroth[8] solve equations (5) and (6) using constant elastic material properties, E and ν. However, in equating stresses and displacements at the boundary between the inner and outer zones it is not necessary to assign the same material properties to both zones. In fact, the inner zone may be subdivided into a number of subzones with different material properties, and boundary conditions (c) and (d) evaluated at each boundary between subzones. The governing equations at each boundary for this analysis are shown in Fig. 3. These equations can be written as a set of simultaneous linear equations with the constants of integration, B and C, for each material as unknowns.

14. A computer program has been developed to solve the governing equations contained in Fig. 3 with as many as 50 different materials in the inner zone. Values of E and ν can be assigned independently to each material. Having solved for the constants of integration, the program calculates the stress changes, displacements and strains as functions of radial offset.

PARAMETRIC STUDIES

15. Parametric studies have been developed to evaluate the sensitivity of the predicted stress changes due to reconsolidation to variations of Young's modulus with radial offset. In these studies, it is assumed the Poisson's ratio does not vary with radial offset. However, the influence of Poisson's ratio on predicted stress changes is considered. Also considered is the influence of the radial extent of the initial excess pore pressure (defined by R/ρ) on predicted stress changes.

16. Figure 4 shows the distributions of Young's

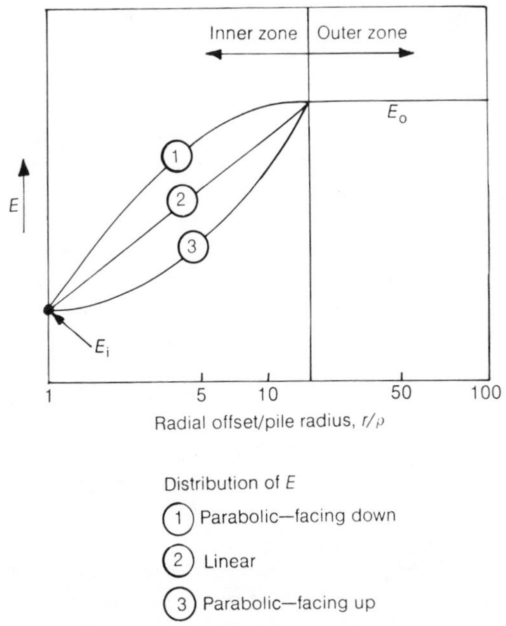

Fig. 4. Distributions of E used in the parametric study: inner zone corresponds to yielded zone during cavity expansion, outer zone corresponds to elastic zone during cavity expansion

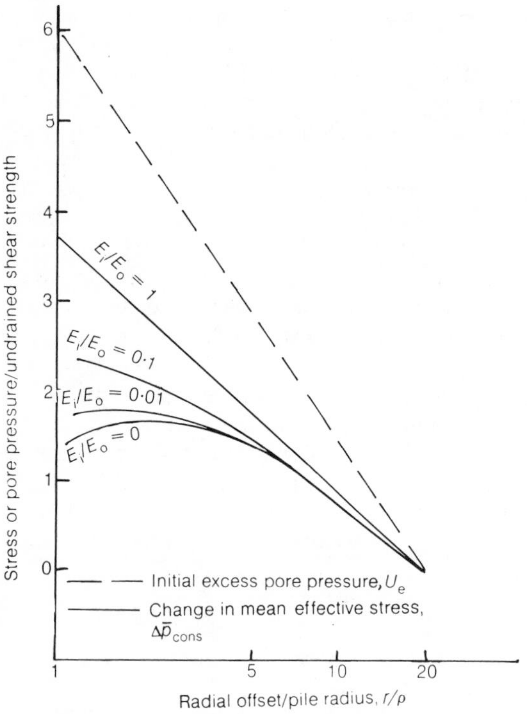

Fig. 5. Influence of magnitude of modulus on change in mean normal effective stress due to reconsolidation after pile driving; log-linear variation of E in the inner zone, $R/\rho = 20$, $\nu = 0.3$, $E_o = E$ in the outer zone, $E_i = E$ at the pile surface

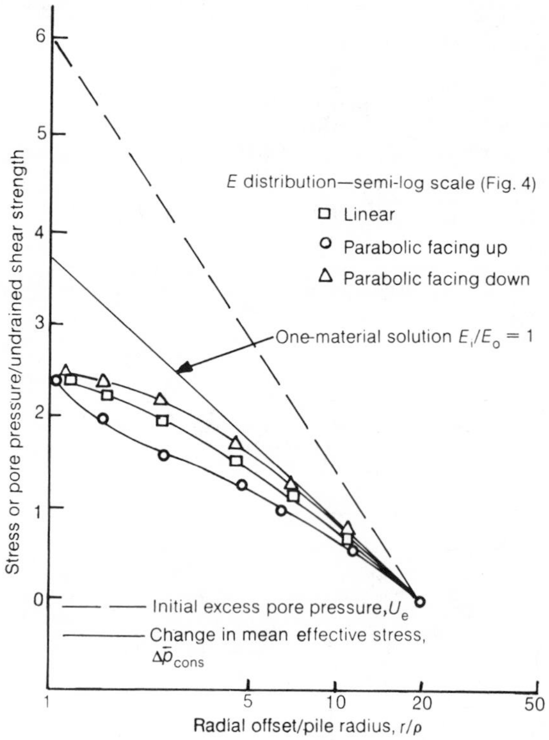

Fig. 6. Influence of distribution of modulus on change in mean normal effective stress due to reconsolidation after pile driving; $R/\rho = 20$, $E_i/E_o = 0.1$, $\nu = 0.3$

modulus used in the parametric studies. The modulus in the outer zone is assumed constant and called E_o. The modulus at the pile/soil interface is called E_i. For the single-material problem, $E_i/E_o = 1$.

17. Analyses of stress changes due to reconsolidation for $R/\rho = 20$, a log-linear distribution of E in the inner zone (Fig. 4), $\nu = 0.3$, and a range of values of E_i/E_o are summarized in Fig. 5. The change in mean normal effective stress at the pile/soil interface (i.e., $r/\rho = 1$) is seen to be about $3.7S_u$ for the one-material solution ($E_i/E_o = 1$). The initial excess pore pressure at the pile/soil interface is $6S_u$. Consequently, the one-material solution suggests that approximately 0.62 of the initial excess pore pressure is locked in as an increase in effective stress during reconsolidation. For values of E_i/E_o of 0.1 and 0, the values of locked in stress (i.e., $\Delta\bar{p}_{cons}/U_e$) at the pile/soil interface are 0.42 and 0.25, respectively. Analyses for values of R/ρ of 20, 10 and 5 suggest that $\Delta\bar{p}_{cons}/U_e$ is independent of R/ρ.

18. Analyses were also made for the three modulus distributions shown in Fig. 4; R/ρ values of 20, 10 and 5; $\nu = 0.3$; and E_i/E_o ranging from 1 to 0.01. The analyses for $R/\rho = 20$ are typical of all values. Results for E_i/E_o values of 0.1 and 0.01 are shown in Figs 6 and 7 respectively. For the modulus distributions and the range of E_i/E_o considered, it appears that the detail of the shape of the modulus distribution does not have a major effect on the mean normal effective stress at the pile/soil interface, except when the modulus ratio approaches zero.

19. The results of the parametric studies are summarized in parts (a)–(c) of Fig. 8. The range in assumed shape of the modulus variation is shown in part (d). The

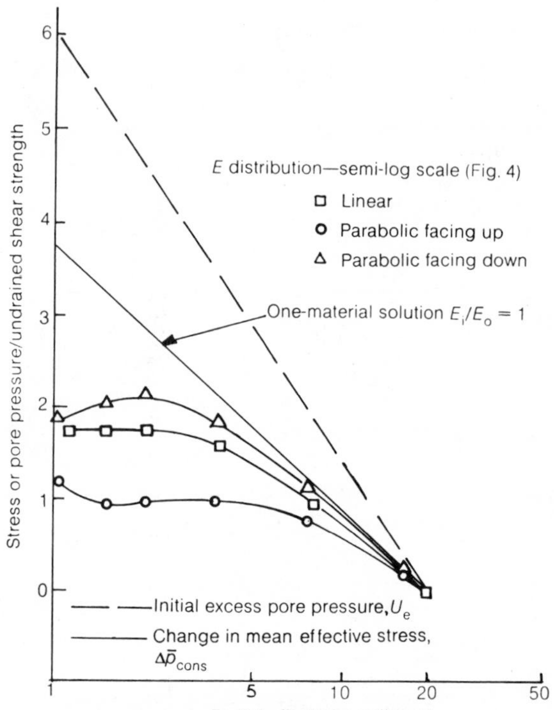

Fig. 7. Influence of distribution of modulus on change in mean normal effective stress due to reconsolidation after pile driving; $R/\rho = 20$, $E_i/E_o = 0.01$, $\nu = 0.3$

ratio of increase in mean normal effective stress to initial excess pore pressure, at the pile/soil interface, is virtually independent of R/ρ and is relatively insensitive to the shape

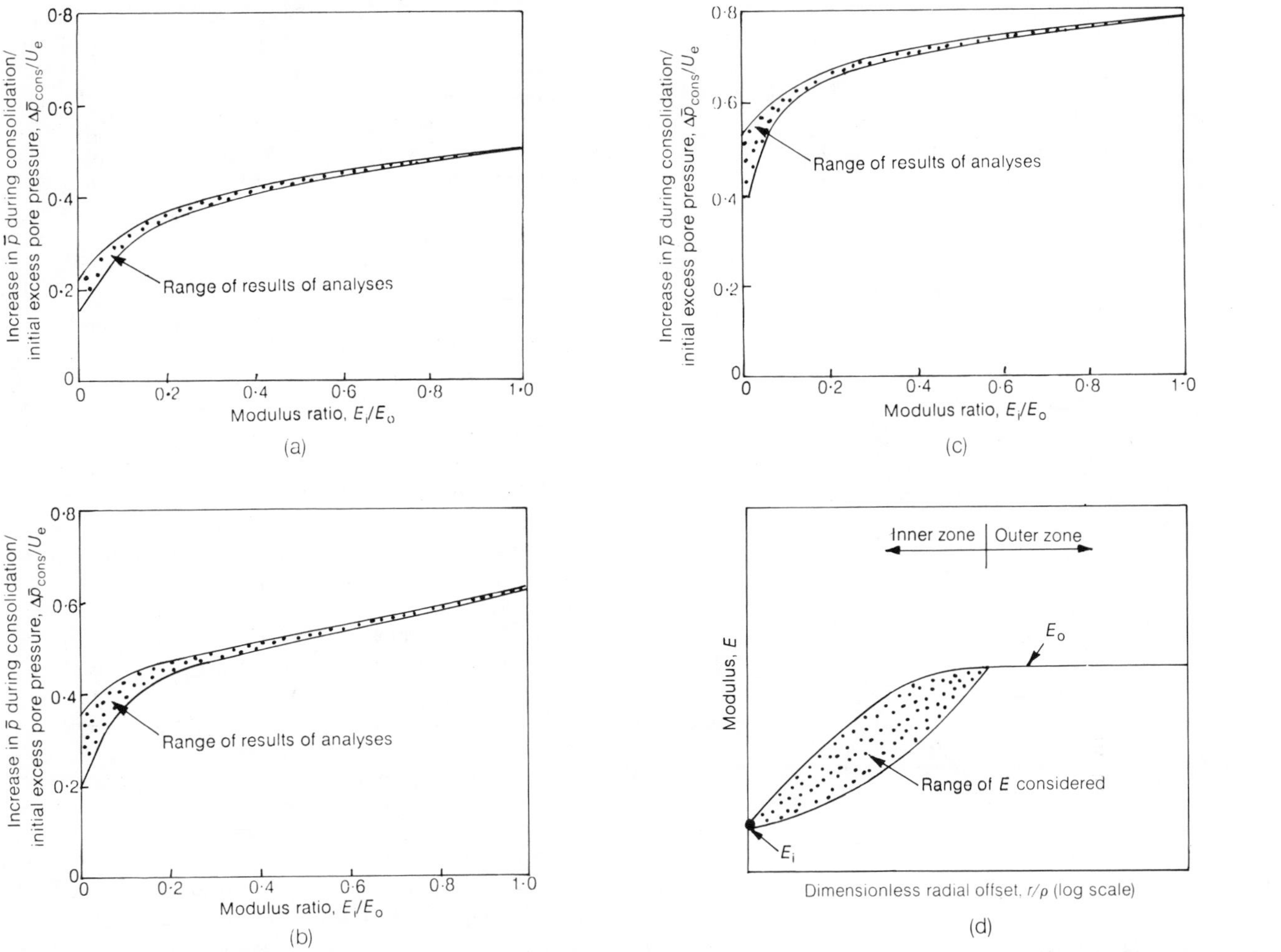

Fig. 8. Predicted increase in mean normal effective stress at pile/soil interface for $R/\rho = 5$, 10 and 20: (a) $\nu = 0.2$; (b) $\nu = 0.3$; (c) $\nu = 0.4$; (d) range of E considered

of the distribution of modulus for E_i/E_o values of 0.1–1.0. For modulus ratios less than about 0.1, the stress change predicted at the interface is found to be sensitive to the shape of the modulus distribution.

20. For $E_i/E_o = 0.1$, the ratio $\Delta\bar{p}_{cons}/U_e$ at the pile/soil interface takes values of approximately 0.3, 0.4 and 0.6 for Poisson's ratios of 0.2, 0.3 and 0.4, respectively. These values of $\Delta\bar{p}_{cons}/U_e$ can be compared with the corresponding values for the one-material solution of 0.51, 0.63 and 0.77. Thus the predictions are seen to be quite sensitive to Poisson's ratio, which is a difficult parameter to evaluate.

APPLICATION OF ELASTIC SOLUTION FOR CHANGE IN STRESS DUE TO RECONSOLIDATION AFTER PILE DRIVING

21. Estimates of an appropriate E_i/E_o ratio and an appropriate Poisson's ratio are necessary to apply the results of the parametric study described previously. Consideration of the stress changes taking place at various locations within the inner and outer zones during pile installation and reconsolidation provides insight toward choosing a value of E_i/E_o.

22. Two typical soil elements are shown schematically in Fig. 9 (part (a)). Element A is adjacent to the pile or expanded cavity; element B is at the boundary between the zone that has yielded during cavity expansion (inner zone) and the zone that has remained elastic during cavity expansion. The effective stresses acting on these elements are represented in e–$\bar{p}$ space (where e is void ratio and $\bar{p}$ is mean effective stress) in part (b) of the figure.

23. Element A, adjacent to the pile surface, has been severely remoulded and stressed to the critical state. Kirby and Wroth[10] discuss the application of critical state soil mechanics to driven piles in clay. The elastic analyses of Randolph and Wroth,[8] as well as physical reasoning suggested by Kirby and Wroth, suggest that the remoulded soil adjacent to the pile consolidates along the critical state line (a line parallel to the virgin compression line in e–log $\bar{p}$ space) and exhibits a compression ratio similar to that of virgin compression. However, this is a controversial issue because consolidation analyses based on a plastic model for soil behaviour[7,11] suggest that the soil elements adjacent to the pile do not consolidate along the critical state line. Nevertheless, it is assumed here that soil element A consolidates along the critical state line.

24. The elastic modulus, E, for virgin compression can be approximated from the compression index, C_c, provided that an estimate of Poisson's ratio is known. A value of Poisson's ratio, ν, of 0.3 is consistent with values of ν backcalculated from a one-dimensional consolidation test, assuming a lateral earth pressure coefficient of 0.5. For

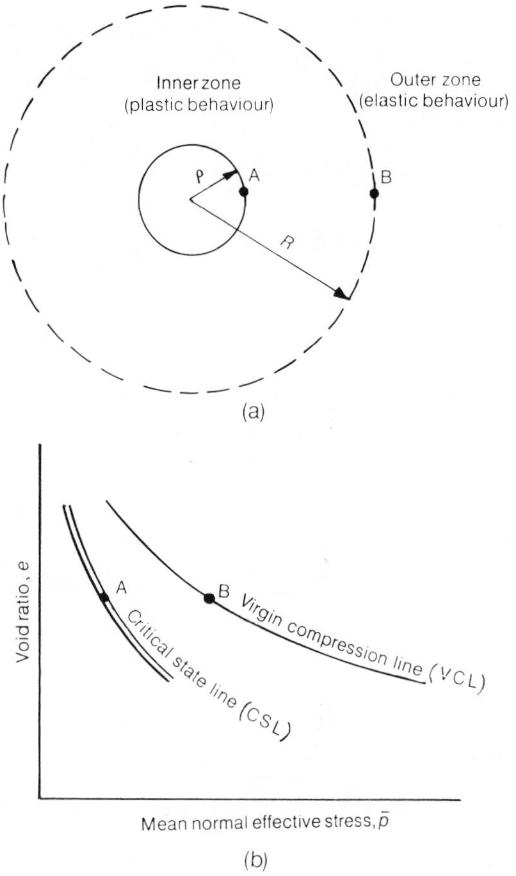

Fig. 9. Location of typical soil elements in e–$\bar{p}$ space after cavity expansion

$\nu = 0.3$, and for virgin compression, a preliminary estimate of E is given by

$$\frac{E}{\bar{p}_o} \approx \frac{2.6}{C_c/(1+e_o)}$$

Typical values of $C_c/(1+e_o)$ for normally consolidated clays range from 0.1 to 0.3. Consequently $E/\bar{p}_o$ ranges from about 9 to 26.

25. Element B, at the elastic/plastic boundary, has been stressed by pile installation, but still behaves elastically. For an elastic material, shear at constant volume requires that no change in mean normal effective stress occur. Furthermore, the consolidation solution indicates that during reconsolidation no volume change takes place in the outer zone. Consequently one would expect the modulus during reconsolidation in the outer zone to be related to the Young's modulus of the soil, appropriate for undrained loading, E_u.

26. D'Appolonia et al.[12] present values of E_u/S_u for normally consolidated and lightly overconsolidated clays ranging from 300 to 1500, with the lower values associated with a higher plasticity index. Estimating $S_u/\bar{p}_o$ as 0.33, $E_u/\bar{p}_o$ ranges from approximately 100 to 500. Converting the values of E_u to values of Young's modulus in terms of effective stress, assuming $\nu = 0.3$, yields values of $E/\bar{p}_o$ for shear at constant volume ranging from approximately 90 to 425.

27. Thus, it appears that typical values of E_i/E_o might range from 0.06 (i.e., 26/425) to 0.10 (i.e., 9/90). If it is assumed that the ratio of E_i/E_o is approximately equal to the ratio of recompression index to virgin compression index (C_r/C_c), values of about 0.1–0.3 are appropriate. A value of E_i/E_o of 0.1–0.2 is probably a reasonable approximation to reality.

28. Estimation of a reasonable value of Poisson's ratio is more difficult, probably because soil does not behave as an elastic material and values of Poisson's ratio backcalculated from laboratory tests vary with the type of test. Nevertheless, a range in values of Poisson's ratio of 0.2–0.4 probably covers most situations.

29. For E_i/E_o in the range 0.1–0.2 and Poisson's ratio 0.2–0.4, the predicted range in values of $\Delta\bar{p}_{cons}/U_e$ (read directly from Fig. 8) is 0.3–0.65. The major source of the uncertainty in these predicted values of $\Delta\bar{p}_{cons}/U_e$ is the range in values of Poisson's ratio.

CONCLUSIONS

30. An elastic solution to the problem of radial consolidation around a driven pile, which considers variation in material properties with radial offset, has been developed. The variation in material properties with radial offset has been accommodated by dividing the area around the pile into a number of zones, each zone having different material properties. A computer program capable of evaluating the multi-material solution for as many as 50 zones (or materials) has been developed.

31. The analysis of stress changes during reconsolidation considering variation in elastic modulus with radial offset indicates stress changes at the pile/soil interface which are substantially less than those obtained with the one-material elastic solution developed by Randolph and Wroth.[8]

32. Parametric studies indicate that the ratio of the increase in mean normal effective stress at the pile/soil interface to the increase in mean normal total stress during cavity expansion is virtually independent of the size of the yielded zone around the pile, expressed as the ratio R/ρ, and is relatively insensitive to the shape of the variation of Young's modulus with radial offset.

33. Considering the properties of the remoulded soil at the pile/soil interface and the elastic behaviour of the soil in the elastic zone, estimates of expected variation of Young's modulus are developed. Based on the estimated modulus values and assumed values of Poisson's ratio, the parametric studies indicate that the predicted change in mean normal effective stress at the pile/soil interface during reconsolidation might vary from 30% to 65% of the initial excess pore pressure. The range in predicted values is shown to be quite sensitive to the assumed value of Poisson's ratio.

ACKNOWLEDGEMENTS

34. The general effective stress analysis for axial capacity of driven piles in clay was developed under a research contract between Woodward–Clyde Consultants, Amoco Production Company and Shell Oil Company. The results of that initial study became part of a larger research study that was organized by Amoco and funded by eleven oil companies. The active participation and enthusiasm of Benton S. Murphy of Amoco Production Company, who served as Project Administrator, contributed significantly to the success of the work. Without Mr Murphy and Mr Robert G. Bea, formerly of Shell Oil Company, this study would not have been possible.

REFERENCES

1. BURLAND J. B. Shaft friction of piles in clay – a simple fundamental approach. Ground Engng, 1973, 6, 30–42.
2. ESRIG M. I. and KIRBY R. C. Soil capacity for supporting deep foundation members in clay. Symposium on Behavior of Deep Foundations, Boston Mass., 1978. American Society for Testing and Materials.
3. CHANDLER R. J. The shaft friction of piles in cohesive soils in terms of effective stress. Civ. Engng Pub. Wks Rev., 1968, 63, 196–228.
4. LADD C. C. et al. State of the art report: stress deformation and strength characteristics. Proc. 9th Int. Conf. Soil Mech. Tokyo, 1977, vol. 2, 421–494.
5. ESRIG M. I. and KIRBY R. C. Implications of gas content for predicting the stability of submarine slopes. Marine Geotech. 1977, 2, 81–100.
6. KIRBY R. C. and ESRIG M. I. Further development of a general effective stress method for prediction of axial capacity for driven piles in clay. Recent developments in the design and construction of piles. Institution of Civil Engineers, London, 1979.
7. WROTH C. P. et al. Stress changes around a pile driven into cohesive soil. Recent developments in the design and construction of piles. Institution of Civil Engineers, London, 1979.
8. RANDOLPH M. F. and WROTH C. P. An analytical solution for the consolidation around a driven pile. Cambridge University, 1978, research report CUED/c-SOILS/TR50.
9. COOKE R. W. and PRICE G. Strains and displacements around friction piles. Proc. 8th Int. Conf. Soil Mech., Moscow, 1973, vol. 2.1, 53–60.
10. KIRBY R. C., and WROTH C. P. Application of critical state soil mechanics to the prediction of axial capacity for driven piles in clay. Proc. 1977 Offshore Technology Conf., Houston, Texas, 483–494, OTC paper 2942.
11. MILLER T. W. et al. Critical state soil mechanics model of soil consolidation stresses around a driven pile. Proc. 1978 Offshore Technology Conf., Houston, Texas, 2237–2242, OTC paper 3307.
12. D'APPOLONIA D. J. et al. Initial settlement of structures on clay. J. Soil Mech. Fdns Div. Am. Soc. Civ. Engrs, 1971, 97, SM10, 1359–1377.

20. Some aspects of the performance of open- and closed-ended piles

J. P. CARTER, BE, PhD (Research Assistant), M. F. RANDOLPH, BA, PhD (Assistant Lecturer) and C. P. WROTH, MA, PhD (Reader; Engineering Department, University of Cambridge)

The shaft capacity of a driven pile will depend not only on the in situ stresses and strength of a soil deposit, but also on the stress changes which occur due to installation of the pile. For a given soil, the stress changes will depend on the type of pile which is driven. In particular, for tubular steel piles, there are likely to be differences in the long term shaft capacity of open- and closed-ended piles. The Paper describes a theoretical study of the stress and strength changes in clay due to the installation of driven tubular piles with open and closed ends. The clay has been idealized as a work hardening elastoplastic material. Pile installation has been modelled as the expansion of a long cylindrical cavity. Numerical results are presented showing how the stress and strength changes in the soil are affected by the consolidation history of the soil prior to pile driving, and by the amount of soil displaced by the pile. It is shown that the final shaft capacity of a closed-ended pile driven into a particular clay deposit may be of the order of 25% greater than that for a similar but open-ended pile.

INTRODUCTION

The construction of offshore oil platforms has necessitated the use of increasingly large friction piles for the foundations, in order to withstand the large forces due to wave loading. The piles, fabricated out of steel tubular sections, are typically 1–2 m in diameter and are often embedded to depths greater than 50 m. Such piles are many times larger than those normally used on land and special high energy hammers have had to be designed to install the piles. It is current practice to drive the piles as open-ended or 'cookie-cutter' piles on the assumption that these are much easier to drive than closed-ended piles.

2. The argument in favour of the open-ended (partial displacement) pile is that the pile has to displace much less soil than a closed-ended (full displacement) pile, a plug of soil being free to move inside the pile. It is possible that this argument may be justified for piles driven into sand where the end bearing resistance of a closed-ended pile is likely to be very high. However, for piles driven into clay, the end bearing resistance is a much smaller proportion of the total resistance to driving. In addition, the presence of a plug of soil inside the open-ended pile may lead to additional damping of the driving pulse and thus a less efficient penetration. Recent evidence suggests that in clay there is relatively little difference in the driving records of open- and closed-ended piles driven on the same site,[1] with, if anything, the closed-ended piles being slightly easier to drive.

3. Besides comparing the driving resistance of open- and closed-ended piles, it is clearly necessary to consider the effect of the type of pile on the long term static capacity. The greater disturbance caused by the driving of a full displacement pile is an *advantage* in terms of final static capacity. The greater disturbance results in the generation of higher excess pore pressures in the soil near the pile; ultimately, after consolidation, this will lead to higher effective stresses, a stronger soil immediately around the pile and a greater shaft capacity. This advantage is likely to outweigh any potential disadvantage which may be associated with a closed-ended pile due to a greater driving effort being required.

4. Recently, analytical techniques have been developed, based on effective stress concepts, which allow estimates to be made of the stress changes in clay soils due to pile driving (modelled as the creation of a cylindrical cavity) and subsequent consolidation.[2,3] In particular, it is possible to make some estimates of the strength of the soil surrounding the pile at any time after driving. So far studies have been made of the effects of installing impermeable closed-ended piles into clay soils.[4–6] The effect of having a permeable pile material (e.g., concrete or timber) where drainage is permitted at the pile face has also been considered.[7] It is the aim of this Paper to investigate the changes in the stress state and the strength of the soil surrounding impermeable hollow piles with open ends and to compare these changes with those for the case of a closed-ended pile.

MODELLING PILE INSTALLATION

5. Elsewhere[3] a numerical procedure has been described which allows predictions to be made of the stress changes in the soil surrounding a driven pile. The method assumes

that the pile is of the full displacement type (i.e., either it is solid or it is hollow with a closed end) and is based on the assumption that a long cylindrical cavity is created under conditions of plane strain. This assumption has been justified experimentally,[8,9] at least for that length of pile which is some distance from both the pile tip and the soil surface. During cavity creation the soil is assumed to deform at constant volume, moving radially outwards from the pile axis; during consolidation the soil skeleton moves radially inwards with flow of pore water away from the pile (i.e., the pile is impermeable). A condition of zero radial displacement is imposed at the pile face during the consolidation phase (i.e., the pile is assumed to be essentially rigid).

6. This analytical method can be extended to cover the case of a hollow pile with an open end, but an additional assumption is required concerning what happens during driving to the soil which is inside the pile. As driving commences this soil moves upwards relative to the pile, though perhaps downwards relative to the mud line. The question is at what stage, if any, does the plug of soil stop moving relatively to the pile and the pile thus start behaving as if it were closed-ended. The assumption adopted here is that the soil inside remains stationary (in space) as the pile continuously penetrates downwards (i.e., the plug does not solidify inside the pile). In accordance with this assumption, it has been further assumed that the soil displaced by the pile wall moves outside the pile, rather than into the plug of soil inside.

7. This situation is depicted schematically in Fig. 1, where r_i and r_o are the inside and outside radii of the pile respectively. The net cross-sectional area of the pile is assumed to be equal to the net volume of soil displaced/unit length and is therefore given by

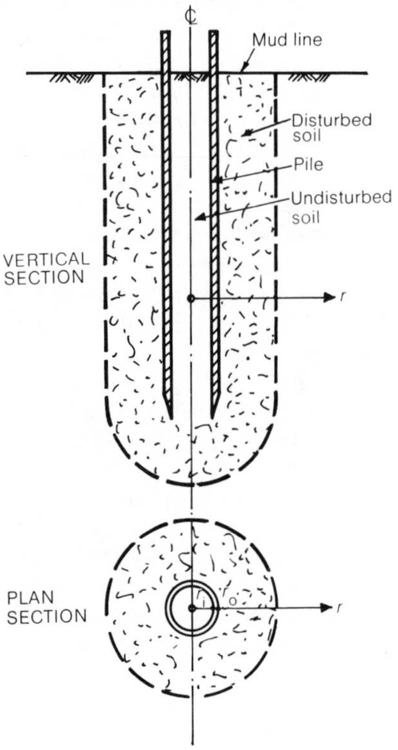

Fig. 1. Model for the installation of an open-ended tubular pile; ρ *= net volume/gross volume =* $1-(r_i/r_o)^2$

$$A_n = \pi(r_o^2 - r_i^2) \quad (1)$$

The gross cross-sectional area is simply

$$A_g = \pi r_o^2 \quad (2)$$

so the relative displaced area ρ is

$$\rho = A_n/A_g = 1 - (r_i/r_o)^2 \quad (3)$$

8. The assumption that the open-ended pile acts as a giant sampling tube, with the plug of soil rising (relative to the pile) continuously within the pile, is a simple and extreme view of the deformation behaviour of the soil. However, it is likely to be a conservative view, providing lower bound predictions of the stress and strength changes in the soil compared with the case where the plug solidifies and the pile acts as a closed-ended pile. Inside the pile the stress regime and the soil strength will be unaltered by driving, while around the outside of the pile the stress and soil strength are changed as a result of the creation of an annular cavity to accommodate the pile section. Thus the soil in contact with the outside of the pile has experienced an outward radial movement which just equals the wall thickness of the pile.

9. This method of analysis neglects any shear stresses which develop at the soil/pile interface during driving. While for a closed-ended pile this may be a reasonable approximation, since most of the effort in driving is directed towards creating a cavity of radius r_o by outward radial movement of soil, the same may not be true for an open-ended pile, where far less radial movement occurs. The manner in which the shear stresses acting on the pile face alter the response of the soil during driving must be the subject of future research. However, it is considered that the conclusions reached here are likely to be qualitatively correct.

MODELLING THE SOIL BEHAVIOUR

10. The soil surrounding the pile is assumed to be fully saturated. The flow of water through the soil is governed by Darcy's law, and the pore water is considered to be incompressible when compared with the skeleton material. Since the soil close to the pile is disturbed by pile installation (with the amount of disturbance depending on ρ), it is likely to be in a remoulded state immediately after driving. Thus it is likely (at least for insensitive materials) that the behaviour of the soil surrounding the pile may be well represented by a model which is based on critical state concepts.[10,11] Such a model couples together, in a realistic manner, consolidation, deformation and strength properties of a soil.

11. In this work a model based on modified Cam clay[12] is used. It has the following important features: the stress–strain behaviour of the soil skeleton is expressed in terms of effective stresses; both elastic and plastic strains are accounted for; undrained conditions with subsequent consolidation can be simulated; and the value of undrained shear strength changes with the current value of mean effective stress as consolidation proceeds. Six parameters are required to define material behaviour, and values for all of them may be readily obtained from standard oedometer

Table 1. Details of numerical computations for Boston Blue clay ($\lambda = 0.15$, $\kappa = 0.03$, $e_{cs} = 1.744$, $M = 1.2$)

Case	OCR	K_0	$G/c_u(0)$	$G/\sigma_z'(0)$	$G/p'(0)$
A	1	0.55	70	24	34
B	2	0.72	78	48	58
C	4	1.07	85	94	90
D	8	1.38	93	187	149
E	32	2.83	108	740	333

and triaxial compression tests. A list of these parameters and their significance is given in earlier papers.[4–6]

TYPICAL RESULTS

12. In order to be specific, a set of parameters was chosen so that the work hardening model would simulate a deposit of soil like Boston Blue clay. The numerical values of these parameters are given with Table 1. The soil is considered to have been initially one-dimensionally consolidated in the field with a value of $K_0 = 0.55$ and then to have been allowed to swell back after removal of overburden stress. Various starting conditions have been assumed just prior to pile installation. Each case corresponds to an overconsolidation ratio, OCR: values are given in Table 1, together with values of K_0 and G. Reasons for the selection of the values are given elsewhere.[4] Values of undrained strength quoted in this Paper are relevant to conditions of plane strain; thus they may differ from values obtained from conventional triaxial tests.

13. Previous studies have looked at the response of this ideal material to the installation of closed-ended piles.[4-7] In the present study, hollow piles with open ends are considered and results are compared with those for closed-ended piles. The features that are of interest include the excess pore pressure generated in the clay by the driving process, the rate of decay of these pore pressures, the change in strength (or water content) of the soil surrounding the pile as consolidation occurs, and the final effective stress acting on the pile face once consolidation is complete.

Effect of OCR

14. Previous studies of solid piles have shown that the stress conditions close to the pile after consolidation are functions of the original undrained shear strength of the soil, $c_u(0)$, and are largely independent of the original value of OCR. Thus the long term strength $c_u(\infty)$ of the soil adjacent to a closed-ended driven pile is greater than the original strength by a factor which is independent of the original value of OCR. However, the excess pore pressure generated during driving showed a slight dependence on OCR.

15. Calculations have been performed to demonstrate the effect of the value of OCR on the results of pile driving for an open-ended pile with $\rho = 0.1$. These results are plotted in Figs 2–4 and are compared with those for a closed-ended pile (i.e. $\rho = 1$). In Fig. 2 It can be seen that in both cases the value of OCR has an effect on the excess pore pressure generated next to the pile, u_{max}, with a proportionally greater effect in the case of an open-ended pile. However, Fig. 3 shows that in both cases the final radial effective stress on the pile face, $\sigma_r'(\infty)$, is practically independent of OCR, although the open-ended pile shows a slight tendency for $\sigma_r'(\infty)$ to decrease with OCR. For a closed-ended pile, $\sigma_r'(\infty)$ has a value of about $5c_u(0)$

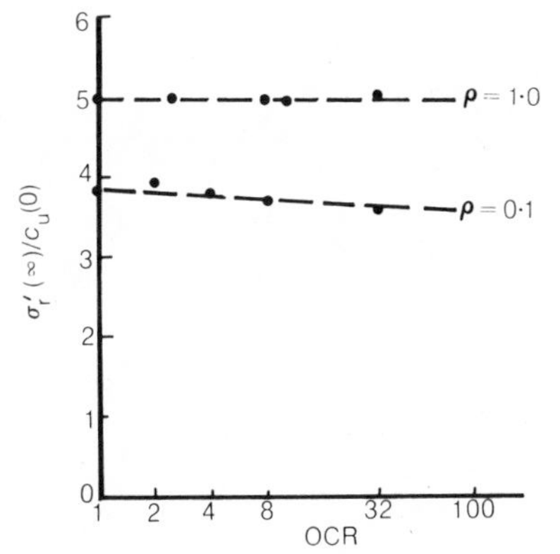

Fig.3. Boston Blue clay: effect of OCR on the radial effective stress acting on the pile after consolidation

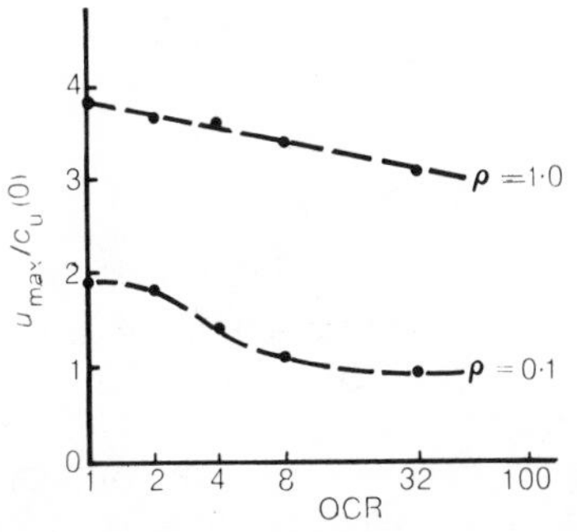

Fig.2. Boston Blue clay: effect of OCR on the maximum excess pore pressure generated next to the pile

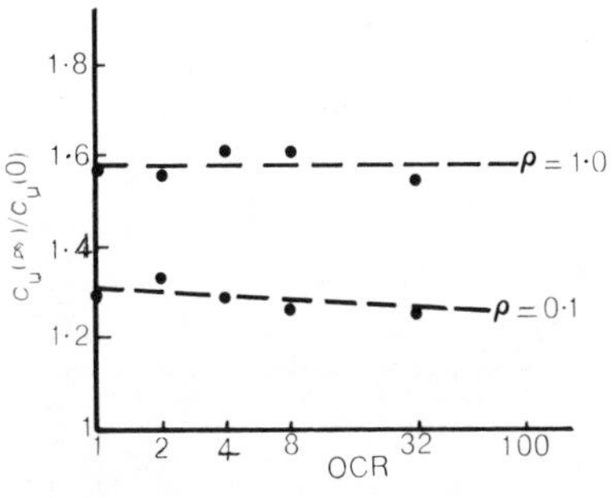

Fig.4. Boston Blue clay: effect of OCR on the undrained strength next to the pile after consolidation

and for an open pile with $\rho = 0.1$ it has a value of about $3.75c_u(0)$. Similar conclusions hold for the final undrained strength of the soil (after consolidation) next to the pile: in Fig. 4, $c_u(\infty) \sim 1.6c_u(0)$ for $\rho = 1$ and $c_u(\infty) \sim 1.3c_u(0)$ for $\rho = 0.1$.

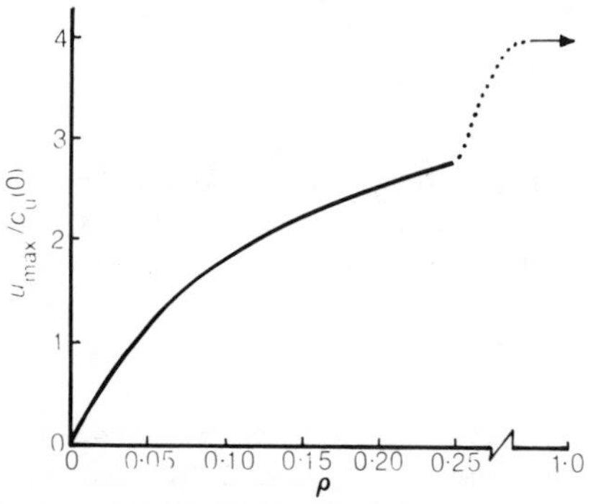

Fig.5. Effect of the wall thickness of the pile on the maximum excess pore pressure generated next to the pile; Boston Blue clay, initial OCR = 2

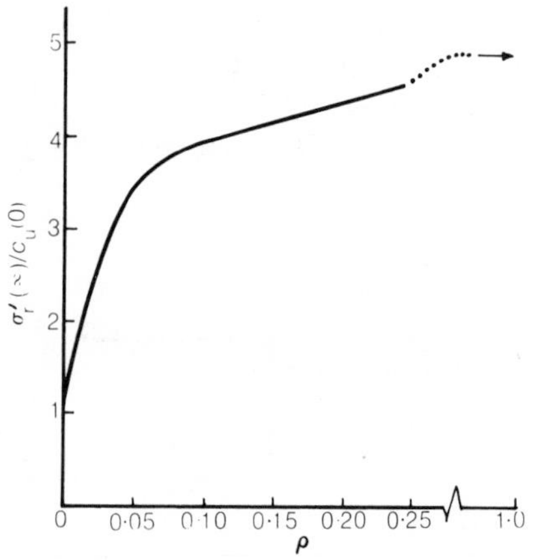

Fig.6. Effect of the wall thickness of the pile on the radial effective stress acting on the pile after consolidation; Boston Blue clay, initial OCR = 2

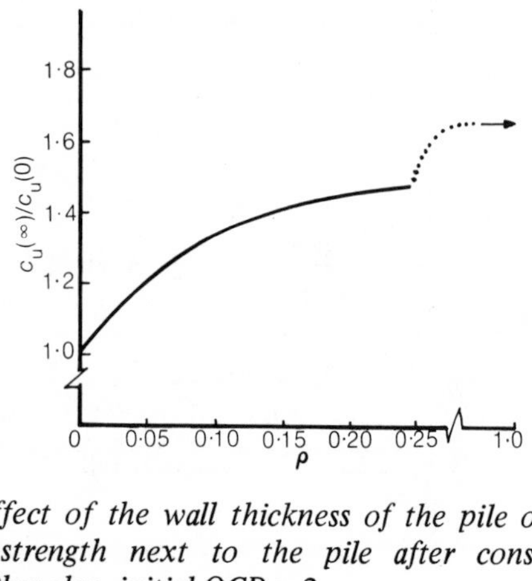

Fig.7. Effect of the wall thickness of the pile on the undrained strength next to the pile after consolidation; Boston Blue clay, initial OCR = 2

16. Thus the work hardening soil model predicts that the final stress state and the final strength of the soil around a driven pile are practically independent of the value of OCR, even for hollow piles with open ends. Previous one-dimensional consolidation of the soil is important only in that it will determine the relative values of the undrained shear strength $c_u(0)$ and the current effective overburden stress.

Effect of ρ on stress changes

17. Although results in paragraph 15 are for only two vlaues of ρ, it is assumed here that the effects of OCR will be generally similar for other values of ρ. The following results show how the value of ρ affects the stress changes in the soil during and after driving. For convenience, a soil with an initial value of OCR = 2 is considered.

18. Figure 5 shows the variation with ρ of the excess pore pressure generated next to the pile. The greater the net volume of soil displaced, the greater is the excess pore pressure generated. The largest pore pressure is due to driving a closed-ended pile ($\rho = 1$) and has a value of about $4c_u(0)$. Fig. 6 shows the variation with ρ of the final radial effective stress acting on the pile face. This quantity also increases with ρ to a value of about $5c_u(0)$ when $\rho = 1$. A similar trend is shown for the final undrained strength of the soil next to the pile (Fig. 7). In this case $c_u(\infty) \sim 1.65c_u(0)$ when $\rho = 1$. The distribution of final undrained strength is shown in Fig. 8 for piles with different values of ρ. As expected this figure shows that the greater the value of ρ, the greater is the disturbance to the soil surrounding the pile. In the case of a closed-ended pile ($\rho = 1$) the disturbance extends to about 10 pile radii while, for hollow piles with open ends, the radius of the disturbed zone and the soil strength next to the pile decrease with ρ.

Effect of ρ on consolidation times

19. Not only does a greater net area of pile result in more soil being disturbed during driving, and greater excess pore pressures being generated, but also it entails a longer consolidation period after driving. In Fig. 9 it can be seen that for a pile with $\rho = 1$, the time taken to reach any given degree of consolidation may be an order of magnitude longer than that for an open-ended pile with $\rho = 0.05$ or 0.1, say. In Fig. 9, time has been non-dimensionalized by the initial value of the undrained soil strength $c_u(0)$, the soil permeability k and the pile radius r_o, as

$$T^* = kc_u(0)t/\gamma_w r_o^2 \tag{4}$$

where γ_w is the unit weight of water. Use of the more conventional time factor[13]

$$T = c_h t/r_o^2 \tag{5}$$

was precluded because this soil model incorporates a variation in the consolidation coefficient c_h as consolidation proceeds.

20. Previous analyses[5] have shown that the times for given degrees of consolidation are relatively unaffected by the choice of soil model. This conclusion enables consolidation times to be estimated with reasonable accuracy from

closed-form solutions based on an elastic soil, with a constant coefficient of consolidation.[14] For a given pile radius and a given consolidation coefficient for the soil, the time for a particular degree of consolidation to occur will be a function of the original degree of disturbance of the soil due to pile driving (i.e., of the original size of excess pore pressure generated as a function of the soil strength). Fig. 10 shows how the times for 50% and 90% consolidation (T_{50} and T_{90}) are affected by the value of $u_{max}/c_u(0)$, where u_{max} is the initial excess pore pressure at the pile face immediately after pile driving.

CONCLUSIONS

21. An investigation has been made of the stress and strength changes in clay due to the installation of a driven pile. The clay has been modelled as a work hardening elastoplastic material; and pile installation has been modelled as the expansion, under undrained conditions, of a long cylindrical cavity. The conclusions outlined below should be viewed in the light of the simplifications implicit in the above methods of modelling the interaction of pile and soil during driving of the pile.

22. The manner in which the previous consolidation history affects the behaviour of a particular soil has been investigated. It has been found that the initial overconsolidation ratio is relatively unimportant. The magnitude of stress changes due to pile driving and subsequent consolidation of the soil depend solely on the initial undrained strength of the undisturbed material, irrespective of the past consolidation history. For partial displacement piles, there is a tendency for the stress and strength changes in the soil to be marginally smaller as the value of OCR increases.

23. A study has been made of the effect of the ratio of net to gross cross-sectional area of the pile on the stress changes in the soil during driving and subsequent consolidation. In particular, it has been shown that the size of the disturbed region of soil increases as the net cross-sectional area of the pile increases (for a given pile diameter). The magnitude of the excess pore pressure generated during driving increases the more the soil is disturbed. This leads to greater stress and strength changes in the soil adjacent to the pile shaft as the area ratio of the pile increases. A comparison of the cases of a closed-ended pile and an open-ended pile with ρ = 0.1 shows that the final strength of the soil adjacent to the pile may be some 25% higher in the former case.

24. The higher the excess pore pressures generated, the longer will the pile take to achieve its full shaft capacity. Some indication is given of the times required for the excess pore pressures to decay to 50% and 10% of their initial values immediately after driving. Typically, an open-ended pile may be expected to achieve its full capacity an order of magnitude faster than a closed-ended pile of the same external diameter.

REFERENCES

1. RIGDEN W.J. et al. Developments in piling for offshore structures. BOSS 1979, Conf. Behaviour of Offshore Structures, London, 1979. British Hydraulics Research Association, Bedford.

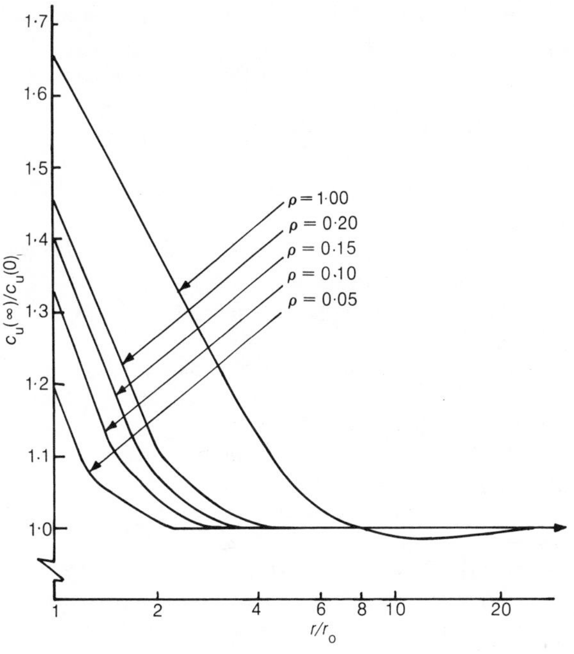

Fig.8. Distributions of undrained strength around the pile after consolidation; Boston Blue clay, initial OCR = 2

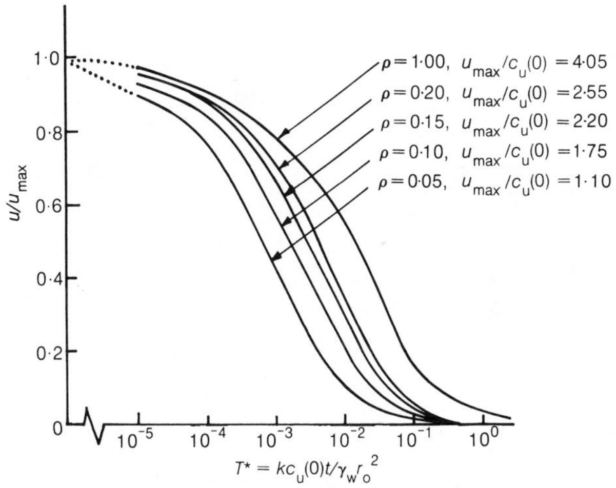

Fig.9. Decay of excess pore pressure in the soil adjacent to the pile; Boston Blue clay, initial OCR = 2

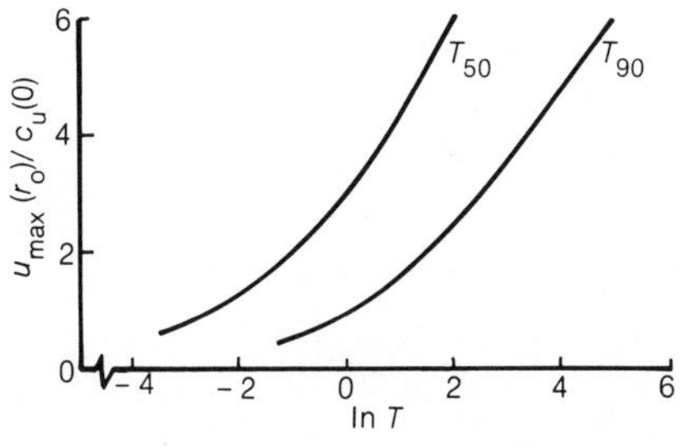

Fig.10. Times for 50% and 90% dissipation of the excess pore pressure adjacent to the pile in an elastic soil

2. KIRBY R.C. and WROTH C.P. Application of critical state soil mechanics to the prediction of axial capacity for driven piles in clay. Proc. Offshore Technology Conf., Houston, 1977.
3. CARTER J.P. et al. Stress and pore pressure changes in clay during and after the expansion of a cylindrical cavity. Int. J. Numer. Anal. Meth. Geomech., 1979, 3, No. 4, 305–322.
4. RANDOLPH M.F. et al. Driven piles in clay: (I) Installation, modelled as the expansion of a cylindrical cavity. Cambridge University Engineering Dept, 1978, tech. report SOILS/53.
5. RANDOLPH M.F. et al. Driven piles in clay: (II) Consolidation after driving. Cambridge University Engineering Dept. 1978, tech. report SOILS/54.
6. WROTH C.P. et al. Stress changes around a pile driven into cohesive soil. Recent developments in the design and construction of piles. Institution of Civil Engineers, London, 1979.
7. RANDOLPH M.F. and CARTER J.P. The effect of pile permeability on the stress changes around a pile driven into clay. Proc. 3rd Int. Conf. Numerical Methods Geomechanics, Aachen, 1979.
8. CLARK J.I. and MEYERHOF G.G. The behaviour of driven piles in clay: (I) An investigation of soil stress and pore water pressures as related to soil properties. Canadian Geotech. J., 1972, 9, No. 4, 351–373.
9. RANDOLPH M.F. et al. The effect of pile type on design parameters for driven piles. In: Design parameters in geotechnical engineering: proc. 7th Eur. conf. soil mech., Brighton, 1979. British Geotechnical Society, London, 1979, Vol. 2, 107–114, Paper c18.
10. SCHOFIELD A.N. and WROTH C.P. Critical state soil mechanics. McGraw Hill, London, 1968.
11. ATKINSON J.H. and BRANSBY P.L. The mechanics of soils: an introduction to critical state soil mechanics. McGraw Hill, Maidenhead, 1978.
12. ROSCOE K.H. and BURLAND J.B. (HEYMAN J. and LECKIE F.A. (eds)). On the generalised behaviour of 'wet' clay. In: Engineering plasticity. Cambridge University Press, 1968.
13. SODERBERG L.O. Consolidation theory applied to foundation time effects. Géotechnique, 1962, 12, No. 3, 217–225.
14. RANDOLPH M.F. and WROTH C.P. An analytical solution for the consolidation around a driven pile. Int. J. Numer. Anal. Meth. Geomech. 1979, 3, No. 3, 217–229.

Discussion on survey paper and drivability (papers 1–7)

Mr F. B. J. BARENDS (Delft Soil Mechanics Laboratory)
It is not certain whether the finite element method, which is suitable for elliptical problems, will solve the problems of internal friction materials so easily. It may be that the method of characteristics should be taken into account, and I would like to know Dr Smith's opinion of this method.

Dr SMITH (Paper 1)
To take account of wedges of soil sliding over other wedges of soil, one would need a very complicated finite element computation. I certainly agree that there is a place for the method of characteristics, but there is nothing intrinsic in the finite element procedure which would prevent it solving hyperbolic problems or any other kind of problem. There is, however, a definite difficulty with discontinuities.

Dr OMAR & PROFESSOR POSKITT (Paper 3)
The Paper uses classical stability theory to examine the circumstances under which piles wander during driving. This is associated with a loss of lateral stability. In the analysis of laterally loaded piles a similar loss of stability can arise if the assumed p–y curves are not properly chosen.

4. Laterally loaded piles are conventionally dealt with using beam-column theory as for example in equation (1) of Paper 17:

$$EIy'''' + P_X y'' - p = 0 \quad (1)$$

The perturbed form of this equation used for stability investigations is given by equation (10) of the Paper:

$$EIh'''' + P_X h'' + \beta_2 h = -\frac{w}{g}\ddot{h} \quad (2)$$

5. For any arbitrarily specified set of p–y data equation (1) satisfies equilibrium and compatibility. However the equilibrium will only be stable if $\beta_2(x,y)$ $[=(\partial p/\partial y)_X]$ (Fig. 1) is such that equation (2) has a bounded oscillatory solution. This is therefore the stability postulate for a laterally loaded pile. It imposes a restriction on the shape of the p–y curves. This, coupled with the inherently unstable nature of the beam-column equation, requires careful consideration during the design stage of piling work.

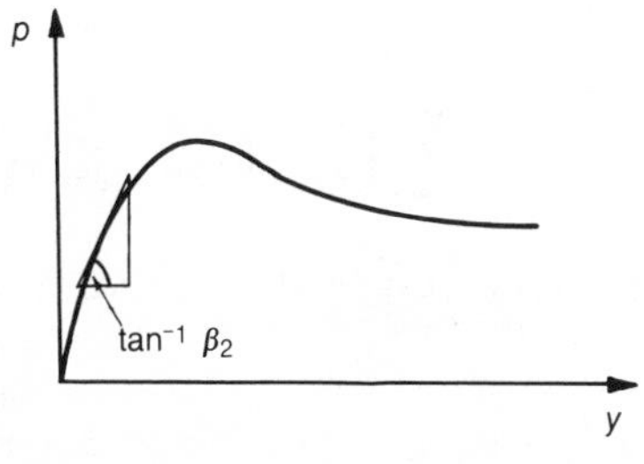

Fig. 1. Typical p–y curve

Dr J.-P. MIZIKOS (Societé Nationale Elf Aquitaine (Production), Paris)
At SNEA(P) we have had experience of driving two curved conductor pipes in very soft mud, softer than in the Gulf of Mexico, and we found that the driving curve is very similar on average to the driving curve of a straight pipe. The curvature was about 2° per 10 m and it would drive up to 100 m. The only difference was that after driving, our curvature was lower than its initial value, so our curved conductor pipe was straighter than expected.

Mr G. E. J. S. L. VOITUS VAN HAMME (Hollandsche Beton Groep n.v.)
Drivability studies are generally made with computer programs, based on the wave equation. This wave equation is a partial differential equation which for the drivability programs is modified by a set of finite difference equations. These are solved for a number of consecutive time steps. In these programs a mathematical model of the pile is used consisting of a number of discrete masses connected by springs. The resistances are modelled by springs and dashpots. Such programs give acceptable results if the number of discrete masses and the time step are chosen with care.

8. It is, however, astonishing that none of the programs known to the writer is based on a solution of the wave equation with the exception of the HBG piledriving program PILEWAVE designed by the writer.

9. The wave equation for stress waves in rods in the absence of friction has a solution which is due to d'Alembert (1717–83). In its simplest form the partial differential equation reads

$$c^2 \frac{\partial^2 u}{\partial x^2} = \frac{\partial^2 u}{\partial t^2}$$

where x is distance along the rod, t is time, u is displacement of a point of the rod and $c^2 = E/\rho$, E and ρ being Young's modulus and the density of the material of the rod; c has the dimension of a velocity, in fact the velocity of propagation of stress waves in the rod.

10. The general solution of this equation reads

$$u(x, t) = f(x - ct) + g(x + ct)$$

Each of the functions $f(x - ct)$ and $g(x + ct)$ depends on a single variable, namely $\xi = x - ct$ and $\eta = x + ct$. The functions $f(x - ct)$ and $g(x + ct)$ represent waves propagating with velocity c; for the former they are propagating in the positive direction and for the latter in opposite direction.

11. From this solution the forces and particle velocities can be derived as follows:

$$F(x, t) = -EA\frac{\partial u}{\partial x}$$

and

$$v(x, t) = \frac{\partial u}{\partial t}$$

This leads to

$$F(x, t) = -EA\left\{\frac{\partial f(\xi)}{\partial x} + \frac{\partial g(\eta)}{\partial x}\right\}$$

$$= -EA\left\{\frac{\partial \xi}{\partial x}\frac{df}{d\xi} + \frac{\partial \eta}{\partial x}\frac{dg}{d\eta}\right\}$$

$$= -EA\left\{\frac{df}{d\xi} + \frac{dg}{d\eta}\right\}$$

$$v(x, t) = \frac{\partial f(\xi)}{\partial t} + \frac{\partial g(\eta)}{\partial t}$$

$$= \frac{\partial \xi}{\partial t}\frac{df}{d\xi} + \frac{\partial \eta}{\partial t}\frac{dg}{d\eta}$$

$$= -c\frac{df}{d\xi} + c\frac{dg}{d\eta}$$

12. These equations can be modified to a form which appeals to engineers by writing

$$F\downarrow(\xi) = -EA\frac{df}{d\xi} \qquad (\xi = x - ct)$$

and

$$F\uparrow(\eta) = -EA\frac{dg}{d\eta} \qquad (\eta = x + ct)$$

This gives

$$F(x, t) = F\downarrow(\xi) + F\uparrow(\eta)$$

and

$$v(x, t) = \frac{c}{EA}\,[F\downarrow(\xi) - F\uparrow(\eta)]$$

The symbols $F\downarrow(\xi)$ and $F\uparrow(\eta)$ represent force waves propagating along the rod in the positive and negative directions respectively, with velocity c. The sign is chosen such that positive force components are compressive. For driven piles the positive x direction is downward; so $F\downarrow$ is a downward-moving force wave and $F\uparrow$ an upward force wave. The quantity $Z = EA/c = c\rho A$ is a property of the pile (the rod) and is called impedancy.

13. As long as there is no side friction the waves $F\downarrow$ and $F\uparrow$ retain their values as $\xi = x - ct$ and $\eta = x + ct$ are constants. So for a point x at time t the total force is equal to $F(x, t) = F\downarrow(x - ct) + F\uparrow(x + ct)$ and its velocity is equal to $v(x, t) = [F\downarrow(x - ct) - F\uparrow(x + ct)]/Z$.

14. For the calculations $F\downarrow$ and $F\uparrow$ must always be considered along the respective wavepaths, namely $x - ct$ = constant and $x + ct$ = another constant.

15. The downward wave $F\downarrow$ came from a point $x - \Delta x$, which it passed at $t - \Delta t$, such that $\Delta x = c\Delta t$, and the upward wave came from point $x + \Delta x$, which it passed at the same time $t - \Delta t$. So the calculations by the computer program are performed for a sequence of times with an interval Δt. The forces and velocities are calculated for 'grid points' along the pile at distances $\Delta x = c\Delta t$.

16. The above applies to the case without skin friction. If skin friction is present, the differential equation contains also a friction term and a solution in closed form is only to be found for particular cases (e.g., for the case of friction proportional to velocity). This case generally does not exist and thus an approximation must be found. The obvious approximation lies in the common engineering practice of replacing the continuously distributed friction by a set of discrete friction forces acting at the grid points. Then the parts of the rod between grid points are without friction and thus the above formulae are valid. Only when the waves pass the grid points their intensity is modified.

17. Assume that at grid point x_i a downward wave $F\downarrow_1$ arrives at a certain instant t simultaneously with an upward wave $F\uparrow_2$. These waves have passed the adjacent grid points $x_{i-1} = x_i - \Delta x$ and $x_{i+1} = x_i + \Delta x$ at time $t - \Delta t$ and thus are known. Simultaneously upward ($F\uparrow_1$) and downward ($F\downarrow_2$) waves start at point x_i.

18. With Fr_i (the friction force at x_i), acting upward (assuming the velocity $v(x, t)$ to be positive, i.e., downward), we have

$$F_1 = F\downarrow_1 + F\uparrow_1 \qquad v_1 = (F\downarrow_1 - F\uparrow_1)/Z$$

$$F_2 = F\uparrow_2 + F\downarrow_2 \qquad v_2 = (F\downarrow_2 - F\uparrow_2)/Z$$

$$F_1 = F_2 + \mathrm{Fr}_i$$

$$v = v_1 = v_2$$

From these equations we find

$$F\uparrow_1 = F\uparrow_2 + \frac{1}{2}\mathrm{Fr}_i$$

$$F\downarrow_2 = F\downarrow_1 - \frac{1}{2}\mathrm{Fr}_i$$

This result is valid irrespective of the laws on which Fr_i depends!

19. The above constitutes the essential features of the piledriving program PILEWAVE. It has important advantage over piledriving programs based on concentrated masses interconnected by springs. First, force and velocity are always calculated for the same points (the grid points, at

intervals of the order of 1 ft), whereas with conventional programs the forces in the pile are calculated for the springs and the velocity for the concentrated masses. Secondly, phenomena which occur at places where no traction can be sustained (e.g., between a pile and an add-on) can be assessed accurately: the time when a gap occurs is found, and how the gap increases and eventually decreases until the parts come into contact again. Thirdly, the piledriving hammer, even a rather complicated hammer such as the Hydroblok (with a built-in gas buffer), and the pile cap, with cushions if these are used, can easily be incorporated in the system. Fourthly, this 'solution of the wave equation' theory not only leads to a simple computer program but also provides a much better understanding of what really happens during piledriving.

Dr H. D. St JOHN (Building Research Establishment)
In practice, what is the magnitude of the problem of instability in offshore piles?

Mr W. J. RIGDEN (British Petroleum Ltd)
By modern standards the Forties piles are quite small – 54 in in diameter and 100 m long. They were, according to Professor Poskitt, highly stable. For a more modern pile which is 70–84 in in diameter and probably a bit shorter, there is less of a problem. Does anyone know of a case where piles have gone unstable offshore? I certainly do not. With conductors it is a different case because the diameters are down to 30 in, but how much of a problem is it with the piles?

Mr F. de KLERK (Nederlandse Aardolie Maatschappij b.v.)
We have had some experience with conductor driving in the southern part of the North Sea. Conductors of 30 in o.d. were driven to a depth of 60–65 m. After driving and removing the soil plug from the conductors we got variations in the order of 3° from the vertical. Particularly in the last 10 m there was quite an arbitrary deviation from the vertical. That is the only experience we have. We have no experience with piles because normally we do not remove the plug and run a survey to determine the deviation.

A SPEAKER
The flutter type of instability of a driven pile, which is dealt with in Paper 2, is a phenomenon which is very similar to that experienced with aerofoils. I wonder whether this type of instability can really be properly treated, in view of the medium, because if the pile undergoes a flutter motion, so does the medium. This means that the inertia of the medium will come into friction and waves will propagate from the driven pile, generating a tremendous amount of damping. I suspect that if the medium were modelled to incorporate this propagation, the flutter would not occur at all.

Mr RIGDEN
Have the Authors of papers 2 and 3 looked at the API recommendations on pile straightness etc., to see if these are great enough to affect the analysis?

PROFESSOR R. BUTTERFIELD & Dr R. A. DOUGLAS (Southampton University)
The following experimental evidence relates to model- and full-scale measurements of pile directional instability (wandering) and is, we believe, relevant to the following questions:

(a) Does Burgess's analysis[1] provide a useful indication of the risk of wandering during installation of driven piles?
(b) Do large piles behave in the manner observed in small models?

26. We have carried out a considerable number of tests[2,3] on small piles of different moduli (E) and cross-sectional properties (I) (from 100 mm to 250 mm long) in a variety of clay beds ($32 < C_u < 260$ kN/m²) in order to cover a reasonably wide range of the main parameters.

27. The key results obtained were as follows. First, even when supported adequately above ground, the very flexible model piles wandered from the start of driving, but if the lateral displacements were plotted to an exaggerated scale (Fig. 2) a notional critical length (L_c) could be defined. Secondly, if all such measurements (around 60) were summarized on a plot of L_c and C_u, and compared with Burgess's prediction of L_c (uniform soil and Skempton's α factor ≈ 1) the analysis fitted the experiments remarkably well (Fig. 3). Thirdly, the deflected profiles (which are not predicted by the analyses), measured both on piles abutting

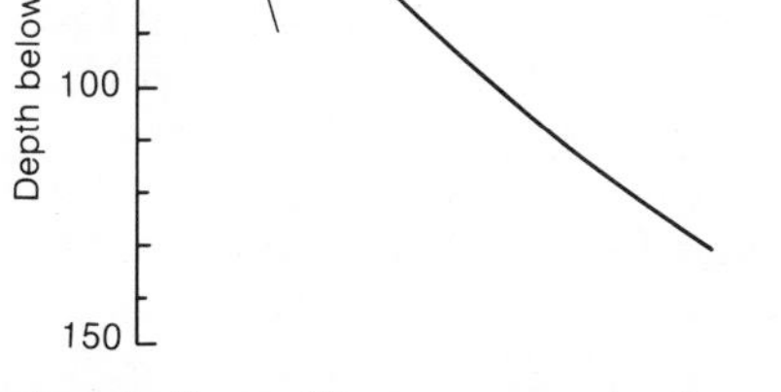

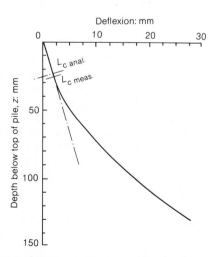

Fig. 2. Displaced pile profile versus penetration

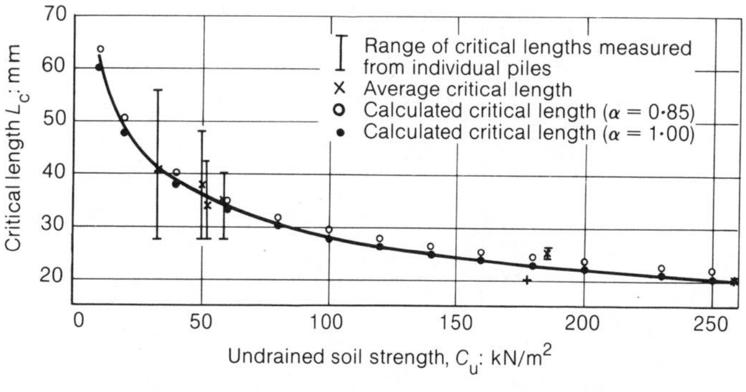

Fig. 3. Critical length versus undrained strength

a glass-sided box and from X ray plates of piles driven centrally in a 100 mm wide box,[2,3] scattered quite widely about the mean line as might be expected in very simple stability experiments of this kind. Fig. 4 shows, in dimensionless form, the mean curve obtained and the range of measured pile centre-line profiles.

28. We have also been fortunate enough to have access

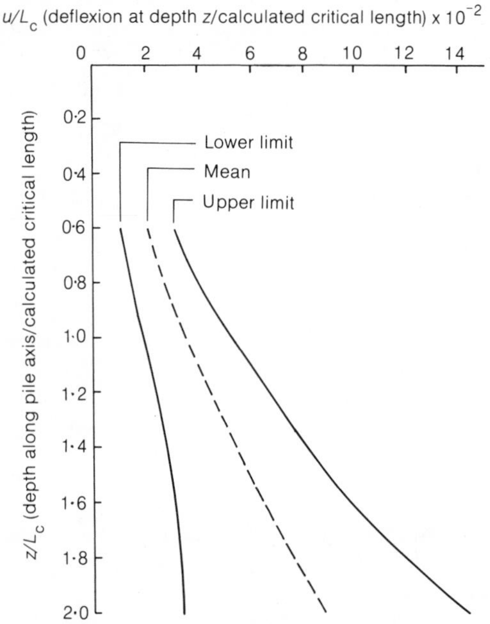

Fig. 4. Dimensionless displaced pile profile (mean and limiting shapes) versus pile penetration

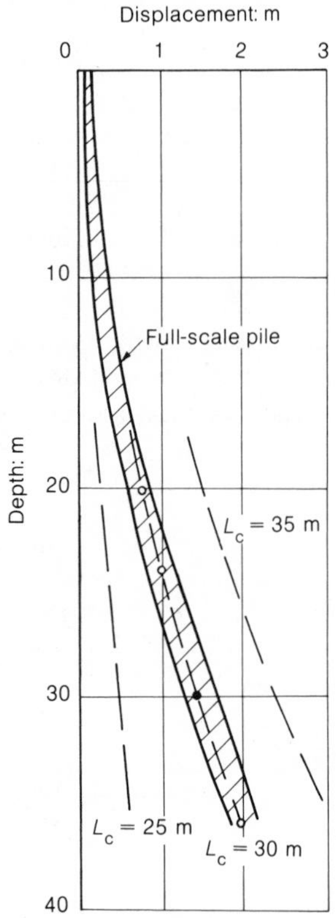

Fig. 5. Full-scale pile; measured profile

to sets of accurate, inclinometer-tube measurements on a large, in situ pile (a concrete cylinder 35 m long, 0.44 m o.d. and 0.31 m i.d., driven by an internal mandrel into a uniform soft silty soil, mean $C_u \approx 20$ kN/m^2). These results are plotted in Fig. 5, together with profiles estimated from the mean line in Fig. 4 ($\alpha = 1$). The L_c values used ($25 \leqslant L_c \leqslant 35$ m) were calculated from Burgess's curves using different EI values for the pile. The alternative EI values arose from different assessments of the equivalent bending stiffness of the composite 'pile + mandrel' cross-section. However, the agreement between the mean ($L_c = 30$ m) profile extrapolated from the model tests and the full-scale pile measurements is extremely good (Fig. 5).

29. For most practical 'pile/homogeneous soil' systems Burgess's results can be summarized in terms of the critical piledriving force (P_c) at the critical depth (L_c) as[2]

$$P_c \approx 40EI/L_c^2 \qquad (3)$$

It is tempting to suggest that equation (3) might be recast, at least as a mnemonic, as

$$P_c \approx 4\pi^2 EI/L_c^2$$

Dr BURGESS & Mr TANG (Paper 2)

Dr Mizikos has described the installation of curved conductor pipes in very soft mud, in which the final curvature was lower than the initial. This behaviour is interesting in that it runs against one's intuitive reasoning of the way in which a curved pile will behave when pushed, but not against the reasoning of the flutter hypothesis. In normal structural terms the initial curvature constitutes a geometric imperfection, and this would be expected to amplify itself considerably during installation. A close analogy to the initially curved pile would be a curved line of articulated trucks pushed tangentially by a tractor at one end. It is easily appreciated that, even if the initial curvature of the line is very small, there is considerable deviation as the motion progresses. The flutter hypothesis is in no way diminished by the overall straightening actually experienced with the conductors, since flutter is unaffected by geometric imperfections. The wandering could therefore take place in any direction for a cylindrical pile, and might well cause an overall straightening even when superimposed on the expected amplification of the initial geometric imperfection.

31. A speaker has pointed to the incomplete nature of the analytical treatment at present, and has raised the question of the effect of damping on flutter instability. On the first point, the Authors feel that to attempt an analysis of a new problem with many independent parameters without attempting to gain some experience of their relative importance would be foolish. We have therefore deliberately avoided the early inclusion of soil mass, continuum elasticity and damping although they do not present very great analytical problems. On the question of the effect of radiation damping on flutter instability, it has been found that damping does not in general have the effect of removing flutter instability and that certain positive damping types can actually reduce critical loads. This will shortly be investigated as part of the study, but it is not expected that any considerable change of behaviour will be caused by damping.

32. Professor Butterfield and Dr Douglas have produced

extremely valuable experimental results in a field where very few results either of laboratory work or of field tests have been published. The Authors are very grateful for these results, which seem to give some support to the theoretical work, at least for the case of uniform clays. It is to be hoped that more results will be published so that the theoretical predictions can be properly tested and given a practical structure.

Dr OMAR & PROFESSOR POSKITT

In reply to Dr St John and Mr Rigden it is worth noticing that when piles for an offshore structure have been installed there is no practical means of determining their actual location in the soil. The layout of the group is such that if the piles drive in the direction intended there is no danger of their making contact in the ground. However, if a pile is initially bent or if the pile–soil system is slightly unstable, a situation occurs where contact could be made. For piles spaced at 3–5 diameters apart in the sleeves it requires very little deformation of a pile to enable the tip to make contact with an adjacent pile. The result might then be that the pile would refuse to drive further.

34. The problem is more critical with conductors as the comments of Mr Rigden and Mr de Klerk indicate. The deviations observed by Mr de Klerk are typical of what would be expected if a conductor suffered a loss of stability.

35. For engineers concerned with using numerical methods to study the lateral behaviour of piles, stability is of the greatest importance. A tendency to a physical instability will lead to a magnification of the errors in numerical routines used to determine deflexions and stresses. The converse is also true, that when a numerical routine is found to be unstable it may be the result of a physical instability.

Dr M. APPENDINO (ENEL, Italy)

I wish to point out a simple experimental procedure which I have used successfully to determine soil tip resistance and driving energy. The procedure is not general because it can be employed only when the following two conditions are satisfied:

(a) the pile is long in comparison with the time required by the incident force pulse to reach the tip and then to go back to the head;
(b) the soil resistance is concentrated at pile tip.

These conditions allow the model of the pile–soil system to be simplified as shown in Fig. 6. Then the analysis of pile-driving reduces to that of propagating waves in a long bar with a yielding support.

37. The procedure is based on the splitting of incident and reflected force pulses, which occurs when the force is measured at distances L_1 from pile tip and L_2 from pile head in such a way the following relationships are satisfied:

$$L_1 \geqslant V_c \Delta t_I$$
$$L_2 \geqslant 0.3\ V_c \Delta t_I$$

where V_c is the velocity of the compressive wave and Δt_I is the duration of the incident force pulse measured at 50% of peak value.

38. The simple inspection of the incident force pulse shape makes it possible to check the efficiency of the driving equipment. The energy L_I, which is introduced into the pile, may be computed by the following expression:

$$L_I = \frac{1}{\rho V_c A} \int [F_I(t)]^2 \, dt$$

where ρ is the pile specific mass, A is the pile cross-section, t is time and F_I is the incident force. The inspection of the reflected force pulse shape gives information on soil stiffness and resistance.[4,5] From both incident and reflected forces we obtain the soil resistance, pile tip velocity and displacement by using the following expressions:

$$F_2(t) = F_I(t) + F_R(t - 2L_1/V_c)$$

$$\dot{x}_2(t) = \frac{F_I(t) - F_R(t - 2L_1/V_c)}{V_c A}$$

$$x_2(t) = \int \dot{x}_2(t) \, dt$$

where F_2 is the soil reaction at the pile tip, and $\dot{x}_2$ and x_2 are the tip velocity and displacement.

39. Although this method is not general, it is worth consideration when the required conditions apply, for the following advantages: first, it requires only a force measurement, which can be easily performed with extensometers stuck or welded on the pile (therefore it eliminates the necessity of integrating accelerations – which can raise some difficulties – and it eliminates also the necessity of operating algebraically on two physical quantities measured by different instruments (i.e., extensometers and accelerometers)); secondly, it gives immediately the perception of how things are going on with the hammer as well as with the soil, as shown by the examples in Figs 7–9. There are also shortcomings in this method: the need to extrapolate the incident force when Δt_1 (Fig. 7) is not long enough; and the necessity to protect the measuring devices against soil friction and water because they must be installed at a distance L_2 from the pile head which is usually below ground or water level at the end of driving.

40. A comparison of Figs 7 and 8 shows how the same permanent set measured on two equal piles, driven with equal hammers (steam hammers) and the same technique,

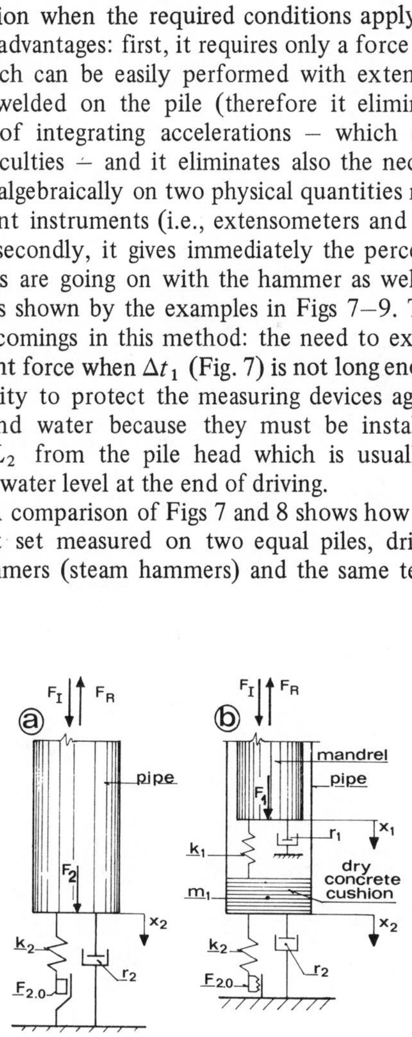

Fig. 6. Pile–soil interaction models for head-driven and mandrel-driven piles

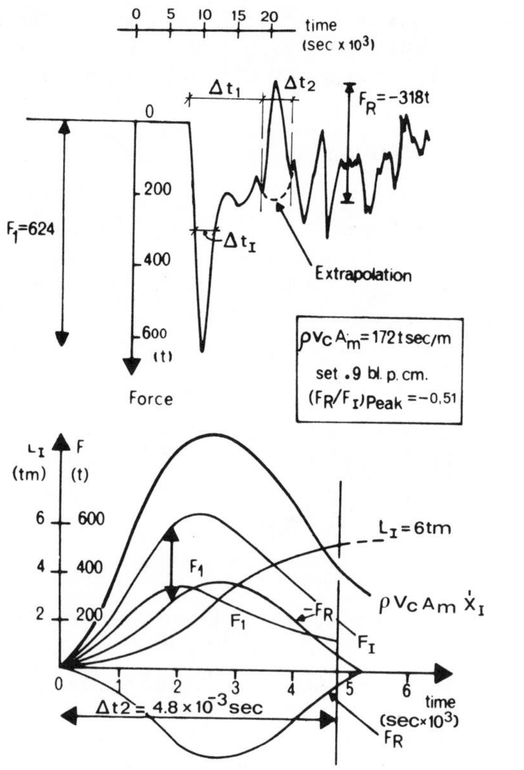

Fig. 7. Force measurement and interpretation procedure: low driving efficiency and low soil resistance (Porto Tolle pile)

may correspond to two very different soil resistances. Fig. 7 shows an initial sharp force pulse, followed by a few lower peaks, revealing that the cushion was too stiff. The energy transmitted to the pile in the first impact is only a low fraction of the hammer nominal energy. The shape of the incident force pulse in Fig. 8, and the ratio of transmitted energy to nominal energy (driving efficiency), indicate that the cushion was chosen properly.

41. Figure 9 shows the incident force measured in an analogous pile driven with the same technique, but with a diesel hammer. The more irregular shape of the force pulse is typical and the driving efficiency is very low. This information proves that the hard driving encountered by using this hammer was a consequence of the equipment and not of the soil resistance.

42. I should like the Authors to comment on the difficulties in integrating accelerometer outputs.

PROFESSOR POSKITT

I think Paper 5 throws some useful light on the behaviour of the soil plug by treating it as a wave transmission problem coupled to the primary stress wave within the pile. This is more realistic than assuming the soil plug moves with the pile wall. For typical driving conditions where accelerations may be of the order of $300g$ the inertia of the soil plug is so great that in order for the soil plug to move with the pile wall a shear stress of about 2000 kN/m^2 would be required between the soil and the pile wall.

44. A second factor which discourages plugging is due to the radial expansion of the pile. As the stress wave travels down, radial strain in the pile momentarily produces a small gap between the soil plug and the pile wall.

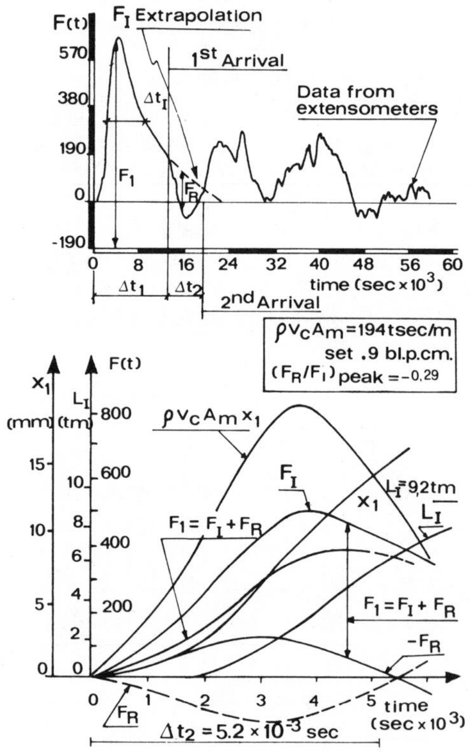

Fig. 8. Force measurement and interpretation procedure: high driving efficiency and soil resistance conforming to design requirements (Porto Tolle pile)

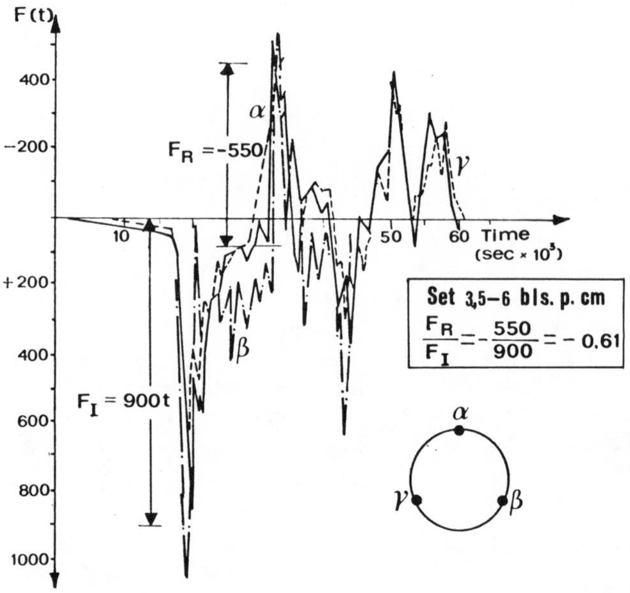

Fig. 9. Force measurement and interpretation procedure: very low driving efficiency and low soil resistance (Porto Tolle pile)

Dr GOBLE (Paper 4)
With reference to Paper 5, there has been an extensive study by Holloway et al.[6] of residual stress effects, I think for values of relatively small area relative to height, and residual stresses can be important in affecting the blow count.

Mr de JONG (Paper 5)
I should like to put a point to Dr Goble and Dr Rausche. It occurred to us that the blow count is affected not so much by the initial stresses in the pile as by the initial stresses in the soil. Those are much more important, I think.

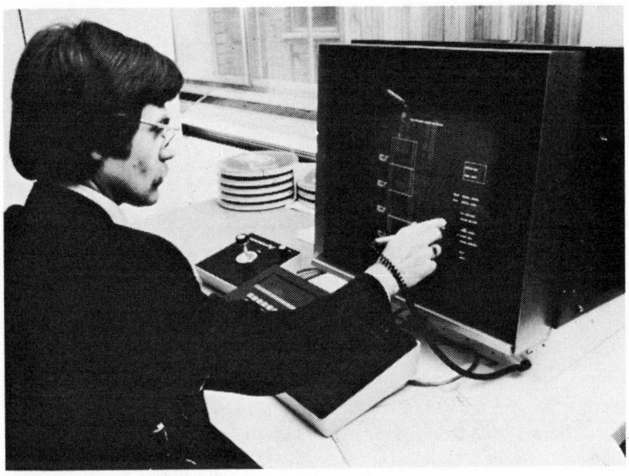

Fig. 10.

Dr R. HOBBS (Lloyd's Register)
I would like to discuss the use of computer graphical techniques for predicting drivability, in particular using a visual display unit.

48. Using a wave equation program and a PDP mini-computer, data can be quickly set up by interaction with the visual display – using light pen, joystick and keyboard (Fig. 10). First the pile make-up is input by selection of options on the right hand side of the tv screen using the light pen and by typing in the required data. In the same way, curves of dynamic skin friction, end bearing, damping values, quake and so on can be fed in. The results themselves can be displayed on the screen in a number of ways.

49. Figure 11, for example, is a plot of the variation of velocity with time at a point near the head of the pile as indicated on the sketch on the left. Also shown is the mud line, just above the toe, so this is a fairly early stage in the driving operation. The hammer hits the top of the pile at time zero, and the wave soon reaches the point under consideration. Later on, at about 0.06 s, the pile starts moving upwards (shown by negative velocity) as the wave returns from the tip.

50. There is a facility which enables plots for different points to be displayed together, and similar graphs of displacement, stress and force with time can also be drawn.

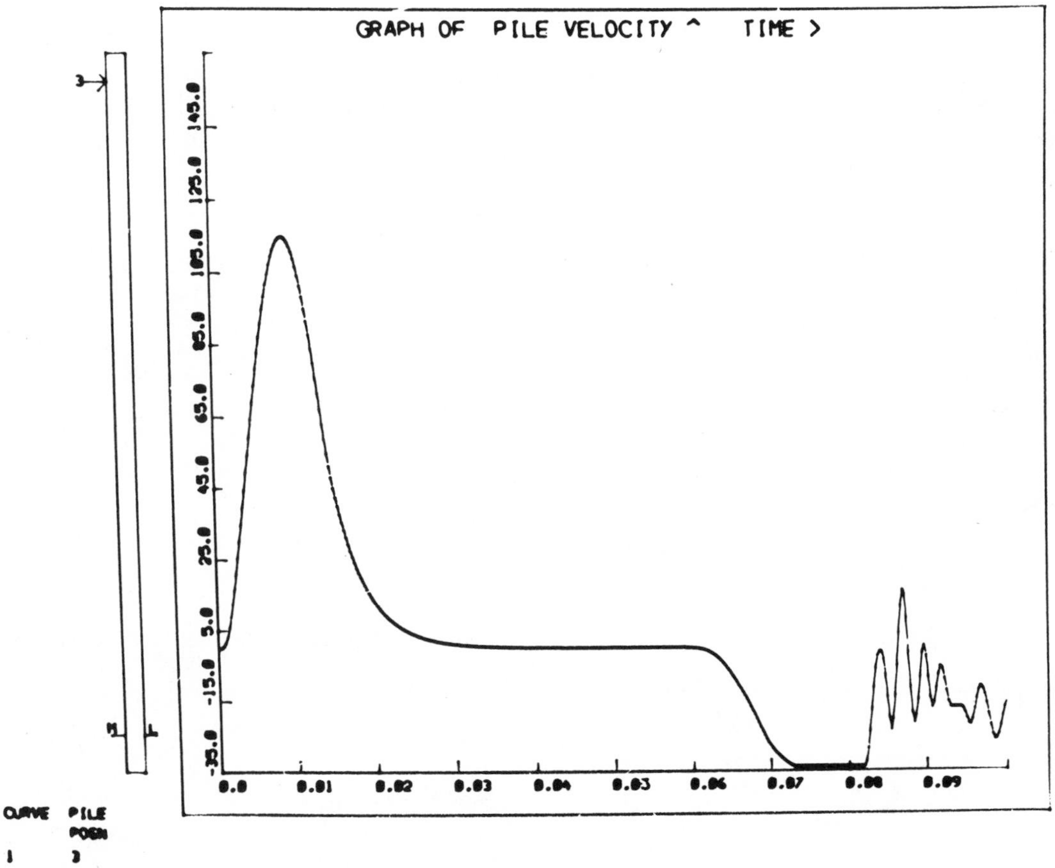

Fig. 11.

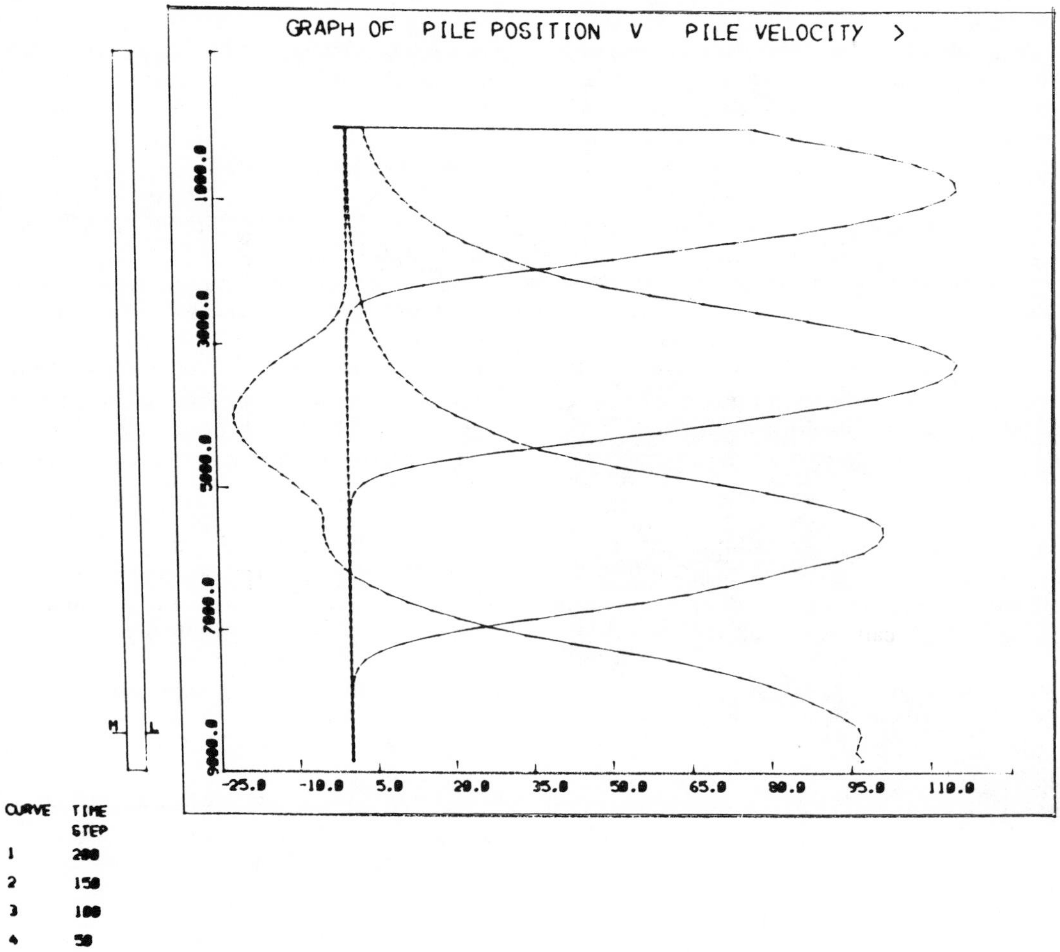

Fig. 12.

51. Another way of presenting the output is in the form of with-depth plots. Fig. 12 shows velocities at four different times. For the last time-step plotted, part of the pile is moving upward.

52. Maximum stresses in the pile over the full driving period that one considers in the analysis can also be drawn. Copies of the pictures can be obtained by taking a photograph or by outputting the data to a graph-plotter.

53. The advantages of the program are apparent. One can select and plot data very quickly and this can aid the engineer's assimilation of the results. The program also allows the engineer to do parametric studies with a minimum amount of new input because the model is stored in the computer. The use of a visual display unit means that data can be checked before the computer run is allowed to proceed.

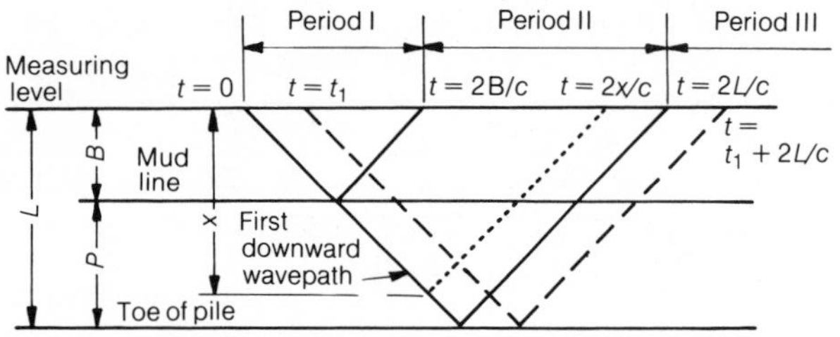

Fig. 13.

Mr VOITUS VAN HAMME

Assessing the driving resistance encountered during piledriving by analysis of measurements of the force–time histories and acceleration–time histories obtained for a certain level near the pile top is an important method of obtaining indications of the capacity of the pile.

55. It is known that, during piledriving, downward and upward stress waves are produced. Obviously the information on the resistances is brought to the surface by the upward wave. Thus it is necessary to derive the time history of the upward wave from the measurements.

56. The relation between the magnitude of both waves ($F_{\downarrow}$ and $F_{\uparrow}$) and the measured total force (F) and velocity (v) is given by the following formulae:

$$F = F_{\downarrow} + F_{\uparrow}$$

$$vEA/c = F_{\downarrow} - F_{\uparrow}$$

From these the magnitude of both waves can be calculated:

$$F_{\downarrow} = \tfrac{1}{2}[F + (EA/c)v]$$

$$F_{\uparrow} = \tfrac{1}{2}[F - (EA/c)v]$$

Here the following conventions are used: compressive forces are positive; the positive direction in the pile is downward and thus downward particle velocity is positive; the variable quantities F, v, $F_{\downarrow}$ and $F_{\uparrow}$ are functions of place and time;

if referred to the level of the measuring devices the place is invariable.

57. Three periods can be distinguished in the upward wave (Fig. 13). The origin of time $t = 0$ is chosen at the instant that the first wave passes the measuring level. In period I, from $t = 0$ to $t = 2B/c$, $F\uparrow = 0$. In period II, from $t = 2B/c$ to $t = 2L/c$, $F\uparrow$ increases from zero to half the value of the total dynamic skin friction. At time $t = 2x/c$ the value of $F\uparrow$ is equal to half of the total dynamic skin friction between mud line and depth x. In period III, $t > 2L/c$, $F\uparrow(t_1 + 2L/c)$ equals total dynamic resistance minus $F\downarrow(t_1)$.

58. So this analysis provides the magnitude of both skin friction and skin friction plus toe resistance as actually encountered along the various wave paths. However, the resistances are not constant but are functions of time (and place). Resistances vary due to damping and thus depend on the local particle velocity $v(x,t)$. The skin friction changes sign when $v(x,t)$ changes sign, which certainly occurs towards the end of the movement produced by the blow which is being analysed.

59. These influences cannot be derived from the measurements as referred to by Dr Goble. It may be possible to gain some insight if blows can be analysed for approximately the same penetration but with different hammer energies.

60. Therefore, in order to obtain the 'basic values' of the resistances it is necessary to know how the resistances really depend on the velocity and to estimate the mean value of the velocities along the various wave paths. Then the 'reduced' resistances can be input to a piledriving computer program which should yield a better estimate. So in fact some iterations are necessary. However, if the first trial is based on the resistances as measured directly, a few iterations should suffice.

61. The success of the method depends on how near the soil model is to reality. Is the damping really proportional to velocity? Is quake as postulated by most authors a reality? The writer believes that neither is true. 'Static' quake (i.e., a certain displacement) in his opinion should be replaced by a 'dynamic' quake (i.e., a certain small velocity, below which the soil does not yet show plastic flow). Probably the damping factor decreases with increasing velocity. Based on considerations of plastic flow and on the elastic response at low velocities the writer has developed a tentative model for skin friction given by the following formula for the shear stress (τ) exerted in the pile:

$$\tau = \tau_0 + K \left\{[1 + 2(Z_s v - \tau_0)/K]^{1/2} - 1\right\} \qquad \text{for } v > \tau_0/Z_s$$

$$\tau = Z_s v \qquad \text{for } v < \tau_0/Z_s$$

where

τ_0 is the basic value of the shear stress

K is a factor with the dimension of a stress

Z_s is the 'soil impedancy' against shear stresses (dimensions for instance kN s/m^3)

v is particle velocity of the pile.

These are tentative formulae, not yet tested sufficiently by measurements. Some confirmation is found by analysis of laboratory measurements performed by Heerema.[7]

62. It is (in theory) possible to perform measurements of local friction during driving of piles with appropriate instrumentation. Such measurements have been performed at piledriving tests at Hoogzand-Oostermeer, however, only with limited success: the average local friction found in these tests seems to be acceptable, but the friction–velocity relation could not be found. These measurements require very sensitive strain gauges and accelerometers, and electrodynamic influences on the devices and the leads should be avoided (if that is possible!).

63. Concerning the numerical results obtained by the Authors of Paper 4, the distribution of the basic values of the skin friction and of the distribution of the damping factors is not consistent if the results for the three penetrations are compared: there are enormous differences in the damping factors of the same soil layers found for pile penetrations of 46.4 m, 50.6 m and 54.9 m. Moreover due to 'soil fatigue', the skin friction exerted by a certain layer decreases as the penetration increases; otherwise, fairly constant blow counts during the penetration of, say, 10 m (see Fig. 16 of Paper 4) could not be obtained. The calculations referred to in the Paper seem to give results contrary to this fact.

Dr GOBLE & Dr RAUSCHE (Paper 4)

In reply to Dr Appendino, soil measurements are not particularly easy, but they can be accomplished and we think they are absolutely necessary if one wants to understand what is happening in pile driving. One must have a motion, and acceleration is the most attractive motion to measure because one has a simple reference frame.

65. We agree with Mr de Jong concerning the importance of the initial stresses in the soil.

66. Getting a match of a computed force with a measured force for the first $2L/c$ time is generally simple and quick convergence can be expected. At the end of matching these two force curves, the computed and the measured, or with the first $2L/c$ time, one only has a resistance and one cannot separate it into that which is proportional to the velocity or damping and that which is proportional to the displacement according to the model which we use. Therefore, another $2L/c$ time period is used in order to divide the resistance into displacement-related resistance and velocity-related resistance.

67. If the measurements and the model are correct, one can match these two records quite quickly for a $4L/c$ time period. The model that we use is quite reliable for dealing with skin friction. The problems arise from the reflections coming directly from the tip, and they tend to be more difficult with closed-end piles than with open-end piles. In some cases it is impossible to achieve as desirable a match as one would like to have. We are generally ascribing that to problems with the soil model. There is clearly work that could be done to improve the model by dealing with problems of the model at the pile tip.

68. With reference to the results presented in our Paper, there was an interruption of driving between the time when the analysis was made and the time when the later one was made. This would, of course, set up effects which would affect the resistance.

Mr HEEREMA & Mr de JONG (Paper 5)

In answer to Professor Poskitt, it is indeed true that the plug does not tend to move down with the pile wall, and that a large shear stress is needed to change this tendency. However, a shear stress of 2000 kN/m^2 would not be needed, provided the plug gets enough time to make its settlement; which it does, as it makes its final settlement considerably later than the pile tip itself.

70. The gap basically does discourage plugging, but it should be kept in mind that this radial expansion is of the order of 0.1 mm only and therefore is not too effective. It also appears in practice that inside friction during driving can be very high, often higher than outside friction.

Dr APPENDINO

I should like to refer to again the simplified pile–soil interaction model described in my contribution to the discussion on Paper 4 (Fig. 6), which can be applied when the pile is long, and the soil resistance is concentrated at the tip. This simplified model is very suitable for generating sets of parametric solutions, which may be useful for determining the role of each parameter, and interpreting measured data by a fitting technique.

72. Figure 14 is an example of the use of parametric solutions, where L_I is the incident energy, F_I is the peak value of the incident force, r_2 is the lumped damping, k_2 is the static lumped stiffness and $F_{2,0}$ is the static soil resistance. The curves were obtained by letting the parameters vary, one at a time, from the values which were obtained by interpretation of data from an instrumented pile at the final depth.[4,8] They show the effect of every parameter on the blow count. Table 1 illustrates the effect of a ±25% variation of every parameter on the blow count. This kind of elaboration gives the way to determine the precision required for every quantity so as to achieve a given precision in the computation. The strong non-linearity of the curves suggests that the precision may vary widely from case to case.

73. Figure 15 shows how a parametric solution may be used to interpret measured data. All curves, with the exception of the permanent set $x_{2,0}$, represent peak values. It is possible to note how the convergence of measured and computed values falls in correspondence with the central

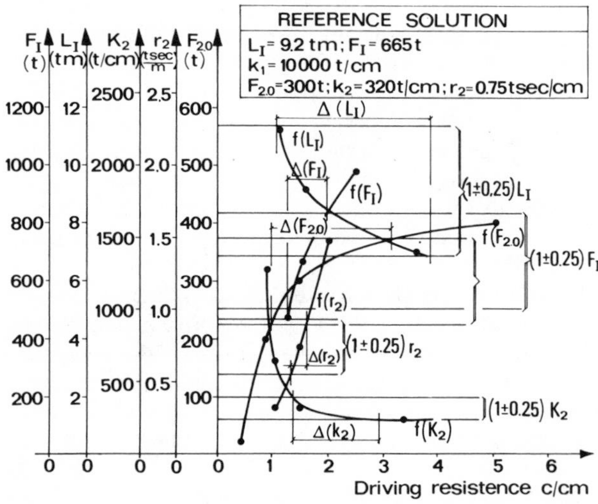

Fig. 14. Influence of the variability of every parameter controlling pile–soil interaction

Table 1. Effect of a ±25% variation of every parameter around the value corresponding to the solution L_I = 9.2 t m, F_I = 665 t, $F_{2,0}$ = 300 t, k_2 = 320 t/cm, r_2 = 0.75 t s/cm, set = 1.5 blows/cm

Parameter	Permanent set, blows/cm		Set variation between ±25%, blows/cm
	Parameter 0.75 of reference value	Parameter 1.25 of reference value	
L_I	1.1	3.9	2.8
F_I	1.25	1.95	0.70
r_2	1.33	1.60	0.27
k_2	1.35	2.40	1.05
$F_{2,0}$	1.45	3.10	1.65

net of curves, built assuming soil lumped parameters as follows: static resistance $F_{2,0}$ = 350 t, static stiffness k_2 = 350 t/cm and damping constant r_2 = 0.50 t s/cm.

74. A different physical interpretation of soil lumped parameters (Fig. 6) may contribute to the understanding of soil interaction. Static stiffness (which is the ratio of static force to quake) may be defined by the relationships given for deep foundations embedded in an elastic medium:

$$k_2 = \alpha \frac{GA^{1/2}}{1-\mu} \qquad (4)$$

where G is the soil shear modulus, μ is the soil Poisson coefficient, A is the pile tip surface and α is a coefficient which may be determined from the formula given by Janbu et al.[9]

75. Static resistance, corresponding to soil reaction when pile tip velocity is zero, may be defined using the Vesić[10] static solution based on the expanded cavity theory. When the pile tip penetrates a sand layer and no variation of pore pressure occurs, the bearing capacity is given by

$$F_{2,0} = AF_q q'_{\text{oct}} = Aq'_{\text{oct}} \frac{3(1+\sin\varphi')}{3-\sin\varphi'} [I_{RR}]^{\frac{4\sin\varphi'}{3(1+\sin\varphi')}} \qquad (5)$$

where

$$I_{RR} = \frac{I_R}{1+\Delta I_R}$$

$$I_R = \frac{G}{q'_{\text{oct}}\tan\varphi'} \qquad (6)$$

q'_{oct} being the effective octahedral confining pressure, φ' the effective friction angle, I_{RR} the rigidity index and Δ the volume deformation in the plastic zone.

76. Equation (4) may be used to express G in equation (6) as a function of k_2. Thereby expression (5) also becomes a function of k_2. This means that lumped stiffness and static resistance are not independent quantities.

77. It is possible to construct the curves shown in Fig.

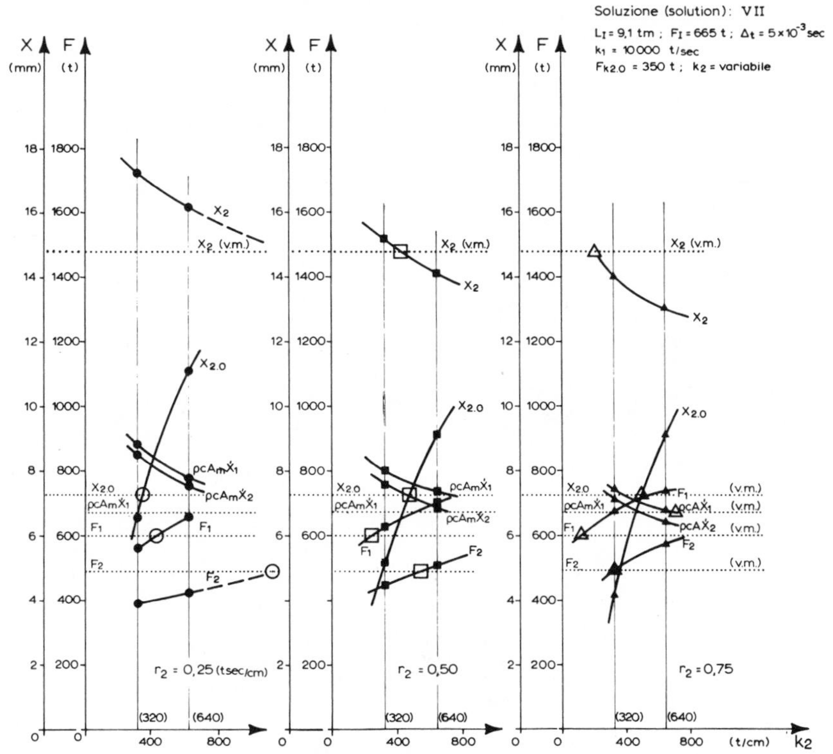

Fig. 15. Example of the use of parametric solutions to interpret measured data

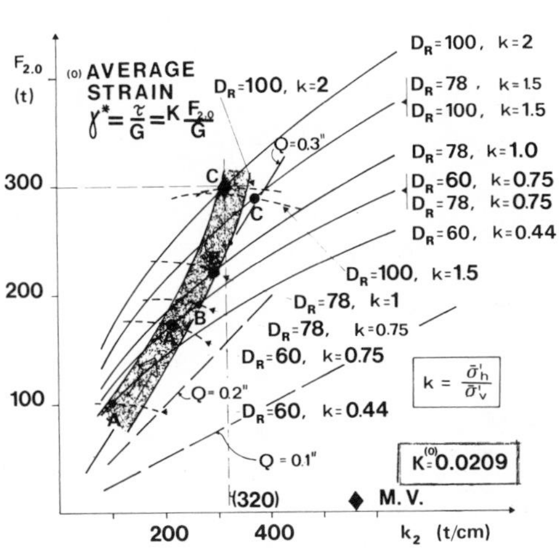

Fig. 16. Resistance dependency from lumped stiffness at different relative densities and confining pressures

16 for different soil densities (D_R = relative density) and ratios k of horizontal to vertical effective soil pressure. If we impose compatibility between the average soil deformation and the average soil modulus (dashed curves) we determine a point on every curve. It represents the unique combination of stiffness and resistance values that may be attributed to a soil with given properties.

78. The curve passing through every point represents how the static stiffness and resistance vary as the soil becomes denser and more compressed, owing to the driving of a number of piles (group effect).

79. The use of the expanded cavity solution allows one to imagine that the resistance at the pile tip is the effect of the friction occurring in the plasticized zone, on which an elastic force proportional to the soil stiffness is acting. Under dynamic conditions the static stiffness is incremented by the dynamic stiffness deriving from inertial effects (elastic waves radiation) at the elastic boundary.

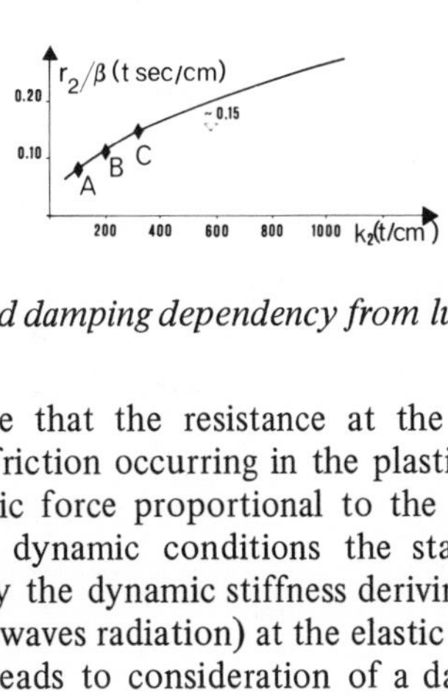

Fig. 17. Lumped damping dependency from lumped stiffness

80. This leads to consideration of a damping lumped parameter as representative of the stiffening effect of wave radiation. It results:

$$r_2 = \beta \frac{AG^{1/2}}{1-\mu}$$

where β is an unknown coefficient.

81. It is evident that damping also may be expressed as a function of soil static stiffness as in Fig. 17, points A, B and C corresponding to soil conditions defined in Fig. 16. If we define a dynamic stiffness $k_2^*(t) = F_2(t)/x_2(t)$, we obtain $k_2^*(t) = k_2 + r_2 x_2'(t)/x_2(t)$. Values of $k_2^*(t)$ may be introduced in Fig. 16 to obtain the resistance $F_2(t)$ (Fig. 18). A check with the measured values shows a fair agreement between computed and measured values.

82. This approach allows a link to be established

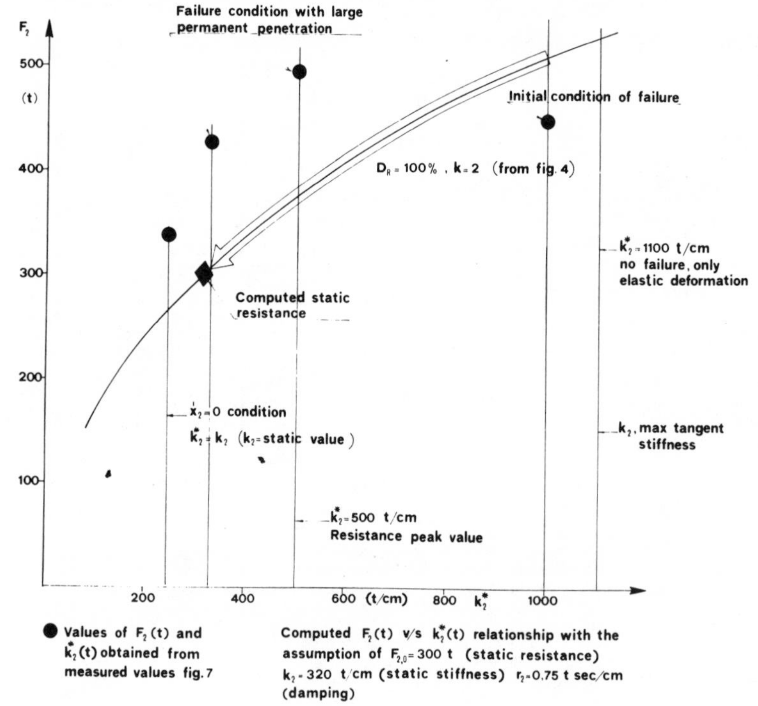

Fig. 18. Dynamic resistance dependency from dynamic lumped stiffness defined by $F_2(t)/x_2(t)$

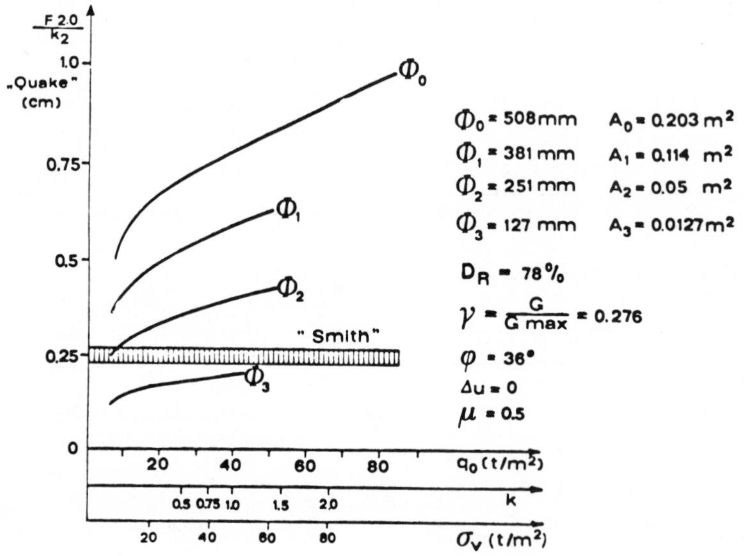

Fig. 19. Effect of pile diameter and soil confining pressure on the quake

between soil lumped parameters and soil mechanics properties and pile dimensions. In Fig. 19, for instance, it is shown how the quake is strongly dependent on pile diameter and – in a lower degree – on the octahedral effective pressure. Therefore the use of a constant quake of 2.5 mm is a good approximation only for small diameter piles driven to shallow depths.

83. These relationships also give a procedure for computing soil lumped parameters corresponding to a static loading from lumped parameters obtained from driving data. The procedure[5] is based on the ratio between static and dynamic soil properties and takes into consideration the different loading conditions occurring at the pile tip as a consequence of different rates of loading.

Mr J. BEERS (Netherlands Offshore Company)
I would like Mr van Luipen to give his definition of mechanical efficiency, and say what the value is when the pile has a batter. Our experience is that the total efficiency of the Menck 3000, with a good working hammer, is about 80%.

85. The pile driver impact velocity is $2^{1/2}$ × efficiency × acceleration of gravity × drop height. Taking 80% as

efficiency factor gives a good relationship between the computed blow counts and the actual pile driving records.

Dr HOBBS

Figure 8 of Paper 6 shows results obtained with blow counts as high as 350 and 380 for 0.25 m. What limit do Menck put on blow count for a given distance?

87. On a recent installation that I attended, the offshore contractor was using an asbestos cushion sandwich. The 3 mm asbestos cushion above and below the Bongossi cushion appeared to reduce the life of the Bongossi cushion and perhaps increase the ability to drive the pile and reduce the blow count. What are your comments on using that type of cushion?

PROFESSOR BUTTERFIELD & Dr DOUGLAS

Interest has been expressed in the question of how the stiffness contribution of each pile in a closely spaced group decreases as the number of piles in the group is increased. Such information is of considerable practical importance, in that the design of large pile groups is more usually governed by their load–displacement behaviour than by their ultimate load capacity (i.e., the design is based on acceptable displacement limits); and, by analogy with the well established analyses of the dynamic load response of machine bases and footings, it is likely that the key, low frequency, dynamic stiffness of pile groups will also be essentially identical to the static stiffness.

89. The following three points are extracted from a forthcoming publication[11] relating to stiffness analysis of pile groups.

90. It is probably helpful practically to discuss the stiffness of an N pile group in terms of a stiffness reduction factor (ρ) rather than the alternative 'group settlement ratio'. If the vertical stiffness of the N pile group is K_N, and K_S is that of a similar, isolated, single pile, then the stiffness reduction factor interrelates them via

$$K_N = \rho N K_S \tag{7}$$

(i.e., the group stiffness is obtained by multiplying the number of piles in the group by the 'reduced' (ρK_S) stiffness of each pile).

91. Figure 20 presents a summary of reduction factor values obtained using the Department of the Environment PGROUP elastic pile analysis program for a practical range of $\lambda = E_{pile}/G_{soil}$ values, with the piles in groups at spacing-to-diameter ratio (S/D) 2.5, embedded in an infinitely deep elastic layer ($H/L = \infty$). The curves relate strictly to piles in square arrays ($N \leqslant 100$) but the added points for 4 × 2 and 8 × 2 groups, similarly spaced, also lie very close to them. The ρ values in Fig. 20 are conservative in the sense that ρ increases if either a rigid layer occurs at any depth below the pile toe or the pile spacing is increased (e.g., for $S/D = 5$, $H/L = 3$, $N = 40$ and $\lambda = 1000$, $\rho = 0.2$ rather than the Fig. 20 value of $\rho \approx 0.1$). Rather surprisingly the ρ values proved to be relatively insensitive to the pile length-to-diameter ratio ($15 \leqslant L/D \leqslant 50$) and the results were averaged over this range with maximum errors in ρ, for large groups, typically within ±35% of the plotted values.

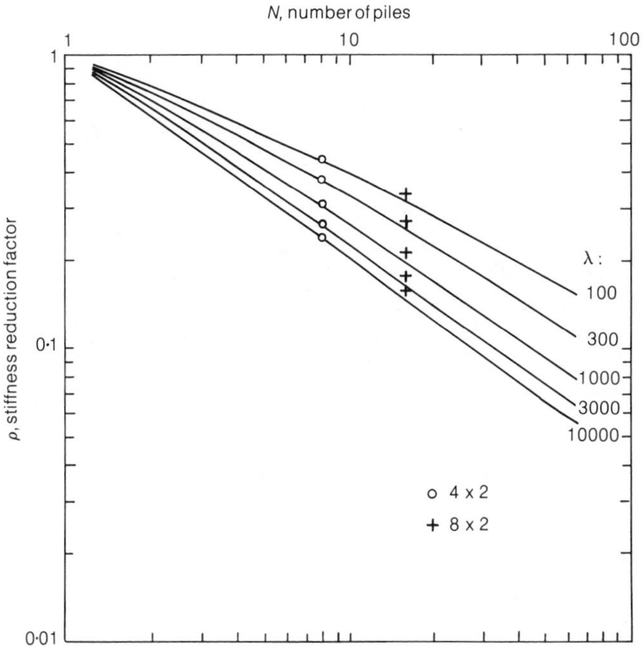

Fig. 20. Stiffness reduction factor versus number of piles; $H/L = \infty$, $S/D = 2.5$

92. In order to use equation (7) the designer needs K_S for the same H/L value. Values are given in Table 2 in the form of a dimensionless vertical stiffness $K_S{}^* = K_S/GD$ ($H/L = \infty$), with the implication that the relevant soil shear modulus (G) might be assessed from, say, the initial, linear portion of the conventional in situ load test on a single pile. (The values of ρ (Fig. 20) are not at all sensitive to the selected value of the modular ratio $\lambda = E/G$.)

Table 2. Dimensionless vertical stiffness ($K_S{}^*$) of single piles ($H/L = \infty$)

$\lambda = E/G$	Length/diameter (L/D)				$K_S{}^* = K_S/GD$ average
	15	25	35	50	
100	11.63	11.82	11.72	11.38	11.64
300	17.24	18.66	18.98	18.94	18.45
1000	22.73	28.01	30.40	31.75	28.22
3000	25.45	34.74	40.98	46.73	36.97
10000	26.67	38.31	48.98	59.88	43.23

Mr VAN LUIPEN & Mr JONKER (Paper 6)

In reply to Mr Beers, we define mechanical efficiency of single acting steam hammers as the ratio of the kinetic energy of the ram at impact and the maximum potential energy of the ram, taking the actual stroke into account. The losses incorporated in the mechanical efficiency are the steam exhaust and the friction on the ram. We present a mechanical efficiency of 96% for a vertical pile in Fig. 8 of the Paper. Due to conditions encountered offshore, including pile batter, we envisage a somewhat lower mechanical efficiency, but this should normally not be lower than 90%. Factors additional to the mechanical efficiency may reduce the maximum rated energy of the hammer; i.e., reduction of the maximum potential energy due to the pile batter (in general less than 1%) and due to shortfall.

94. Mr Beers has not given his definition of total efficiency. We understand that it is the same as wave equation efficiency. We define this as the ratio of the kinetic energy of the ram at impact as used for wave equation computations and the maximum rated energy of the hammer. Incorporated in this wave equation efficiency are – apart from the factors quoted above – losses below the impact level, which are difficult to quantify in other wave equation parameters, and a certain safety margin. Therefore, this wave equation efficiency is much lower than the mechanical efficiency.[12]

95. In reply to Dr Hobbs' question concerning Fig. 8 of the Paper, blow counts of 350 and 380 blows per 0.25 m were indeed given. As a general rule we state a driving limit of 200 blows per foot. If all operating details are known, it may be possible to deviate from this rule. An example on pile refusal given by API[13] has very high blow count figures. One should be careful in applying these figures for North Sea conditions, which are very different from the driving conditions in the Gulf of Mexico. Under most North Sea circumstances a progressive increase in blow count is experienced above a certain blow count limit. Thus, only a marginal increase in pile penetration is obtained, while the pile driving hammer is highly stressed. The high blow counts in the test with the MRBS 8000 were due to the special test circumstances, where the comparison drive was done at a high soil resistance.

96. Drivability studies with the aid of the wave equation program show that the blow count of a certain pile–hammer–soil combination is not only dependent on the spring constant and the coefficient of restitution of the cushion, but also on the relative height of the soil resistance during driving (SRD). Two wave equation curves showing the relation between SRD and blow count for two different cushions may intersect each other or not. Only a later analysis using the best data possible for such a sandwich cushion could give an answer to the question regarding the drivability. The asbestos sheets above and below the Bongossi could isolate the hardwood, thus keeping the developed heat inside. A shorter lifetime for the Bongossi could then be expected.

REFERENCES

1. BURGESS I. W. Analytical studies of pile wandering during installation. Int. J. Numer. Anal. Meth. Geomech., 1979, 3, 49–62.
2. DOUGLAS R. A. Slender piles. PhD thesis, Southampton University, 1979.
3. KUEH Y. S. The stability of long slender piles during installation. MSc project report, Southampton University, 1978.
4. APPENDINO M. Analysis of data from instrumented driven piles. Proc. 9th Int. Conf. Soil Mech., Tokyo, 1977, 2/1, 359–370.
5. APPENDINO M. et al. Driven piles: measurement and interpretation procedures. Design parameters in geotechnical engineering: proc. 7th Eur. conf. soil mechanics and foundation engineering, Brighton, 1979. British Geotechnical Society, London, 1979, 3, 9–14.
6. HOLLOWAY D. M. et al. The mechanics of soil pile interaction in cohesionless soils. School of Engineering, Duke University, Durham, North Carolina, 1975, soil mechanics series 39.
7. HEEREMA E. Ground Engineering, 1979, Jan.
8. APPENDINO M. Discussion contribution. Recent developments in the design and construction of piles. Institution of Civil Engineers, London, 1980.
9. JANBU N. et al. Publication 16, Norwegian Geotechnical Institute, Oslo, 1956.
10. VESIĆ A. S. Expansion of cavities in indefinite soil mass. J. Soil Mech. Fdns Div. Am. Soc. Civ. Engrs., 1972, 98, Mar., No. SM3, 265–290.
11. BUTTERFIELD R. and DOUGLAS R. A. A parametric study of pile group stiffness matrices. Written for Construction Industry Research and Information Association, London.
12. VAN LUIPEN P. Experience with heavy pile driving hammers in the North Sea. Proc. Offshore Brazil 78 Conference. CELP, Rio de Janeiro, 1978, OB-78.13.
13. AMERICAN PETROLEUM INSTITUTE. API recommended practice for planning, designing and constructing fixed offshore platforms. API, 1977, API RP2A, 8th edn, para. 6.12.h.

Discussion on dynamics (papers 8–10)

Miss Y. BARTON (University of Cambridge)
With respect to the static analysis, I have found that the use of equivalent springs for foundations can be very misleading. It is all very well for the structural engineers to be given a set of equivalent springs, but very often one is just trying to model the effects of the soil rather than modelling the soil and its effects on the structure, and one has to be very careful in the formulation of this type of approach. It is all right having an impedance function, but one has to know what one puts into the impedance function in the analysis. The main problem which I came across when I first analysed the dynamic force on a structure was the problem of the effects of an earthquake sending seismic shear waves from bedrock through the foundations to the structure. The structure can be analysed quite effectively; much is known about structural response; but nothing is known of how the soil mass and the piles themselves attenuate and emphasize the shear waves before the data is used as input for the analysis. I would like to ask the three speakers on dynamics how they actually treat data from the field to attack the actual design problem.

2. May I draw attention to the Mindlin equation. The Mindlin solution is just one way of solving one particular Green's function for an integral equation. Green's function came first so may we concentrate on the fundamental solutions and not just one solution.

Dr H. MATLOCK (Paper 16)
The following is a general process which has recently been used in handling the earthquake problem. The soil motion analysis employed was the random type of analysis such as Professor Roesset has described. The model did have the capability to produce liquefaction, and the time-dependent dissipation of pore pressures was accommodated in the solution. The soil motion description was much as he has described, with the inclusion of the transmitting boundary to take care of reflections from below. The accelerations were introduced at a particular elevation. The pile was modelled as a discrete system in which the soil elements were based on *p–y* curves, but were actually fully hysteretic on gapping near the surface. Furthermore the pore pressures generated were transferred to the *p–y* curves. Complete liquefaction took place. The soil reactions would be eliminated completely. This was a sandy type soil. The pile was modelled representing the complete foundation of about 20 piles. Because of the spread of the foundation, the model effects and the rotational strain effects were input as additional linear strains at the soil surface. The structure itself was modelled as a stick model, so that it all became one stick from the soil up. The motions were transferred to the bases of the springs, and the resulting response of the whole system was carried out. When the liquefaction became severe, there were significant effects, not so much in the first mode but in some of the higher modes. The second and third modes of the structure were affected quite significantly.

4. I was disappointed to find that the problem seemed to be rather insensitive to the normal degradation of *p–y* curves with just a hysteretic degradation. It took a significant amount of liquefaction on top of the hysteretic degradation to produce interesting results.

Dr E. G. PRATER (Federal Institute of Technology, Zurich)
The Authors of Papers 8 and 10 treat in a thorough manner various theories for the lateral vibration of piles embedded in an elastic or viscoelastic medium. They remark, however, that the influence of the inelastic behaviour of the soil may be substantial. Under these conditions the interaction problem is usually handled by using a more simplified structural model, in which the soil continuum is replaced by a series of springs and dashpots acting independently of one another. Such a model is presented in Paper 9. One must, however, go a step further and allow the spring elements to respond non-linearly.

6. In static problems it is well established practice to incorporate a non-linear soil modulus in the calculations, by taking secant values to the *p–y* curves relating lateral pressure and displacement empirically at various depths. This approach is followed, for instance, in Paper 17. In passing it is observed that there is really no compelling reason, with the comparative cheapness of computer time today, for using a secant modulus approach instead of a tangent modulus approach.

7. Under conditions of dynamic excitation not only the loading modulus curve must be defined, but also the unloading modulus and possibly degradation effects under cyclic loading. The use of such models can account for the gap effect (i.e., the separation between the pile and the ground) which is especially important near the ground surface, where larger deformations occur. If the embedded medium comprises cohesionless soil it is likely that the gap will quickly close. In cohesive soil, however, complete closure during a load cycle is unlikely. In this case the effective embedded depth is reduced, with a consequent influence on the vibrational behaviour of the system. Thus the effect of the non-linearity is not only the introduction of hysteretic damping associated with the plasticity of the soil, but also to modify the flexural vibration characteristics of the pile–soil system.

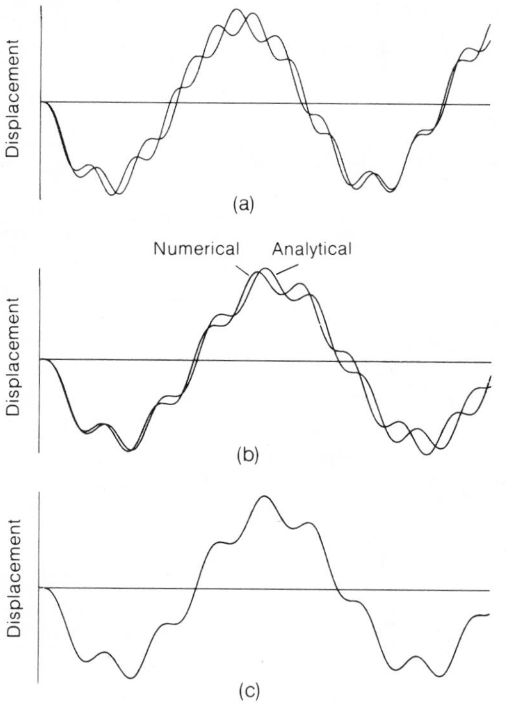

Fig. 1. Comparison of numerical and analytical solutions for beam on elastic foundation; applied load sinusoidal, $\omega = 0.2$ Hz: (a) $\Delta t = 2.1$ ms, $y_{num}/y_{anal} = 0.99$; (b) $\Delta t = 1.3$ ms, $y_{num}/y_{anal} = 1.00$; (c) $\Delta t = 0.13$ ms, $y_{num}/y_{anal} = 1.00$

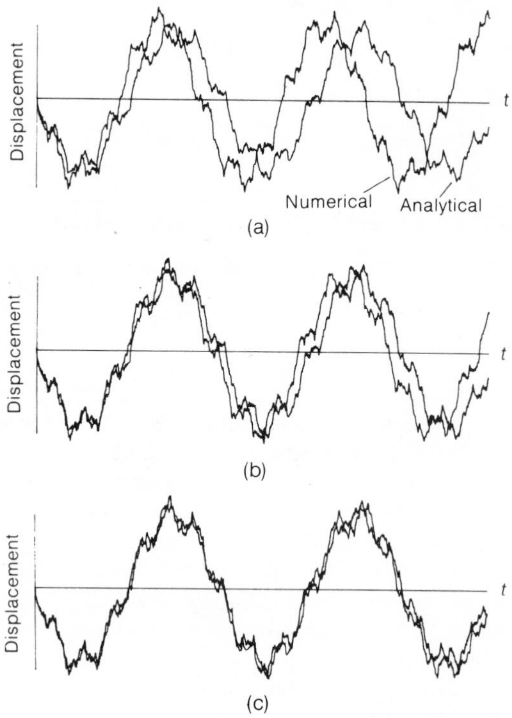

Fig. 2. Comparison of numerical and analytical solutions for beam on elastic foundation; applied load Dirac pulse; (a) $\Delta t = 2.1$ ms, $y_{num}/y_{anal} = 1.15$; (b) $\Delta t = 1.3$ ms, $y_{num}/y_{anal} = 1.06$; (c) $\Delta t = 0.13$ ms, $y_{num}/y_{anal} = 1.04$

8. Two topics concerning the flexural vibration of such a system are the accuracy of explicit numerical procedures, and the model for separation of the pile and the soil.

9. Generally, the most economic solutions to non-linear wave propagation problems are obtained using explicit numerical procedures, so that a system of linear equations does not have to be solved after every time step increment. The writer has presented such a method elsewhere,[1] in which the dynamic relaxation algorithm was used to produce a solution in the time domain. The algorithm is virtually the same as that presented by Richtmyer and Morton.[2] The condition for stability of a simple thin beam was given by Collatz[3] as

$$\Delta t < \frac{\Delta x^2}{2}\left\{\frac{m}{EI}\right\}^{1/2} \qquad (1)$$

where Δt is time step, Δx is discretization in length of beam, m is mass per unit length of beam, and EI is flexural rigidity of beam.

10. If, in addition, the beam is loaded axially a modified form of equation (1) is required.[2] The writer has performed some numerical calculations to compare the results with a known analytical solution. It has been found that due to the parabolic form of the differential equation of flexural vibration the above condition, while ensuring stability, does not give accurate numerical results. The problem investigated was that of a beam on an elastic foundation acted on by a concentrated force applied centrally. Both a sinusoidal and an impulsive type loading were considered. The results are shown in Figs 1 and 2 for the displacement–time response at the centre. The comparison between the analytical solution and the numerical solution is made for three time steps, the largest corresponding to equation (1), the next being a value $\sqrt{3}$ smaller corresponding to a condition given by Stangenberg,[4] and the final time step being very small. The mesh size was kept constant at 1/20 the length of the beam. The time step must be made much smaller than that required for numerical stability in the calculations if high accuracy is desirable. The condition of stability for flexural vibration (equation (1)) already imposes a rather severe restriction on Δt compared with problems involving body wave propagation,

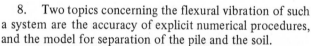

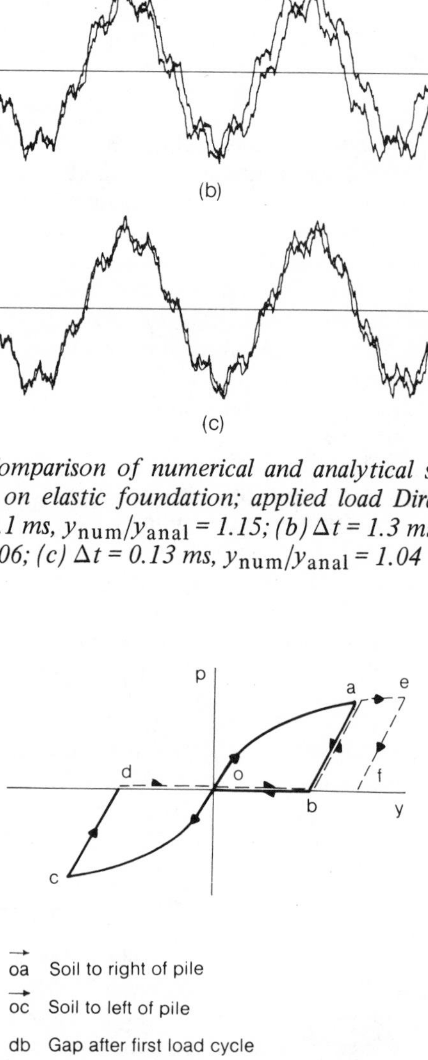

Fig. 3. Simple model for inelastic soil reaction giving separation between soil and pile during unloading

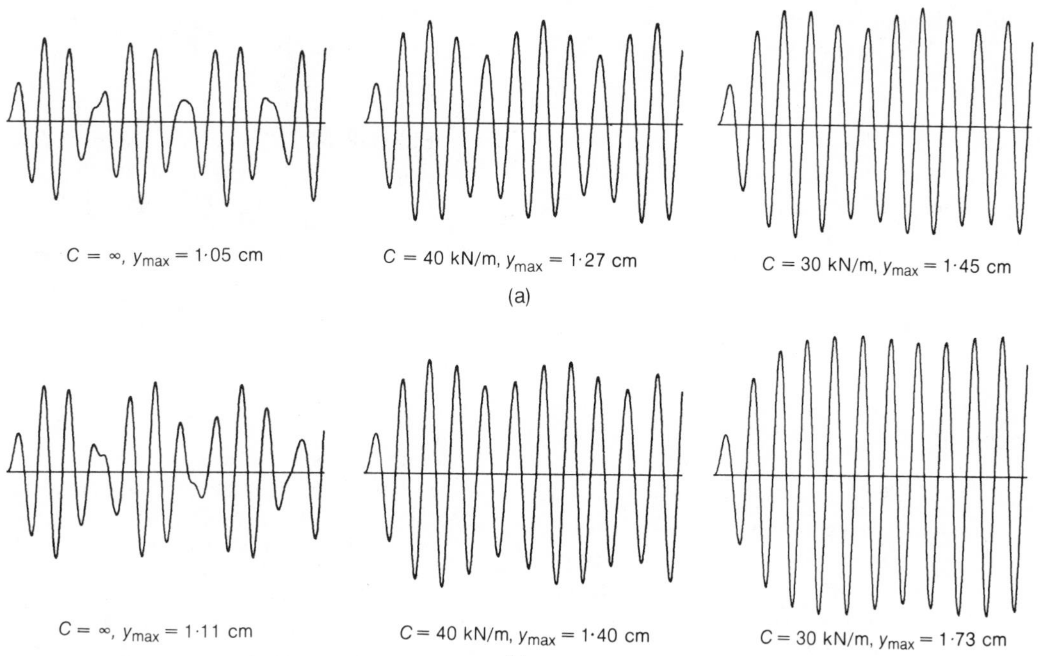

Fig. 4. Displacement–time response at top of pile; sinusoidal loading, frequency $\omega = 10$ Hz: (a) bottom end fixed; (b) bottom end free

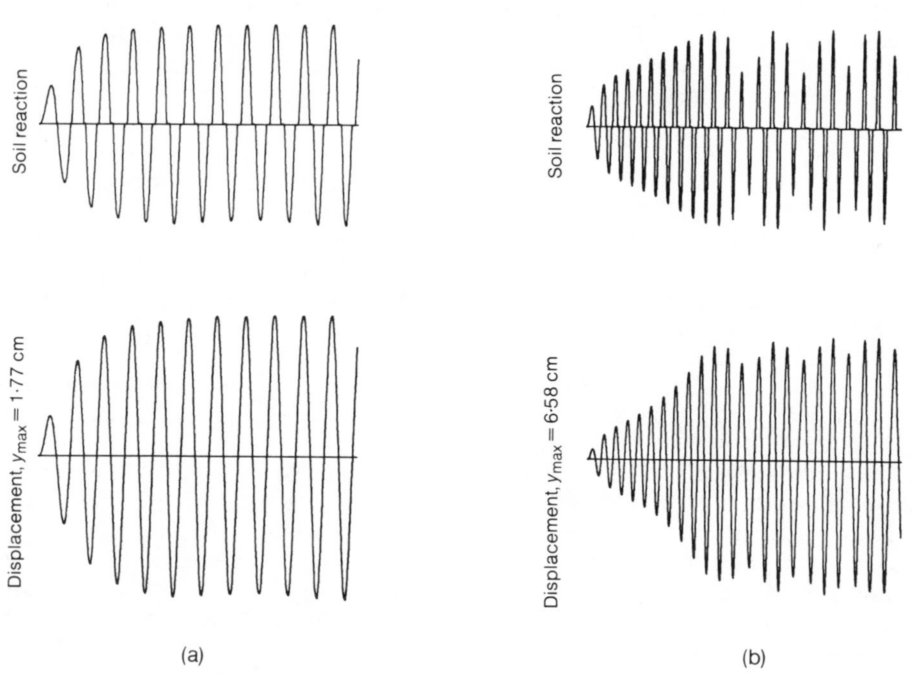

Fig. 5. Displacement and soil reaction time response to sinusoidal loading, $\omega = 10$ Hz; soil constant $C = 25$ kN/m: (a) duration $t \approx 1.1$ s, bottom end fixed; (b) duration $t \approx 2.7$ s, bottom end free

and the explicit method of solution is only practicable in this case due to the one-dimensional structural model for a beam or pile, which requires relatively few degrees of freedom in the discretization.

11. A simple model for simulating the inelastic soil reaction on the pile is shown in Fig. 3. The initial loading portion of the curve may be described by a function (e.g., hyperbolic, exponential) or by discrete values equivalent to a complex system of springs and sliders.[5] Some numerical results are shown in Figs 4 and 5 for the displacement at the top of a pile of length $L = 12$ m, with $m = 0.012$ t/m^2 and $EI = 50\,000$ kN m^2, subjected to a sinusoidal horizontal force $H = 5 \sin 20\pi t$ (kN). The lower boundary condition is for the pile tip fixed or free.

12. The soil resistance is governed by a hyperbolic law

$$p = \frac{y}{\dfrac{1}{E_i} + \dfrac{|y|}{p_u}}$$

in which E_i is assumed for the calculated examples to be constant with depth x and equal to 800 kN/m^2, and p_u varies according to the relation

$$p_u = C[2 - \exp(-x/4L)]$$

where C takes on values infinity, 40 kN/m, 30 kN/m and 25 kN/m. The soil resistance is also shown in Fig. 5 for the case $C = 25$ kN/m. The formation of a gap is clear from this figure.

13. The case for $p_u = \infty$ corresponds to elastic behaviour with the soil modulus varying exponentially with depth. The vibration behaviour of the non-linear system is seen to be modified by the form of the p–y curve, and there may be cases when a quasi-resonant condition may be masked in the analysis if the yielding of the soil is neglected.

PROFESSOR NOVAK (Paper 8)

In reply to Miss Barton, two main types of approach are used in the analysis of structural response to earthquakes. In the first approach, often called the complete soil–structure interaction analysis, a complete finite element model of the soil and the structure is exposed to an input seismic motion usually applied at the top of the bedrock (lower boundary of the model).

15. In the second type of approach, the site amplification analysis is performed first. In this step, stiffness and damping of the structure and its foundation are sometimes included, in which case the site response incorporates what has come to be known as kinematic interaction. In the second step, the response of the structure is investigated for the input motion obtained from the first step for the level of the foundation; the elasticity of the soil can be accounted for by means of the impedance functions. This second step of the analysis is known as analysis of inertial interaction.

16. With pile foundations it is most often assumed that the kinematic interaction is negligible. This assumption is supported by field experiments conducted in Japan.[6] The building, the piles and the soil were all instrumented and monitored. These experiments indicate that slender piles follow the motion of the ground except for their uppermost parts. The difference between the motions of the piles and the soil near the pile heads may be attributed to inertial interaction.

17. Dr Prater reports on his approach in which he accounts for soil reaction non-linearity and pile separation (gapping). These factors are certainly important even though their effect on dynamic response can sometimes be less significant than in the case of static loading.

18. Non-linearity and gapping are complex phenomena that are treated by most authors in terms of a lumped mass model. This model can be particularly successful when the experimental results are known beforehand and the theoretical model can be readjusted to yield an agreement with the experiments. A true prediction of the response on the basis of given soil properties is much more difficult because the non-linear p–y curve has to represent the whole stress field. More research into these problems is needed.

PROFESSOR ROESSET & Mr ANGELIDES (Paper 10)

Miss Barton is quite right in stating that a difference must be made between models used to predict structural response, simulating the effect of the soil, and models which would also allow one to determine the state of stresses and deformations in the soil. This distinction must be made not only for pile foundations but also for all soil–structure interaction problems. It is clear that care must be exercised in the formulation of these models and perhaps more importantly in the interpretation of the results. The difficulty in using at the present time a unique model to depict the complete problem lies in our limitations to obtain reliable three-dimensional non-linear constitutive equations for a specified soil. A great deal of research is being done in this area and significant improvements can be expected in the next few years.

20. If the earthquake motions can be defined in terms of a specific train of waves (whether the most commonly used vertically polarized shear waves or any train of body waves at arbitrary angles, or surface waves) the motion and stresses in a soil mass can now be obtained by a variety of methods (particularly assuming linear soil behaviour). The motions that would occur when a pile is inserted in the soil can also be studied and the filtering of high frequencies by the pile can be investigated (see for instance Blaney's work). The problem at present is not how to solve the mathematical problem but how to decide what trains of waves should be considered. There is very little data to help the designer with this decision. One would need for this purpose arrays of instruments to record earthquake motions not only at various depths but primarily in two orthogonal directions in a horizontal plane. Steps are now being taken to install such instrumental arrays.

21. It is clear finally that Mindlin's solution is just one particular Green's function. I do not see, however, any problem in referring to Mindlin's equations when this particular influence function is used. One could of course state that the Green's function for the problem, as derived by Mindlin, is being used, but this seems unnecessarily long.

22. Dr Prater discusses the need for non-linear solutions which incorporate non-linear soil behaviour, degradation and gapping effects. His points on the accuracy of numerical integration techniques and the need to distinguish between the time step requirements for stability and for accuracy are well made. They apply both to explicit and implicit integration schemes and they may be even more important for the latter if unconditionally stable.

23. A point which deserves some additional consideration is the use of the P–y curves for incremental dynamic analyses. These curves are intended to provide a lower bound for the stiffness of the soil springs when the pile is subjected to a steady state harmonic excitation in a given frequency for a sufficient number of cycles. As such they can be used for design purposes when considering a periodic force at the top of the pile of large duration. Their generalization to other types of excitation is open to question. Their use for incremental analyses in the time domain would be entirely inappropriate.

24. While it is possible to conduct non-linear analyses

such as those shown by Dr Prater using one-dimensional, non-linear Winkler springs, and these studies are of great value to investigate qualitative aspects of behaviour, the numerical results cannot be considered reliable unless they are substantiated by experimental data or by more sophisticated three-dimensional models with appropriate constitutive equations for the soil. It is in this area that a considerable amount of research is needed.

REFERENCES

1. PRATER E. G. Analysis of laterally loaded piles. Proc. 3rd Int. Conf. Numerical Methods in Geomechanics, Aachen, 1979, 1087–1096.
2. RICHTMYER R. D. and MORTON K. W. Difference methods for initial value problems. Interscience, New York, 1967, 2nd edn.
3. COLLATZ L. Zur Stabilität des Differenzenverfahrens bei der Stabschwingungsgleichung. Z. angew. Math. u. Mech., 1951, 31, 392–394.
4. STANGENBERG F. (ZERNA W. (ed.)). Berechnung von Stahlbetonbauteilen für dynamische Beanspruchungen bis zur Tragfähigkeitsgrenze. Konstruktiver Ingenieurbau, Ruhr University, Bochum, 1973, report 16.
5. MATLOCK H. et al. Simulation of lateral pile behavior under earthquake motion. Proc. American Society of Civil Engineers' Speciality Conf. Earthquake Engineering and Soil Dynamics, Pasadena, 1978, 600–618.
6. Proc. 6th Wld Conf. Earthqu. Engng., New Delhi, 1977.

Discussion on axial/lateral behaviour of piles and groups (papers 11–20)

Mr BAGUELIN & Dr FRANK (Paper 11)
Our paper deals with several theoretical studies using the finite element method. As to the mechanism of shaft friction it shows how, from the shaft friction curve obtained with the self-boring friction probe, one can derive the vertical shear law of the soil. There are now three self-boring tests for which, from the stress–strain curve obtained, one can derive elementary shear curves: the pressure-meter test, which corresponds to a horizontal shearing of the vertical planes at 45° to the radial and tangential directions; the shear-meter test, which is a horizontal shearing of the concentric vertical surfaces; and the friction probe test, which is a vertical shearing of the same surfaces.[1] Original reference dates for the three derivations are respectively 1972, 1976 and 1975.

2. These elementary shear curves are used at present for research purposes only. As regards the self-boring pressure-meter (PAF), the practical rules for designing foundations are obtained directly from the expansion curve of the test. These rules have now been published[2] and are relative both to the bearing capacity of shallow foundations and pile tips and to the settlement of shallow foundations.

Mr F. B. J. BARENDS (Delft Soil Mechanics Laboratory)
I feel that papers 11 and 12 discuss in the main the deterministic approaches for construction. One thing that is important in my view is the sensitivity of a model, or the possibility of a probablistic approach to understand something about the safety of the structure and the design.

Dr R. HOBBS (Lloyd's Register of Shipping)
This contribution describes a finite element study of soil–pile interaction for piles in clay which are subjected to quasi-static axial loading. Account has been taken of the constitutive non-linearity of the soil, assumed to be undrained, by idealizing it as an isotropic, linearly elastic/perfectly plastic material obeying the von Mises yield criterion and the Prandtl–Reuss equations.

5. The calculations have been performed using an incremental, iterative, initial stress program FEEPHO, previously employed in the analysis of embankments[3,4] and of offshore gravity structures such as the North Sea Frigg CDP1[5] and Ninian Central platforms as part of the independent design appraisal carried out for certification purposes.

6. Figure 1 shows the axisymmetric finite element mesh for a hypothetical 30 in (0.76 m) dia., 25 mm wall thickness open-ended pile embedded 20 m into very stiff to hard overconsolidated clays, reported[6] to be representative of the West Sole field in the southern North Sea. The mesh

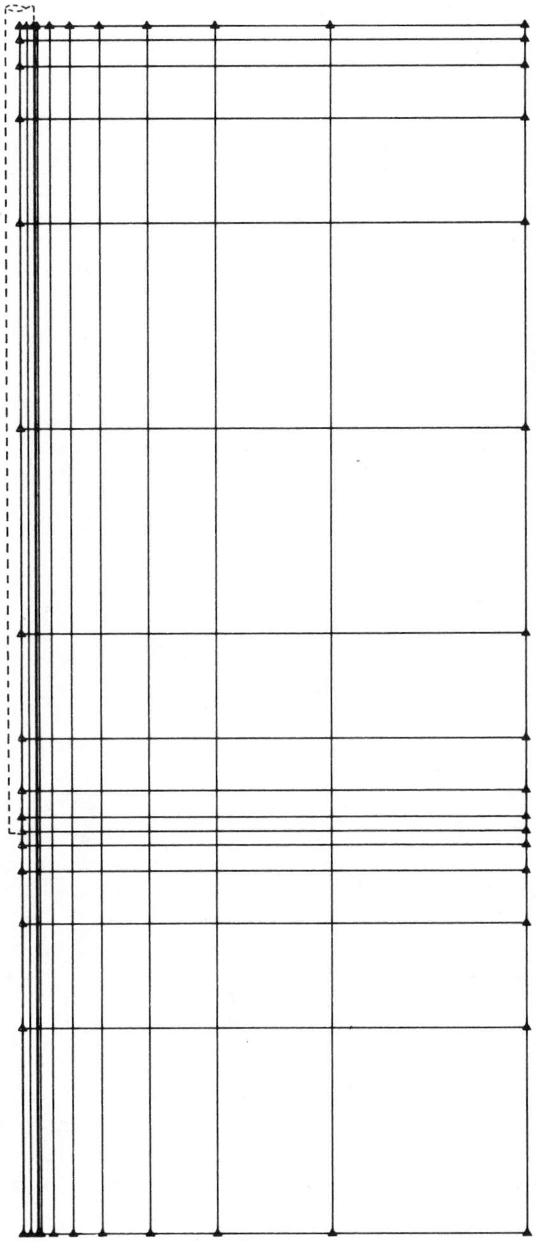

Fig. 1. Finite element idealization

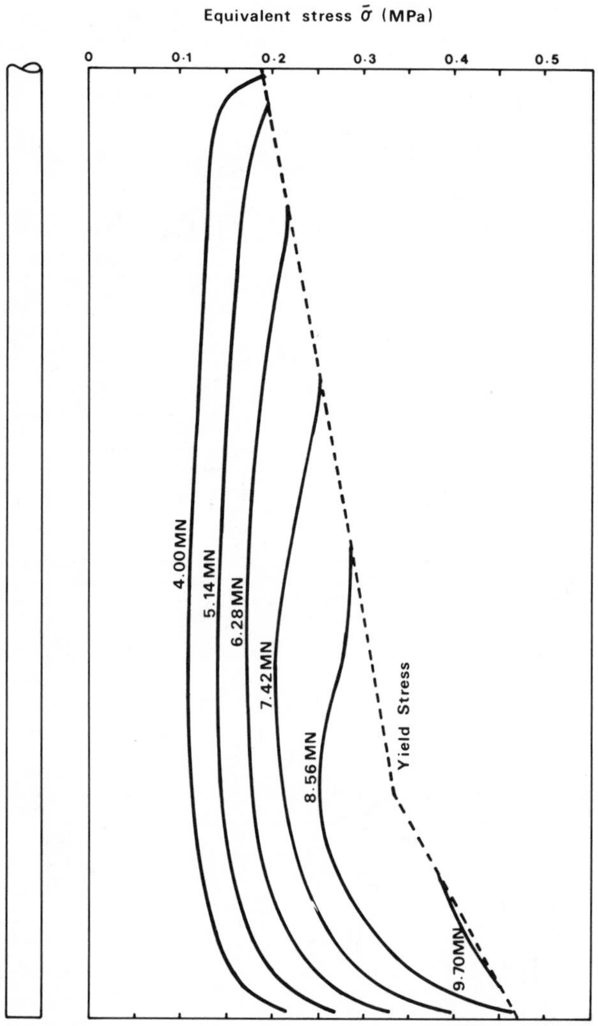

Fig. 2. Development with increasing load of equivalent stresses in soil–pile interface

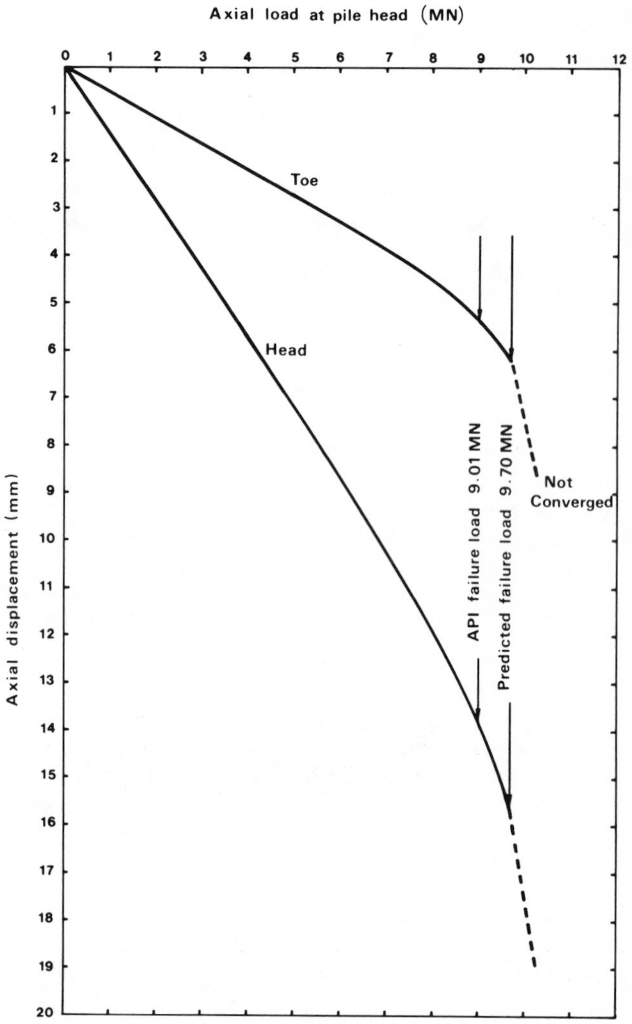

Fig. 3. Load–displacement behaviour

incorporates linear strain isoparametric quadrilateral elements.

7. In an attempt to model interface behaviour, soil within a distance of approximately one twelfth (60 mm) of the pile diameter outside the pile has been assigned modified properties. In this example, the undrained shear strength, C_u, in this zone has been taken as half that in the soil mass, in accordance with the skin friction parameters employed in determining axial capacity of piles in clays by the API α method.[7] No modification has been made to the properties of soil within the pile in the calculations reported here.

8. In the elastic range, Young's modulus, E, has been taken as 250 times the local undrained shear strength throughout the soil. Poisson's ratio, ν, has been taken as 0.48 to model the relative incompressibility of saturated undrained clay. The pile steel has been considered as a linearly elastic material with $E = 2.07 \times 10^5$ MPa and $\nu = 0.3$. In situ soil stresses have been assumed to be isotropic, and thus to have no effect on yield, although this is not a limitation of the program.

9. The computed development of equivalent stress

$$\bar{\sigma} = 2^{-1/2}\,[(\sigma_r - \sigma_v)^2 + (\sigma_v - \sigma_\theta)^2 + (\sigma_\theta - \sigma_r)^2 + 6\tau_{rv}^2 + 6\tau_{v\theta}^2 + 6\tau_{\theta r}^2]^{1/2} \tag{1}$$

(Fig. 2), where subscripts r, v and θ refer to the radial, vertical and hoop directions, indicates the progress of soil yield along the pile/soil interface with increasing axial pile load. At yield, $\bar{\sigma} = 2C_u$.

10. Yielding initiates at the pile head (sea bed) at a load of 4.00 MN. As yielding progresses down the shaft, load is redistributed to deeper soil, as reflected in the shape of the equivalent stress profiles. Yielding is predicted just above the toe at a load of 8.56 MN and, at failure, almost all the interface has yielded.

11. 'Failure' in this context is defined as the last load step before lack of convergence in the iterative, initial stress calculations. This has been found[3,4] to give good agreement with $\phi_u = 0$ calculations for embankments. In the present calculations, a criterion of convergence within 100 iterations has been imposed. This gives a load which corresponds to a stage immediately before a rapid acceleration in the load–settlement curve (Fig. 3), and immediately before complete yielding of the interface and soil just below the toe. Increasing the maximum number of iterations to 150 gives identical results. The computed failure load on this criterion is 9.70 MN while that calculated using API criteria is 7% lower.

12. Load increments of about 6% of the failure load have been employed and, except adjacent to the pile head, the overshoot of the yield surface by $\bar{\sigma}$ is less than 0.05%.

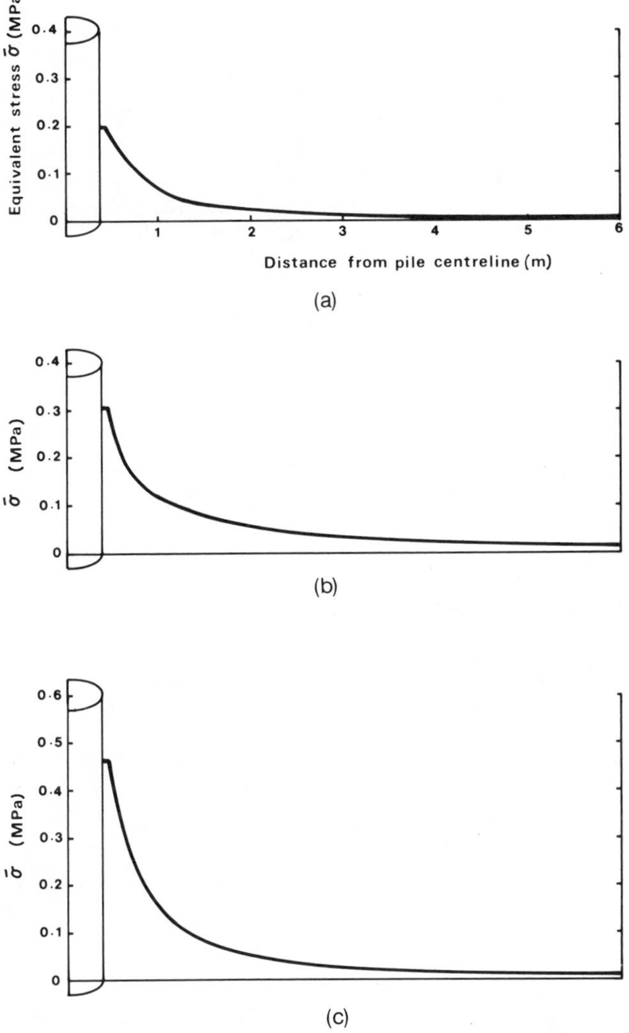

Fig. 4. Equivalent stresses at various horizons at failure: (a) 0.175 m below sea bed; (b) 12.55 m below sea bed; (c) 19.825 m below sea bed

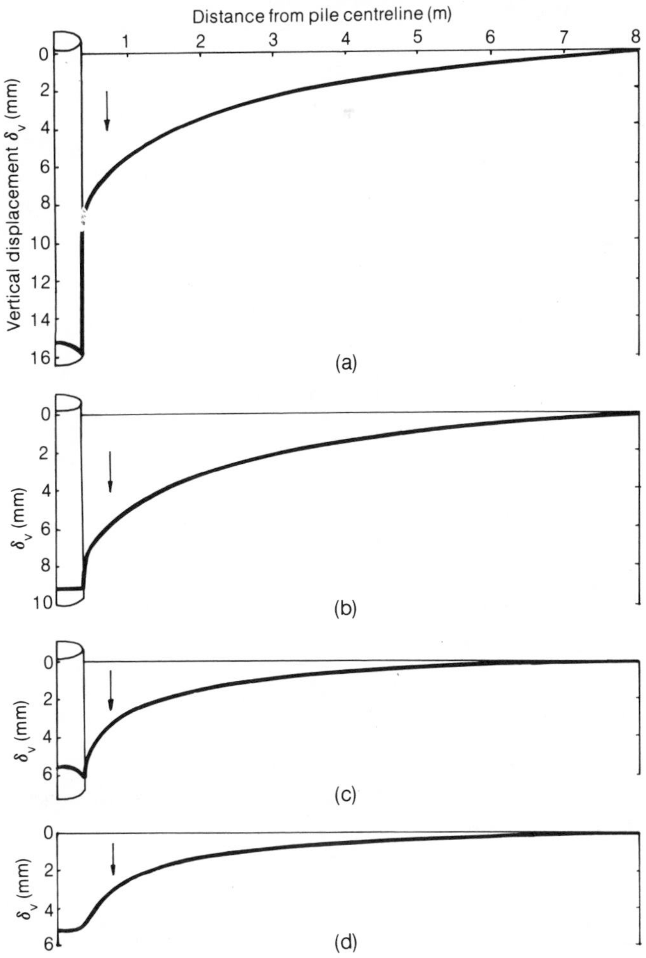

Fig. 5. Vertical movements at various horizons at failure: (a) pile head; (b) pile mid-point; (c) pile toe; (d) 0.35 m below pile toe

Thus, it is not considered that significant changes in failure load would result from the use of smaller load steps.

13. In contrast, some overshoot of C_u by the shear stress, τ_{rv}, is predicted, so that the Mohr–Coulomb failure criterion is violated. Putting $\sigma_r = \sigma_v = \sigma_\theta$ and $\tau_{v\theta} = \tau_{\theta r} = 0$ in equation (1) shows that the shear stress can exceed the undrained shear strength by a maximum of 15.5%.

14. As expected, the calculations indicate that soil failure is limited to a narrow band around the pile (Figs 4 and 5), suggesting that a fine mesh should be used to model the interface between pile and soil. Clearly, mesh refinements are limited by considerations of cost, and by numerical problems engendered by high aspect ratio elements in zones of rapidly changing stresses. In this regard, elements which allow slippage may be appropriate, although it is shown below that good agreement with field tests can be achieved without such elements. Doubling the interface thickness in the present calculations results in only a 5% increase in failure load (not significant for the load increment size used) and almost identical load–deflexion behaviour (in the elastic range, axial displacements are increased by 3%), so a reduced interface thickness might not lead to significantly different results.

15. The distribution of vertical stress changes in the interface, and in soil below the toe (Fig. 6), shows that a discontinuity exists. Just above the toe, a tensile stress change is predicted while, just below it, a large compressive stress change is predicted.

16. It should be remembered that the stresses discussed here are changes relative to in situ conditions. Nevertheless, the model does suggest that absolute tensile stresses may develop in the soil just above the toe, although the soil may be unable to sustain these in practice.

17. Mesh refinements to toe and interface did not relieve these stresses, so they may represent a physical phenomenon rather than a numerical problem caused by the rapid stress change at the toe. Tomlinson[8] reports the measurement of tensile stress above the toe of an instrumented driven pipe pile in stiff London Clay at Stanmore. This was attributed to the formation of a tension crack in the soil at toe level and subsequent slumping of soil immediately above the crack causing local downdrag on the pile.

18. Apart from those at the pile head and toe, the vertical stress changes are relatively small, as are the radial and hoop stress changes.

19. Inspection of stresses below the toe at failure indicates that full end bearing is not mobilized, the mean vertical stress change amounting to about 25% of $9C_u$ and

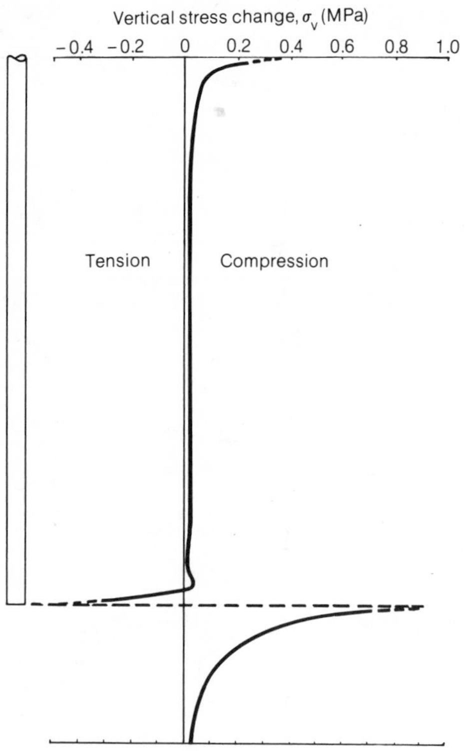

Fig. 6. Vertical stress changes in soil–pile interface and below toe at failure

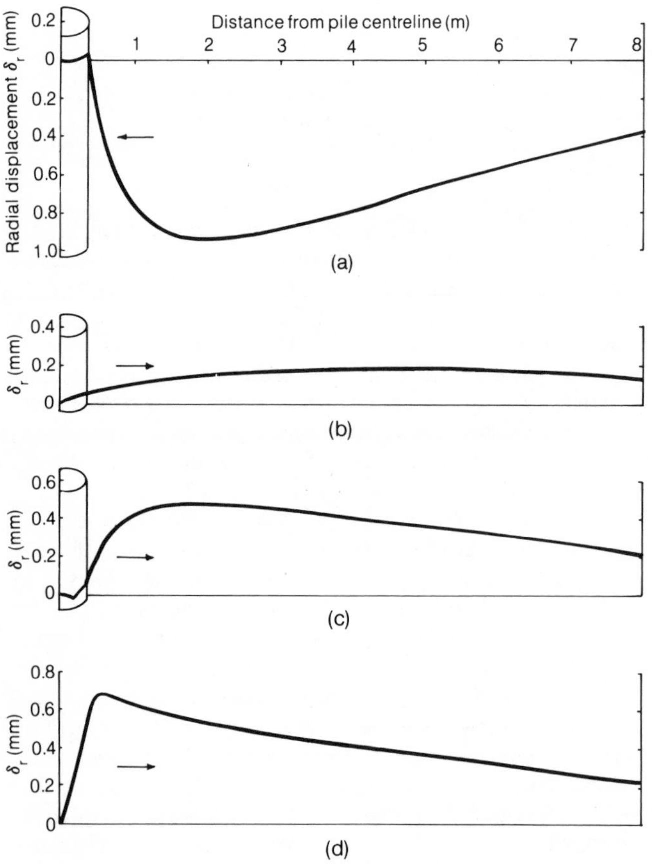

Fig. 7. Radial movements at various horizons at failure: (a) pile head; (b) pile mid-point; (c) pile toe; (d) 0.35 m below pile toe

the mean equivalent stress amounting to about 30% of the yield stress at this horizon.

20. Large vertical movements are computed at the pile/soil interface (Fig. 5) and the sea bed is pulled down to a distance of some 8 m from the pile. No heave is predicted. Within the pile, the soil does not move significantly relative to the wall, indicating that under the quasi-static axial load, a 'plugged' condition pertains for the soil/pile interface model employed within the pile.

21. At the pile head soil moves inwards, towards the pile (Fig. 7), while at and just below the toe it moves outwards. At the mid-point of the pile there is a small outward movement.

22. The above calculations have been repeated for a uniform soil with $C_u = 0.3$ MN/m^2, which is close to the mean value over the 20 m length of the pile. Almost identical load–deformation behaviour is predicted for the pile, although there is less yielding at shallow horizons and more at deep horizons at equivalent stages of loading.

23. In order to validate the model, analyses have been performed of axial load tests[6] carried out at the West Sole C site on 18.3 m long, 30 in (0.76 m) dia. piles. A modified finite element mesh with the same soil parameters as in the previous example has been used save that, in the elastic stress range, all soil stiffnesses have been arbitrarily doubled in order to give better agreement with measured deformations. Thus, in the interface, a Young's modulus of 250 times the undrained shear strength of the soil mass has been adopted. Use of the soil stiffnesses previously employed has no significant effect on failure load, but leads to earlier 'first yield' and to an increase in displacements in the elastic

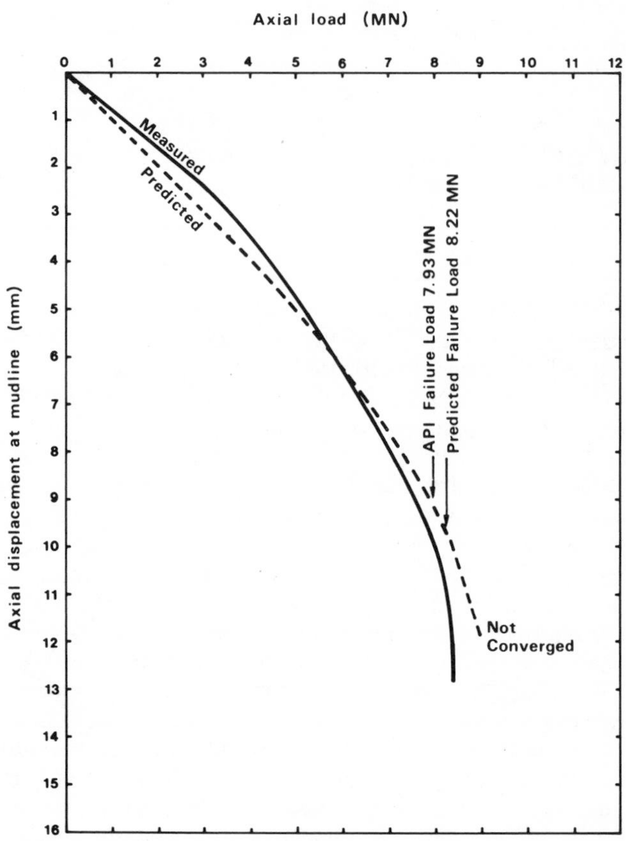

Fig. 8. Predicted and measured load–displacement behaviour of West Sole test piles

range of about 50%, and immediately before failure of about 30%. Fig. 8 compares the computed and measured load–settlement curves for the pile at the mud line: the numerical model gives a good fit to the observed behaviour.

24. While the analysis discussed here is more expensive than that for the widely used *T–z* idealization, it is able to model the coupling between soil strata, treating these as a continuum, and enables a better understanding to be gained of soil behaviour around the pile.

25. The present incremental calculations are force controlled; a more economical solution may result from the use of displacement control (Paper 1). Lateral, or combined lateral and axial, loading of single piles can be modelled by using Fourier series expansions (Paper 1), and by incorporating superposed bending elements into the finite element mesh. Group analyses are expensive since, in general, they require true three-dimensional idealizations; thus, it is not forseen that elasto-plastic calculations of the type reported here will be used for this purpose in the immediate future, other than for research purposes.

26. Cyclic degradation of clay can be simply modelled by reducing the soil yield stress at each cycle by some predetermined amount, or by reference to calculated stress levels and laboratory test data. By keeping the elastic properties constant, the stiffness matrix in the initial stress calculations need not be reformed for each load cycle. Analyses of this kind are expensive, however, and the comments made above for pile groups again apply. Pore pressures and dynamic effects have also to be considered.

27. The calculations presented show that, using a relatively simple elasto-plastic constitutive model and a relatively crude finite element idealization, good agreement can be found with the results of simple hand calculations and of pile load tests for axially loaded piles in clay. It is not suggested that they should replace simple hand calculations, however; rather, that they can give an insight into possible mechanisms of behaviour of the soil–pile system.

28. I wish to thank the Committee of Lloyd's Register for permission to publish this contribution and British Petroleum for permission to publish data from the West Sole pile tests.

Mr M. CHIN (Manchester University)

I should like to ask Mr Baguelin and Dr Frank whether they have done any work on the analysis of a pile in sand, and whether they have observed any significant increase in skin friction at the base. The results I have obtained show a significant increase in skin friction at the side of the pile during the mobilization phase of the shaft friction.

Mr BAGUELIN & Dr FRANK

In all our calculations we found that the values for the shaft friction at the pile head or the pile base were far higher than the values along the main part of the shaft (Fig. 4 of Paper 11). These higher values did not seem realistic to us, according to the results we commonly had from full-scale instrumented pile tests, and we concentrated only on the values along the main part of the pile. (However, during pile tests, we cannot usually measure the shaft friction exactly at the tip of the pile. We rather measure a mean value on the last few decimetres of the pile.)

31. In our theoretical studies we made calculations with dilatancy inside the soil mass. The results were surprising. We thought we would obtain an increase of the radial stresses, normal to the pile's shaft, in comparison with the elastic case. We found no increase, neither of the radial normal stress, nor of the vertical tangential stress. This was during the mobilization phase of the shaft friction (with no relative pile–soil slipping).

32. About ultimate values of skin friction, they depend, of course, directly on the ultimate friction characteristics and the initial stress state one introduces oneself at the pile/soil interface for one's calculations. It is, thus, not something one will get entirely by the finite element analyses. To allow for dilatancy of sands, one should introduce, for instance, higher values for the angle of friction than in non-dilatant soils. Overall shaft friction resistance will therefore increase because one increases the angle of friction. According to our calculations, it will not increase because of an increase of the normal stresses along the shaft of the pile.

Dr M. APPENDINO (ENEL, Italy)

When we model numerically the pile–soil interaction we must consider soil properties as they are modified by the installation technique. The importance of this is shown by an example deriving from a vertical load test performed on a driven pile. The pile belongs to a large group where the driving disturbed heavily the soft, normally consolidated, cohesive soil above the sand bearing layer, as pore pressure and soil displacement measurements confirmed.[9,10] Fig. 9 shows the vertical load distribution, and Fig. 10 shows the

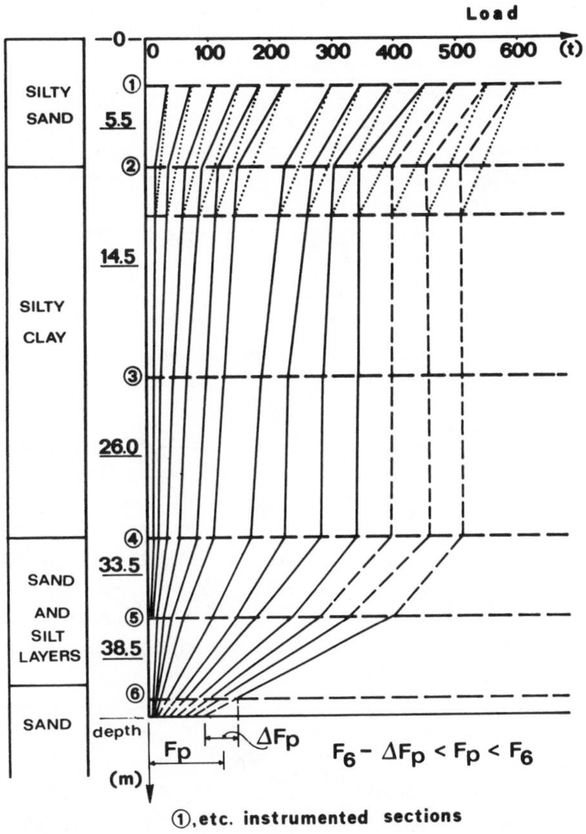

Fig. 9. Vertical load distribution curves due to static loading: Porto Tolle instrumented pile

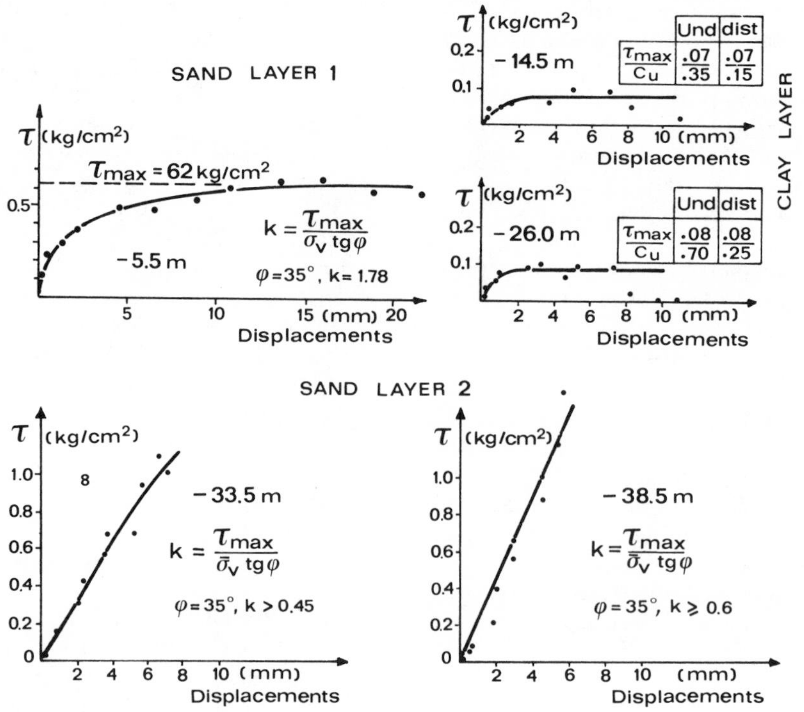

Fig. 10. Load transfer curves at different shaft depths: Porto Tolle instrumented pile

load transfer along the pile shaft. The values of shear resistance in the cohesive soil can be seen to be even lower than the remoulded values. At the same time the shaft resistance in the granular soil is greater than might be expected in the undisturbed sand. This is more evident in the upper sandy layer where the load was sufficient to produce the failure, but it appears also in the tip bearing layer where the transfer curve is linear up to a shear load requiring a ratio of horizontal to vertical effective soil pressure larger than 0.6. These data had not been back-analysed in terms of load–displacement behaviour. However, the static stiffness resulting from the interpretation of driving data shows that the sand deformability is lower than that existing before piles were driven. (Similar considerations have been pointed out before.[11–13])

34. Up to now we have not a suitable numerical procedure to pass from the original soil properties to soil properties modified by pile penetration.

35. As long as we perform a back-analysis it is possible to repeat a campaign of investigations to determine final soil properties. The procedure is expensive and moreover it can only produce data of limited precision. In fact the driving makes the soil properties and stresses at rest vary sensibly with the distance from pile axis, whereas it is practically impossible to establish the exact location of the point where properties are measured or the sample is taken when piles are long.

36. Some information may be derived from the interpretation of driven data, if the interpretation is performed to estimate soil deformability in addition to soil resistance. But in a design analysis, we can only apply qualitative information obtained from the literature. I hope that some effort will be made to develop procedures to simulate the effect of pile installation on soil properties. That will enable us to pass from the original soil properties to the ones which do exist when computing the pile behaviour under different loading conditions. This step is of primary importance.

Dr BANERJEE & Dr DAVIES (Paper 13)

It is probably not widely known that the boundary element method is a completely general method for obtaining numerical solutions to virtually all linear and non-linear problems of engineering science. The method is already well developed and has been applied to steady state and transient problems of solid mechanics, fluid mechanics, acoustics, electromagnetism etc. A comprehensive account of these developments has been given recently.[14,15] For three-dimensional problems the method is probably the only reliable means of getting adequately detailed results at a reasonable cost.

38. The necessary requirements for developing a boundary element formulation are

(a) a reciprocal identity between two solutions (one may be called real and the other virtual) of the governing differential equations of the problem;
(b) a fundamental solution of the governing differential equation for unit excitations.

Thus for the solution of an elastic or elasto-plastic problem one has to use Maxwell-Betti's reciprocal work theorem and a point force solution such as the Kelvin solution or the Mindlin solution as outlined in Paper 13.

39. Fundamental solutions (point force solutions) which are used as virtual solutions in the reciprocal identity play the same role as the shape functions do in a finite element formulation. Computational work can often be substantially reduced by choice of the most advantageous point force solution for each problem. Some of these may be exact solutions (e.g., Kelvin's solution, Mindlin's solution)

while others may be approximate (e.g., Banerjee and Davies[16]) or numerically constructed as shown by Wilson and Cruse.[17]

40. The boundary element method can also be combined with other numerical methods such as the finite element or the finite difference method to provide an efficient hybrid solution algorithm as outlined in Paper 13.

Dr M. F. RANDOLPH (Paper 20)

The axial stiffness of pile groups is customarily estimated by first estimating the load–settlement response of an isolated pile and then allowing for the group effect by means of interaction factors calculated from elastic analyses (Paper 15). The configuration of offshore pile groups is one where the piles are located evenly round the legs of the superstructure. The symmetry of the group entails that, under axial loading, each of the piles will carry the same share of the load. For such groups, the methods of analysis proposed by Randolph and Wroth for single piles[18] and for pile groups[19] may be combined to give a closed-form solution for the load–settlement behaviour of the pile group.

42. The solution for a single pile is derived in detail by Randolph and Wroth.[18] The resulting load–settlement ratio is

$$\frac{P_t}{G_l r_o w_t} = \frac{\dfrac{4}{\eta(1-\nu)} + \dfrac{2\pi}{\zeta}\rho\dfrac{l}{r_o}\dfrac{\tanh(\mu l)}{\mu l}}{1 + \dfrac{4}{(1-\nu)}\dfrac{1}{\pi\lambda}\dfrac{l}{r_o}\dfrac{\tanh(\mu l)}{\mu l}} \quad (2)$$

where

P_t = load at mud line
w_t = settlement at mud line
l = pile length
r_o = pile radius
G_l = soil shear modulus at level of pile base
ν = Poisson's ratio for the soil
λ = pile stiffness ratio, E_p/G_l
ρ = ratio of shear modulus at pile mid-depth to that at pile base, $G_{l/2}/G_l$ (a linear variation of soil stiffness with depth is assumed)
ζ = $\ln(r_m/r_o)$, where r_m is the maximum radius of influence of the pile, given by $r_m = 2.5\rho l(1-\nu)$
μl = $(2/\zeta\lambda)^{1/2}\, l/r_o$
η = 1 for the solution for a single straight-shafted pile

The compression of the pile may be estimated from the formula giving the ratio of the mud-line settlement w_t to the base settlement w_b:

$$w_t/w_b \approx \cosh(\mu l)$$

43. Following the methods described by Randolph and Wroth,[19] the interaction of piles within a group entails that the constants η (relating to the pile base load–settlement behaviour) and ζ (relating to the shaft load–settlement behaviour) should be replaced, for the analysis of a symmetrical pile group, by

$$\eta = 1 + \frac{2}{\pi}\sum_{i=2}^{n} r_o/s_i \quad (3)$$

where S_i is the spacing of the ith pile from pile 1, and

$$\zeta = \ln(r_m/r_o) + \sum_{i=2}^{n}\ln(r_m/s_i) \quad (4)$$

To account for the slightly larger radius of influence of the group of piles compared with a single pile, r_m should be supplemented by the radius of the pile group, r_g, to give

$$r_m = 2.5\rho l(1-\nu) + r_g$$

With these modified factors, equation (2) may be used to calculate the load–settlement ratio of each pile within the pile group.

44. As an example of the method, the pile group discussed by Poulos (Paper 15) is analysed. The pile group consisted of eleven piles evenly spaced around a circle of diameter 10.8 m. Each pile was of embedded length 72 m, radius 0.685 m and wall thickness 50 mm. The last figure enables the equivalent pile stiffness to be calculated as

$$E_p = E_{steel}(1 - r_i^2/r_o^2) \approx 30 \times 10^3 \text{ MPa}$$

Adopting a shear modulus of $G = 150c_u$, the profile of soil shear strength given by Poulos would indicate a value for G_l of 30 MPa and a ratio $\rho = G_{l/2}/G_l$ of about 0.8. These values give a load–settlement ratio for a single pile (see equation (2)) of $P_t/G_l r_o w_t = 52.5$, giving

$$P_t/w_t = 1080 \text{ kN/mm} \quad (5)$$

The relative settlement of the top and base of the pile is calculated to be $w_t/w_b \approx 4.5$.

45. For the pile group, the factors η and ζ may be calculated from equations (3) and (4) to be $\eta = 1.71$ and $\zeta = 28.9$. The resulting load–settlement ratio for each pile within the group is thus $P_t/G_l r_o w_t = 17.2$, giving

$$P_t/w_t = 354 \text{ kN/mm} \quad (6)$$

The compression of each pile may again be estimated from the ratio of the settlements of the top and base of the pile, calculated to be $w_t/w_b \approx 1.4$. From equations (5) and (6), the settlement ratio for the group is 1080/354 = 3.05. This value is lower than the values commonly quoted for settlement ratios of offshore pile groups. One of the reasons for this is that the method of analysis outlined above takes account of the increased proportion of the load which is taken by the pile base due to the presence of similarly loaded neighbouring piles. The interaction between piles near the level of the bases is less than that along the shafts, leading to overall lower values of group settlement ratios (i.e., less interaction) than predicted by methods which fail to take the redistribution of load into account. Another interesting feature of the analysis is the lower ratio of w_t/w_b calculated for the piles in the group compared with the isolated pile. This has important consequences in the manner in which the pile mobilizes its full shaft capacity. For an isolated pile, the full shaft adhesion will be mobilized near the mud line well before that near the pile base. On the other hand, for piles within the group, although the absolute compression of the pile increases for a given load level, the ratio w_t/w_b is reduced, thus giving a more uniform mobilization of shaft adhesion.

PROFESSOR M. BOULON (Institut de Mécanique, Grenoble)

I present here the ideas, the numerical model and some results of our research team in the University of Grenoble; some of these ideas are held in common with the soil mechanics section of the Institut Francais du Pétrole. Our philosophy differs significantly from the approaches of the various authors in this conference.

47. An idealized numerical model for piles under each kind of load, treated by the finite element method or any other numerical method, requires

(a) knowledge of the mechanical properties of the materials (grains of soil, water, pile) and their connection (interface);
(b) the ability to treat the soil–structure system by algorithms representing the various kinds of behaviour;
(c) reproduction of the history of loading for each piece of soil, from the initial state.

48. The soil must be represented by a functional rheological relationship including directional dependence, which means an incremental stress–strain matrix depending on the parameters (e.g., strain ratios ϵ_1/ϵ_3 and ϵ_2/ϵ_3) of the loading increments. The pore water can be taken as an incompressible viscous (Newtonian) fluid. The structure can be considered as linear elastic. The interface between soil and pile (structure) knows a localized failure and must be represented by a criterion for the beginning of the relative sliding, and a law for the residual friction.

49. Three main behaviour algorithms are necessary: research of the local stress–strain matrix, soil–structure interaction and soil–fluid coupling.

50. In research of the local intergranular stress–strain relationship, if we assume a directional dependence of the response to a loading path, we must search the local available incremental stiffness matrix (unknown) before each increment of loading; thus many more equations (or inequalities) than those of the discretized equilibrium have to be verified. Special attention has to be given for cyclic loading applied to offshore piling: in the case of gravity structures and also in the case of floating structures, the loading due to a storm consists of many thousands of cycles. As it is impossible to compute incrementally such a loading because of the volume of the calculations, a simplifying assumption is necessary.

51. The soil–structure interaction is more complex than a heterogeneity, in the stiffness, because relative displacements occur between soil and structure, especially in the case of piles, which modify consequently the load transfer by limiting the shear stresses at the contact.

52. For soil–fluid coupling an effective stress analysis must be developed, using only the intergranular mechanical properties of the soil. This analysis is efficient for a quick loading (an undrained step corresponding to no change in local volume); quickness is not absolute, but is relative with respect to the permeability of the skeleton of the soil. This analysis must also be efficient in periods of consolidation.

53. The soil behaviour is a functional relation between stress path and strain path. Thus the behaviour of a pile depends on the soil loading path. The history of the pile working load is generally not a problem, but the initial state, which is a very important parameter, can only be known by the history of the soil before pile installation (well simulated by an oedometric path), the history of the pile installation (very difficult to take into account), and the change in effective stress between pile installation and pile loading (reconsolidation).

54. Our pile numerical model has the following characteristics, obviously inferior to the last idealized model. It represents a single pile under axial load with an axisymmetric geometry. The calculations are performed with the finite element method. The rheological relationship between stress and strain is a functional relationship with discrete directional effect (eight forms for a stress state); an algorithm allows one to specify the available form by verifying three inequalities in each point where the stresses are known. The shaft friction is taken into account in a Coulombian way: the interface sliding criterion and the law of residual friction are assumed to be identical:

$|\tau| \leqslant \sigma_n \tan \delta$ for granular drained soils

$|\tau| \leqslant \alpha C_u$ for undrained saturated cohesive soils

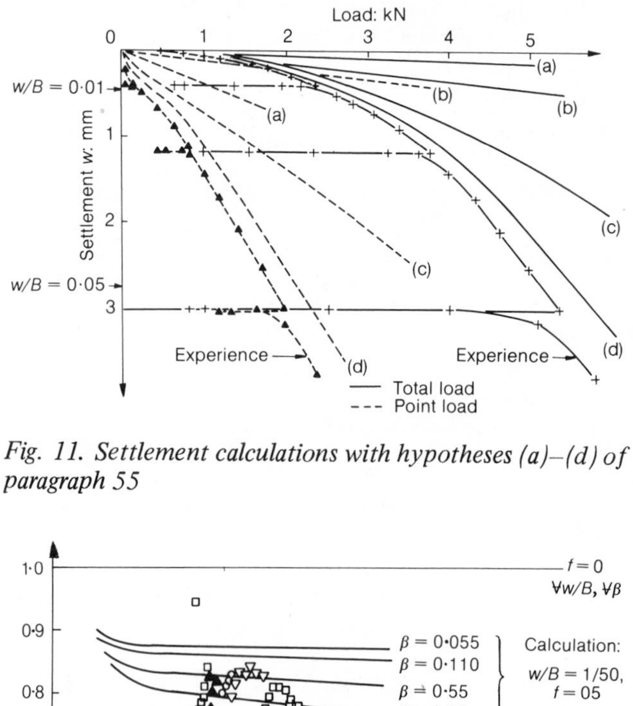

Fig. 11. Settlement calculations with hypotheses (a)–(d) of paragraph 55

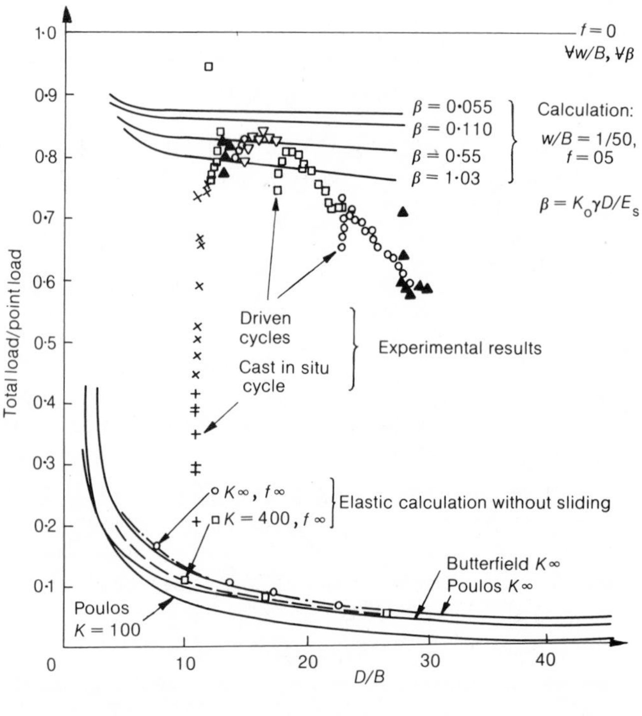

Fig. 12. Settlement and transferred load

where τ is shear stress, σ_n is normal stress, δ is soil–pile friction angle, C_u is undrained cohesion and α is adhesion coefficient. We have developed an effective stress analysis for the undrained problem only; thus, we do not obtain a good solution for the question of initial state, which requires modelling of the reconsolidation after pile installation. For the initial state, we simulate only a K_0 state which represents cast-in-situ piles; we search now for a simple way to model driven piles. We prepare also calculations, according to experience, for introducing the irreversible strains after a high number of cycles by a pseudo-creep from the static state (governing by the static load), as a function of the level and of the shape of the cycles; this assumption permits one to group the cycles by series; thus the first cycle of each series is computed incrementally, and the effects of the following cycles are derived from the effect of the first. We solve the three problems of soil–pile interaction, effective stress analysis, and use of rheological law with directional dependence by our method of influence coefficients which gives direct solutions.

55. Fig. 11 gives the influence of each hypothesis on the result of one calculation for a compression pile in dense sand:

(a) linear elasticity for the soil, without sliding along the shaft;
(b) linear elasticity for the soil, with sliding along the shaft;
(c) directional dependence for the soil law, without sliding along the shaft;
(d) directional dependence for the soil law, with sliding along the shaft.

Calculation (d) is very close to experience, and seems to give the best simulation of the phenomenon.

56. Figure 12 presents these results plotted in the Poulos diagram. The experimental points, which measure the change in the point load to total load ratio, vary from the lower curves (elastic calculation without sliding) to the upper curves (elastic calculation with sliding). At greater depth the experiment shows a decrease of this ratio, corresponding to the limit of shaft friction and point load.

57. Figure 13 is a loading–unloading curve for one cycle, with sliding and linear elasticity: the effect of the sliding appears clearly, with a hysteresis loop. Fig. 14 shows the shear stresses along the shaft during the same simulation: residual stresses can be noted.

58. Figure 15 represents a pile pulling test and the corresponding calculation with the hypotheses (a), (b) and (d) of paragraph 55. The soil is a loose sand ($\gamma = 1.50$ g/cm^3). Hypothesis (b) leads to a limit load, depending only on the initial state. Fig. 16 gives the calculated mechanism of load transfer showing the progressive saturation of sliding.

59. The results show clearly the influence of the

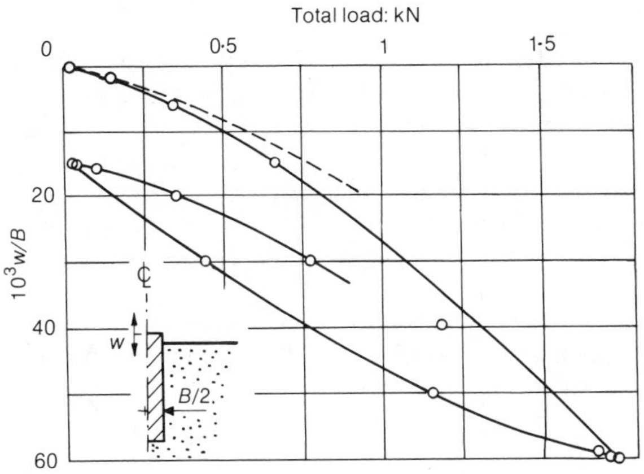

Fig. 13. Load–settlement curve for one cycle (loading, unloading and reloading); D = 25 m, D/B = 50, K = 100

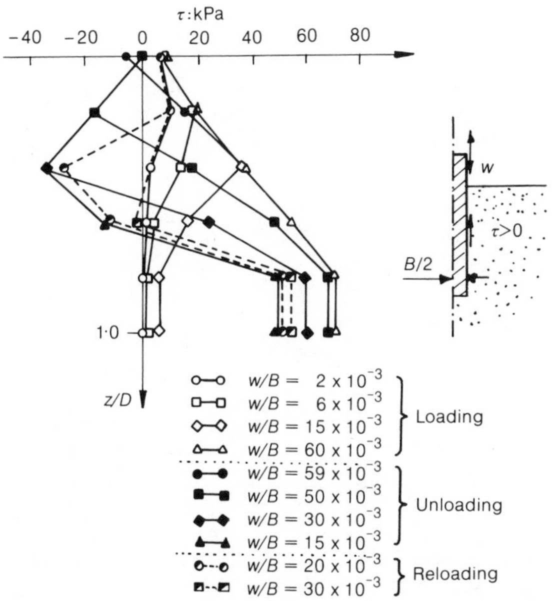

Fig. 14. Change in the shear stresses along the shaft for one cycle

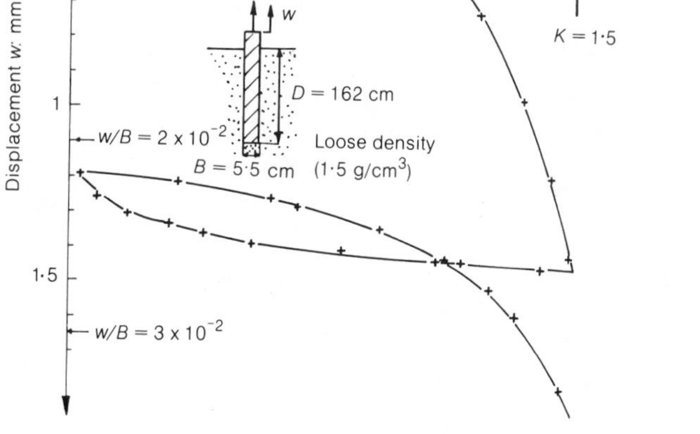

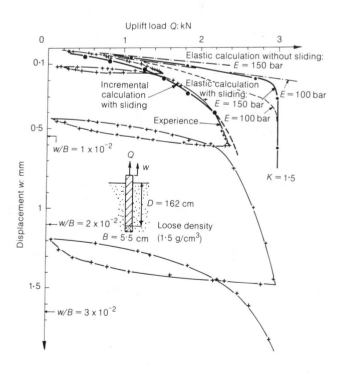

Fig. 15. Comparison between calculation and experiment (uplift)

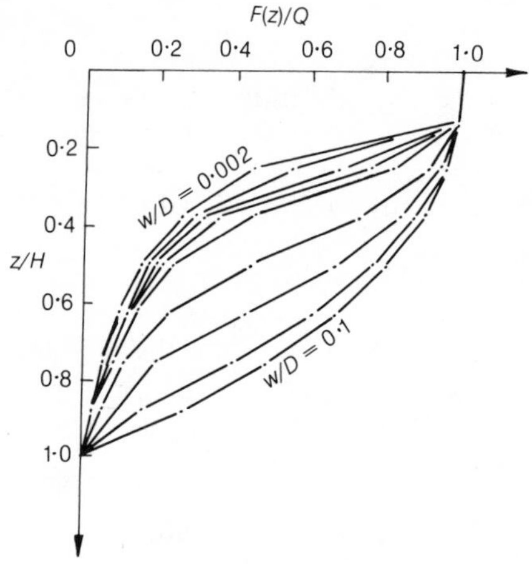

Fig. 16. Change in the load transfer (uplift, calculated)

hypothesis and especially the great effect of taking into account the sliding along the shaft in a numerical model for pile, even for working loads.

Dr R. M. SEMPLE (McClelland Engineers)
A pile group for a North Sea structure might have nine piles, spaced at least two pile diameters apart, with a length-to-diameter ratio of 20–40. The model test data presented in Paper 14 for a 3 × 3 pile group would indicate a settlement ratio of the order of 4–5. This sort of measurement has been made in other model tests[20,21] and is supported by results of analyses using elastic theory.[22] There are, however, field measurements on nine-pile groups that show a settlement ratio at working load of the order of 1.5–2.5.[23–26] In evaluating the axial stiffness of offshore pile groups, it is McClelland Engineers' practice to adopt a settlement ratio in accordance with these field measurements rather than to accept larger values where these are indicated by the results of analyses using elastic theory.

Mr D. J. WILLIAMS (Cambridge University)
I should like to make a comparison of various analytical methods for the lateral load response of single piles. Basically there are two idealizations of the pile–soil system. One is discrete, in which the soil is replaced by a number of independent Winkler springs. In the second, the soil is considered continuous. The methods of solution are dependent on the pile–soil system idealization and the assumed distribution of soil modulus with depth. For the discrete case, taking a constant modulus distribution, there is a closed form solution of the governing differential equation. For a linear modulus – from zero at the surface – the differential equation can be solved by difference equations. The accuracy will be limited by computer precision and the number of springs employed. The least accurate case will be for short, rigid piles. For the continuum case, there are solutions based on the boundary element method and on the finite element method. The boundary element method requires some approximation in dealing with a linear modulus.

62. The methods may be compared by their influence coefficients for lateral pile deflexion at the soil surface. First, a constant soil modulus is considered. For I_H (Fig.

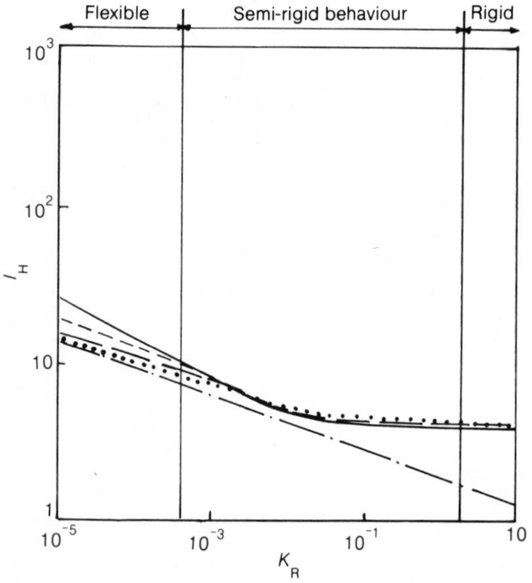

Fig. 17. Constant modulus: influence coefficients for horizontal load, I_H

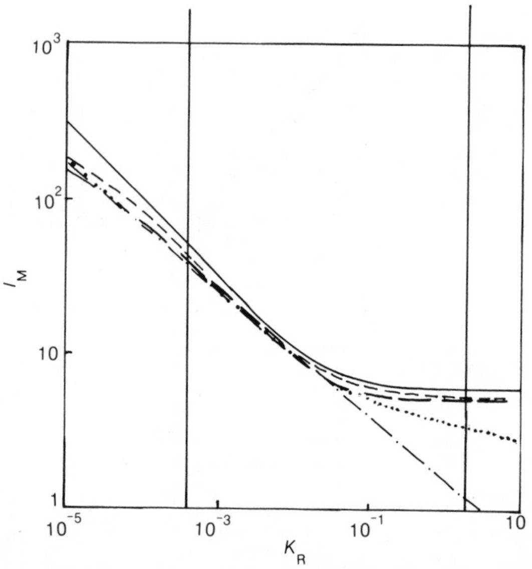

Fig. 18. Constant modulus: influence coefficients for moment, I_M

17), all the methods are in reasonable agreement, with the exception of Kuhlemeyer. The main difference between the discrete and continuum approaches is in the assumed transfer of stress. The discrete approach allows for no transfer of stress and the continuum approach allows for elastic transfer. Real soil behaviour is possibly somewhere between the two approaches. As for I_M (Fig. 18), all the methods are in reasonable agreement, with the exception of Kuhlemeyer for short piles.

63. Secondly, a linear soil modulus is considered. For I'_H (Fig. 19), the agreement between methods is not as good as for a constant modulus, possibly because of the approximations required in the boundary element solutions. As for I'_M (Fig. 20), there is fair agreement between the methods of solution.

64. Concerning the verification of these methods, given a field test most of the methods can be fitted to show good agreement. But what is the rationale behind the choice of soil properties or fitting parameters? Can verification be extended to prediction?

PROFESSOR M. NOVAK (Paper 8)

Considering practical applications it is clear that interaction of piles in a group is most important. For a number of years I have been using the interaction coefficients due to Poulos and felt quite happy with them for small groups. However, when one applies these interaction coefficients based on linear theory and homogeneity to a very large group, say of 200 or 300 piles, one gets a reduction of the stiffness of the order of 15–20. Such a large group effect raises the question of the applicability of the procedure. It appears that the interaction factors based solely on linearity and homogeneity could be exaggerated because the soil is softer (non-linear) around the loaded pile and the soil stiffness often diminishes towards the surface.

66. Also, the application of the interaction coefficients based on the consideration of two piles to a large group may be conservative. The interaction of two piles is affected by the presence of all piles of the group.

67. I would appreciate it if some of those who are engaged in research into group effects could suggest whether they agree that the general practice may be exaggerating the interaction effects.

Dr BANERJEE & Dr DAVIES

Mr Williams presents an interesting comparison between the various methods of analysis for the load–displacement behaviour of laterally loaded piles. The analyses based on the continuum approach (finite element and boundary element) seem to yield very similar results for the homogeneous and non-homogeneous cases. The major differences seem to be between the continuum idealization and the discrete spring approach. For the single pile case the spring constants may be adjusted to fit any set of analytical or practical data. Indeed one of the advantages of this approach is that any form of non-linearity can be included in the analysis by adjustment of the values of the spring constants via a set of p–y curves. However, the real test of the applicability of these discrete methods will be in analysis of the case histories of laterally loaded pile groups as described in Paper 13, where we analysed a number of groups using the same set of soil parameters and demon-

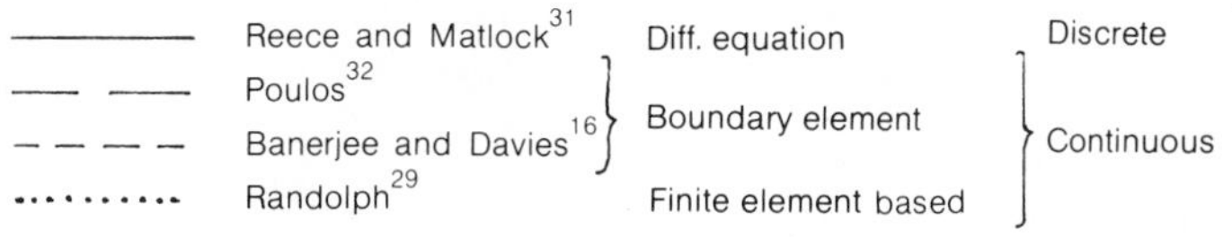

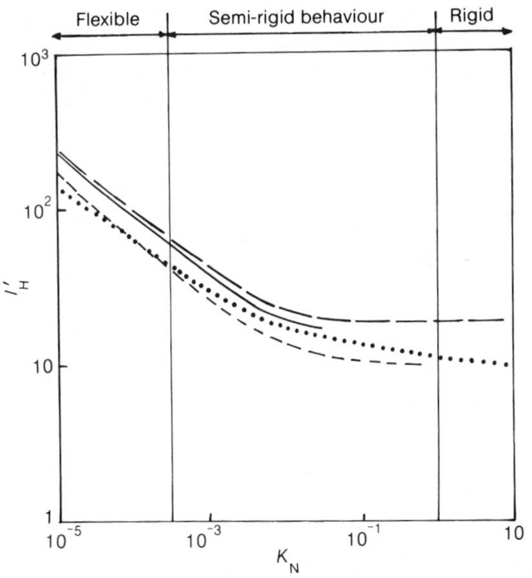

Fig. 19. Linear modulus: influence coefficients for horizontal load, I'_H

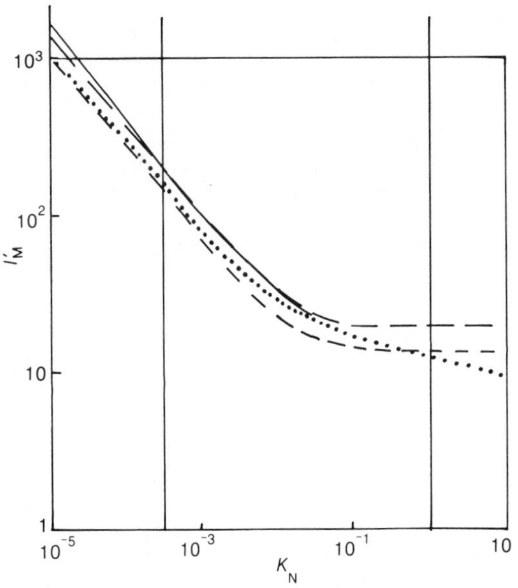

Fig. 20. Linear modulus: influence coefficients for moment, I'_M

strated that in spite of vast differences in geometrical scale, shape and loading conditions, good agreement with the observed data could be achieved. Clearly in only this way can a 'rationale' behind the choice of soil parameters be established.

69. Regarding the use of the approximate point force solution in the boundary element analysis of the non-homogeneous case, there is a parallel approximation in the adoption of a quadratic shape function for displacements as an incomplete virtual state in the stiffness matrix calculations for the $E = mz$ case using the finite element method. Any differences between the two continuum approaches would be primarily due to the discretization or due to the computer implementation of the two formulations.

70. Using our computer programs we have back-analysed nearly 50 reported case histories of axially and laterally loaded piles and pile groups. We found, rather surprisingly, that although the analysis was originally developed for undrained deformation of stiff overconsolidated clays, consistently better results were obtained for piles and pile groups embedded in sand. The elastic moduli required to fit the observed test data in clayey soil ranged from 100 C_u to 300 C_u, where the higher values correspond to stiffer clays. For sand a more consistent range of values of m (for the $E = mz$ model) of 40–50 lbf/in^2 per in seem to fit almost all reported test data. These parameters are appropriate for predicting the behaviour at half the ultimate load by using the elastic analysis.

71. Professor Novak has made some important remarks on the interaction factors. Indeed these factors are always overestimated by the homogeneous elastic analysis as has been shown recently.[33,34] Furthermore, even for the homogeneous case these interaction factors are decreased due to the reinforcing effect of neighbouring piles. For the $E = mz$ case we have found[33] that beyond a spacing-to-diameter ratio of about 15, the interaction factors are negligible.

Table 1

S/D	Theoretical ρ values: $H/L = \infty$	$H/L = 3$
2.5	0.25	0.28
5.0	0.33	0.38

PROFESSOR BUTTERFIELD (Paper 14)

With reference to Professor Novak's query, an answer is given by Butterfield and Douglas in paragraphs 88–92 of the discussion on papers 1–7.

73. Contrary to Dr Semple's statement, my experience is that, correctly interpreted, practically measured settlement ratios (or stiffness reduction factors – see Butterfield and Douglas contribution, paragraphs 88–92 of the discussion on papers 1–7) agree rather well with the elastically predicted values. It is, however, useful to bear in mind that the fully bonded, elastically interacting pile settlement ratio will provide an upper bound value since any one of slippage, plastic deformation or modulus increasing with depth will reduce the pile interaction and therefore the settlement ratio.

74. The following is my interpretation of results of three sets of field measurements[23–25] mentioned by Dr Semple, based on very simple measurements taken directly from the papers. It should be emphasized that

(a) the measurements are taken from incremental load test results;
(b) the relevant stiffnesses are those at working load (i.e., well away from 'failure load displacements');
(c) the soils data is very crude indeed and the best assessment of $\lambda = E_{pile}/G_{soil}$ that can be made is $1000 < \lambda < 3000$, although this value is quite adequate for the present purpose and is used below.

75. The American Railway Engineers Association[24] reports tests on three single, tapered, thin-wall, fluted steel tube piles from 50 ft to 100 ft embedded length. The results are very scattered and the best estimate one can make of the vertical stiffness (K_s) is $6 \times 10^6 \leqslant K_s \leqslant 8 \times 10^6$ lbf/ft. Measurements on a 3 × 3 group ($L \approx 60$ ft, $N = 9$) of such piles spaced at 4 ft centres (i.e., $S/D \approx 4$, since the pile diameter tapered from 16½ in at the head to 8 in at the toe) indicated a stiffness (K_N) of about 20×10^6 lbf/ft. Therefore the measured stiffness reduction factor $\rho = K_N/NK_s$ was $0.29 \leqslant \rho \leqslant 0.38$. The elastically calculated ρ values are approximately as shown in Table 1 for an elastic layer of total thickness H. (There is insufficient information in the paper to assess the value of H but the predicted ρ values are reasonably close to the experimental values.)

76. Schlitt[25] reports on rather similar steel piles (55 ft $\leqslant L \leqslant$ 65 ft and D tapering from 12 in to 8 in), tested both singly and in a 3 × 3 group (S = 3 ft 9 in), again driven into a soft silty-clay soil with C_u scattered widely about a mean value of around 700 lbf/ft. Now $K_S \approx 11 \times 10^6$ lbf/ft for

Table 2

Group size	Vertical displacement, ft, under 30 t/pile	Stiffness per pile, 10^6 lbf/ft	Measured ρ value	Theoretical ρ values ($L/D = 50$; $S/D = 2.5$): $H/L = \infty$, $\lambda = 1000$	$H/L = \infty$, $\lambda = 3000$	$H/L = 3$, $\lambda = 1000$	$H/L = 3$, $\lambda = 3000$	$H/L = 1.5$, $\lambda = 1000$	$H/L = 1.5$, $\lambda = 3000$
2 × 1	0.016	4.3	0.66	0.67	0.64	0.69	0.65	0.70	0.65
2 × 2	0.020	3.4	0.52	0.45	0.40	0.48	0.42	0.50	0.45
4 × 2	0.025	2.7	0.41	0.30	0.25	0.32	0.28	0.35	0.30

the single piles, and, for the group, $K \approx 44 \times 10^6$ lbf/ft; hence $\rho \approx 0.44$. Using $S/D = 5.0$ the elastic results are approximately $\rho = 0.33$ and $\rho = 0.42$ for H/L values of ∞ and 1.5 respectively. Again the calculated values underestimate ρ slightly (i.e., overestimate the interaction) and the effective layer thickness is indeterminate from the borehole log.

77. Perhaps the most interesting results are given by Masters,[23] who presents a large number of tests on timber piles, again driven into soft silty clay but now underlain by a much stiffer sand layer at about 120 ft depth. The single piles, typically 50–60 ft long, tapered from 16 in to 8 in dia., and had a mean vertical stiffness of about 6.5×10^6 lbf/ft under a 40 t load. Masters' results on groups of such piles ($L = 50$ ft, $S = 3$ ft), using his mean vertical displacement measurements under a load of 30 t per pile, are summarized in Table 2. Once more the theoretical ρ values are seen to be able to predict the field group stiffnesses quite adequately from in situ measurements on single vertical piles.

Dr POULOS (Paper 15)

Dr Randolph has devised a very neat analysis for the settlement of pile groups which overcomes some of the shortcomings of the interaction factor approach. In particular, the ability to consider the changing proportion of base load with increasing numbers of piles leads to smaller group settlement ratios. I give an indication of this effect in an earlier paper.[35] For the 11 pile group considered, Dr Randolph's solution gives a settlement ratio of 3.05, whereas the interaction factor approach, for the soil parameters selected in the paper, gives a value of 4.94. However, this is not a fair comparison as the shear modulus values used by Dr Randolph are significantly greater than those used in the original example. There, a constant value of E_s of 18 MPa was used, whereas Dr Randolph has used a value of shear modulus G at the pile tip of 30 MPa. A higher modulus will lead to a decrease in relative pile stiffness and a consequent decrease in interaction between the piles in a group.

79. In order to make a fairer comparison, I have reanalysed the problem, for vertical loading only, using non-uniform soil modulus values compatible with Dr Randolph's. The analysis has been carried out by two different programs. The first is the program DEFPIG, described in the Paper. The second is the program AXPIL5, which does not employ the interaction factor approach, but considers, via simplified boundary element analysis, the effects of all the elements of each individual pile in a symmetrical group: the proper distribution of stress and displacement along each pile in the group should therefore be obtained.

80. The results of the analyses are summarized in Table 3. The agreement between the solutions for single-pile stiffness is quite good (except that the ratio of top to base settlement given by Randolph's analysis is higher). For the groups however, Randolph's analysis gives a considerably higher stiffness, and hence a smaller settlement ratio, than either the DEFPIG or AXPIL5 analyses. While the reason for this is not clear, it could be associated with the assumption of the value of r_m in the Randolph group analysis. It is interesting to observe that the DEFPIG and AXPIL5 analyses give solutions which agree closely with each other, thus confirming the validity of the interaction factor approach, at least for symmetrical groups. The redistribution of base load, and the decreased ratio w_t/w_b, which are emphasized by Randolph, are reproduced by the AXPIL5 analysis. It should be noted that the assumption of an equivalent uniform soil modulus would lead to a slightly higher settlement ratio; e.g., the value from the AXPIL5 analysis increases from 3.80 to 4.05.

81. The discussion contributions from Dr Semple and Professor Novak also deal with interaction between axially loaded piles, and both suggest that the present theories may overestimate group settlement ratios. As mentioned above, the relative stiffness of the pile has a significant effect on interaction, and a conservative estimate of soil modulus will result in a higher stiffness factor and hence a higher settlement ratio (as well as a higher single-pile settlement). Other factors which may contribute to the overestimation of settlement ratios are the following:

(a) the presence of underlying stiffer layers will decrease interaction as compared with a deep soil mass;
(b) interaction between piles in a non-homogeneous mass is less than in a homogeneous mass;
(c) the reinforcing effect of other piles between two interacting piles is ignored in the theory.

Fig. 21 shows a typical example of the effect of soil non-homogeneity on the settlement interaction factor α. Consideration of all the above factors will lead to smaller settlement ratios than if a homogeneous deep soil layer is considered. The differences will increase as the group size increases. Thus, I must agree with Professor Novak that the currently used theories (which assume homogeneous soil) will generally exaggerate the interaction effects, particularly for larger groups of piles. While the applicability of the interaction factor approach has been reasonably well verified for groups of up to 25 piles at relatively close spacings, there have been no comparisons made (by myself, at least) for groups containing large numbers of piles. In such cases, it may be preferable, from a practical point of view, to

Table 3. Eleven-pile group; vertical load only; elastic soil; $\nu_s = 0.3$

Analysis	Single pile		Pile in group		Settlement ratio
	Stiffness, kN/mm	w_t/w_b	Stiffness, kN/mm	w_t/w_b	
Randolph	1080	4.5	354	1.4	3.05
DEFPIG	1042	–	278	–	3.75
AXPIL5	1042	3.6	274	1.36	3.80

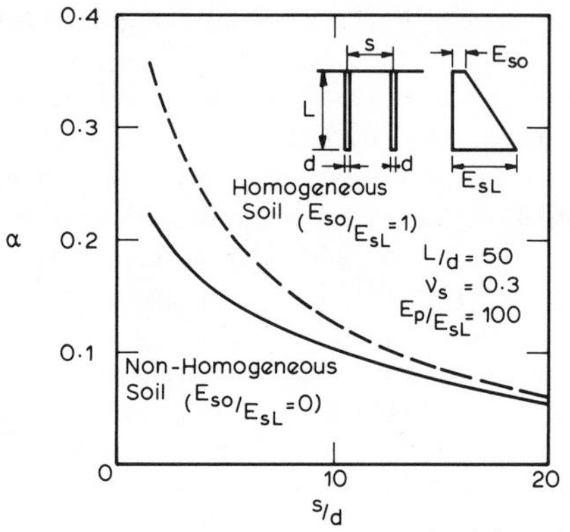

Fig. 21. Effect of soil non-homogeneity on interaction factor α

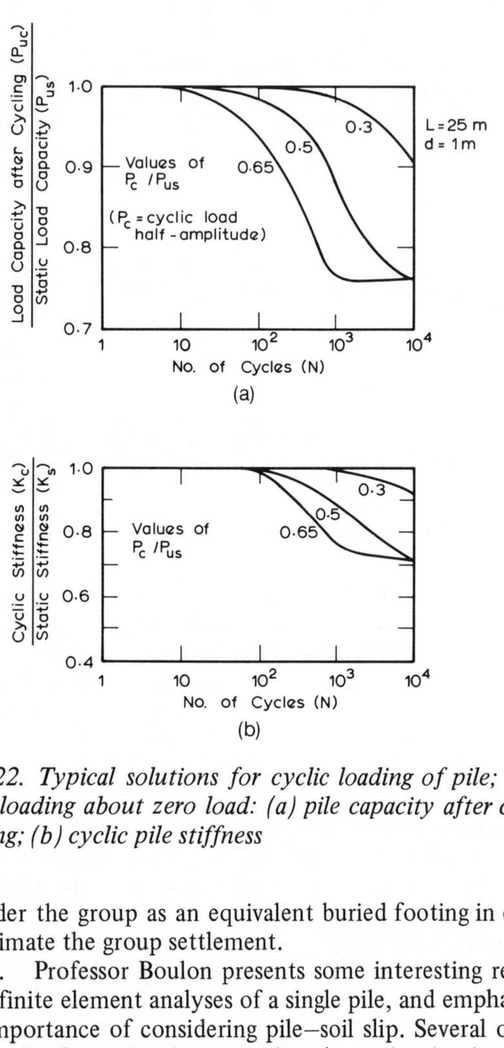

Fig. 22. Typical solutions for cyclic loading of pile; 'two-way' loading about zero load: (a) pile capacity after cyclic loading; (b) cyclic pile stiffness

consider the group as an equivalent buried footing in order to estimate the group settlement.

82. Professor Boulon presents some interesting results from finite element analyses of a single pile, and emphasizes the importance of considering pile–soil slip. Several of the factors Professor Boulon considers (e.g., the development of residual stresses, the occurrence of negative friction, and the response of a pile to cyclic loading) can also be considered in a simplified type of boundary element analysis. Such an analysis has been used[36] to consider the effects of cyclic loading on the ultimate axial capacity and the axial stiffness of a single pile, using an effective stress analysis which considers the effects of pore pressures generated near the pile by the cyclic loading. Fig. 22 shows a typical result for a hypothetical pile, showing the decrease in load capacity and pile stiffness with increasing cyclic load level and numbers of cycles. However, the proper consideration of pore pressure generation and dissipation during loading of a pile does require a finite element analysis, in terms of effective stress, of the type described by Professor Boulon. Such analyses are valuable in aiding our understanding of the mechanisms governing pile behaviour.

83. Mr Williams compares a variety of theoretical approaches for calculating the deflexion of laterally loaded piles and finds that most are in fair agreement. He questions, however, the rationale behind the choice of soil parameters or fitting parameters for these theories. I agree that this is the most difficult task facing the geotechnical engineer. Since we are using simplified theories to model the soil behaviour, the best that we can expect is that we can obtain, for a particular theory, fitting parameters which will enable the same theory to be used subsequently to predict the behaviour of other piles or pile groups. The best way at present of obtaining such fitting parameters is to backfigure them from the results of a pile load test. Alternatively, we can make use of empirical correlations between easily measured soil parameters (e.g., undrained shear strength) and our fitting parameters (e.g., soil modulus). In order to progress from this approach to a more fundamental approach, it is necessary to use more realistic theories of soil behaviour and to model all stages of the history of a pile, from installation onwards. It is encouraging that a start to this type of analysis has already been made (Papers 19 and 20 of this Conference).

Mr SULLIVAN, PROFESSOR REESE & Mr FENSKE (Paper 17)

The approach that is described in the Paper for the analysis of piles subjected to lateral loading is believed to be the best method currently available. While the method has some weaknesses (e.g., the use of the Winkler assumption in describing soil response), the advantages of the method far over-shadow such deficiencies. For example, if one wished to employ complete rigour in describing soil responses by taking into account the possible interaction between adjacent soil layers, it would only be necessary to employ an appropriate family of p–y curves at each position along the pile. Thus the method can be improved in the above and in other respects as additional information is gained from research on soil response.

85. The user should be cautioned, however, that the use of p–y curves in analysing the response of pile foundations to lateral load requires a thorough understanding of the techniques by which the experimental p–y curves were derived. The several experiments that were employed in deriving the various criteria for soil response are referenced in the Paper. The papers describing those experiments should be carefully studied prior to the making of computations.

86. In connection with experiments, the soil response due to cyclic loading is in most instances a lower bound. Furthermore, as indicated by the criteria, the loss of resistance in clay soils is of significant magnitude. The 'after-cyclic' behaviour of a pile foundation could be an important consideration. That aspect of the p–y curve is not discussed in the Paper and the reader is referred to the work of Matlock[37] for a discussion of this significant aspect of soil response.

Mr C. J. F. CLAUSEN (Consulting Engineer Soil Mechanics, Copenhagen)
I should like to report on a computer program that was developed to analyse the interaction between a jacket and its non-linear piled foundation, including pile group effects. In carrying out this project we were able to draw a great deal on the work done by Professor Reese, Dr Matlock and Dr Poulos.

88. The results of an interaction analysis are the displacements and forces at the jacket/pile interface points. The structural designer can then take this solution and make a back-substitution up through his structure. In order to reach a reasonable solution, the following items should be included:

(a) three-dimensional geometry;
(b) actual pile geometry;
(c) pile–jacket interaction;
(d) pile–soil–pile interaction (group effects);
(e) non-linear soil behaviour axially and laterally;
(f) second order moments in the piles.

89. The structural designer supplies the reduced (condensed) jacket stiffness matrix; that is, the relationship between jacket support displacements and jacket support forces. In addition he has to select the load cases to be investigated, and to compute the fixed interface support reactions for these cases.

90. The soil consultant has to advise on required pile penetrations and to supply pile/soil load–displacement data, $p–y$ and $t–z$. Standard geotechnical data, such as modulus of elasticity, ultimate strength values etc., will be needed as well.

91. A typical problem is now solved by the following iterative procedure:

(a) Assume displacements of the pile nodes and the surrounding soil volume.
(b) Compute pile/soil stiffness values from the relative displacements pile/soil, using the given $p–y$ and $t–z$ data.
(c) Compute the 6×6 stiffness matrix and the 6×1 load correction vector at each pile head. The load correction vector is the forces required to prevent displacements at the pile head.
(d) Form the governing system of equations for the jacket/pile interface by adding the 6×6 pile head stiffness matrices to the sub-matrices along the leading diagonal of the jacket reduced stiffness matrix, and adding the load correction vectors to the fixed interface load vector.
(e) Solve this linear system for interface node displacements.
(f) Back-substitute through the piles to find pile node displacements and pile/soil forces.
(g) Check for convergence. If satisfied, terminate.
(h) Compute displacements of the soil volume surrounding the pile nodes by Mindlin interaction values and known pile/soil forces.
(i) Repeat from (b) above.

92. Table 4 shows some results from a study of a 12 pile well head platform, with four three-pile groups, interacting with a jack-up rig. The jack-up rig will be located close to the well head platform for the period of time required to drill the wells. During this period the soils will settle and displace laterally as a result of rig weight and environmental forces, thereby inducing stresses in the piles supporting the well head platform.

Table 4. Importance of pile group effects and influence of jack-up

	Horizontal displacement of piles at mud line, mm	Maximum pile stress, % of yield
Neglecting group effects; neglecting jack-up	13	46
Including group effects; neglecting jack-up	25	59
Including group effects; including jack-up	27	64

PROFESSOR T. J. POSKITT (Paper 3)
With reference to Paper 17, the comments made by Dr Omar and myself in the introduction to discussion on Paper 3 are applicable.

Dr SEMPLE
In Paper 18, the Authors indicate that the $t–z$ method of analysing axial deflexion of offshore piles tends to overestimate axial stiffness. The characteristic pile shaft $t–z$ curve developed by Vijayvergiya[38] is roughly parabolic in shape whereas the 'Curve C′' of Coyle and Reese[39] is linear up to working load levels. In addition to providing a softer axial pile response, the Coyle and Reese $t–z$ curve probably better characterizes the soil behaviour after repeated loading. Due to scale effects, the $t–z$ method may overestimate axial stiffness of single, large-diameter piles. However, as discussed by Banerjee,[40] use of interaction factors for pile groups in homogeneous soils (Poulos and Davis[41]) probably overestimates group settlement ratio. Therefore, two errors are commonly present in the analysis of offshore pile groups that tend to compensate and may lead to reasonable prediction of axial stiffness of the structure foundations.

95. In paragraph 24, the Authors state that soil appears to be stiffer under axial load than under lateral load. I would agree that this effect is related to stress level, there being higher stress levels near the head of the pile, which is the critical zone for lateral loading. Nevertheless, I do not think it is correct to say, as is stated, that the soil along the

lower length of the pile and at the tip controls axial pile-head deflexions. That is true for a rigid pile; however, a 2 m dia. pipe pile with a 65 mm wall is a flexible pile in stiff North Sea clays. In this situation, the soils along the lower length of the pile and at the pile tip have a negligible effect on pile head response to axial loading.

PROFESSOR NOVAK

I would like to clarify the remarks I made earlier about interaction effects.

97. I did not want to express any doubts about the existence of pile interaction. I am perfectly aware that the addition of piles to a group which is already crowded does not change a thing. What I wanted to question is whether the interaction coefficients derived under the assumption of perfect elasticity and homogeneity do not actually represent the upper bound of interaction.

98. There are a few factors that reduce the interaction effects compared to those based on linear elasticity and homogeneity. These are the reduction of soil stiffness towards the ground surface (mud line), soil non-linearity, pile separation from the soil and tip fixity. All of these factors reduce the off-diagonal flexibility coefficients in the group flexibility matrix compared to the diagonal terms and, consequently, the interaction effects. Finally, most of the solutions available are limited to static loading. Dynamic interaction is more complicated and, depending on frequency, may differ from static interaction. For all these reasons, more research into pile interaction is needed.

Dr BANERJEE (Paper 13)

From contributions made in this discussion it appears that some workers may not be aware of developments made in recent years in the analysis of the behaviour of pile groups. The hybrid analysis that Dr Davies and I present in Paper 13 is fully capable of providing interaction factors for homogeneous, non-homogeneous ($E = E_0 + mz$ type) and layered soils, which are necessary for a simple analysis of axially and laterally loaded pile groups. These developments and others have been reported,[33,34,42,43] and a typical set of results for the analysis of axially loaded pile groups embedded in a $E = mz$ soil is shown in Fig. 23.

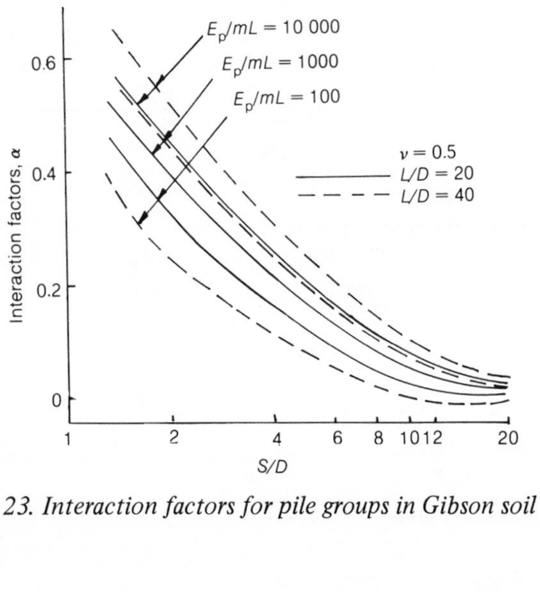

Fig. 23. Interaction factors for pile groups in Gibson soil

100. Furthermore, this method of analysis has been extended[34,44] to include non-linear behaviour such as slipping and yielding of soils by introduction of additional volume discretization over the expected plastic regions, which are likely to be confined to the immediate vicinity of the piles. It is possible to use this analysis and obtain interaction factors at a prescribed fraction of the ultimate load. However, I seriously doubt the wisdom of using these interaction factors in any non-linear analysis of axially or laterally loaded pile groups since they cannot adequately reflect the loading history of individual piles within a particular group configuration. For the non-linear analysis I would strongly advocate the fuller analysis (as described in Paper 13) so that the entire loading history of individual piles within a group can be followed. Experience gained from such a non-linear analysis indicates that quite dramatic redistribution of individual loads in piles can occur.

101. A suite of computer programs which includes the linear elastic analysis of pile groups embedded in homogeneous, non-homogeneous and layered soils has been available for the last two years from HECB Division, Department of Transport, London, virtually free to an established UK user. The non-linear program will also be available for distribution in the UK in the near future. The cost of computing is about £2 per pile for the linear analysis and about £15 a pile for the complete non-linear one – negligible in comparison with the cost of the actual structures.

Mr O. E. HANSTEEN (Norwegian Geotechnical Institute)

What is the exact definition of the p–y curves? My understanding is that each point on the curve represents the cyclic force recorded, after a number of displacement cycles (I do not know how many), with an amplitude corresponding to the abscissa of the point.

103. It appears that some people regard the p–y curve as something that one can use in essentially the same way as a stress–strain curve; that is, as a curve describing the force as a unique function of the displacement. This is not the case, if my understanding of the definition of the curves is correct. One should be very careful to ensure that the p–y curves are not used in analysis in a way that is not fully consistent with the way they are derived.

A SPEAKER

I should like to make two points basically with regard to what people may call empirical and what I call fudge factors.

105. The inclusion of a relatively thin sand stratum within the top 5–10 diameters of the pile length can significantly change the response of a pile to lateral loading, and to use some number times the shear strength as being a typical modulus in that situation or in any situation can be very dangerous.

106. In his contribution to this discussion (paragraphs 4–28), Dr Hobbs uses a shear strength around the pile of 0.5 times the soil shear strength that has been reported in the literature. This 0.5 presumably is the 0.5 that the American Petroleum Institute have suggested in their rules. Back-figuring the soil shear strength data from published information indicates an α of about 0.7. However, we at BP have some new soils data from West Sole and we back-

figured this to an α value of the order of 0.35–0.4, which is appropriate for the pile load tests that we carried out at West Sole, and this supports a recent presentation by Dr Semple that for heavily over-consolidated dilatent soils the α value may well be below 0.5.

Mr S. W. M. KOMÁROMY (C. J. B. Earl & Wright Ltd, London)

In Paper 16, it appears as though the cyclic hysteretic degradation factors are based on the assumption that a stress reversal must take place. If this is the case and the load condition suggested in the Paper is assumed, it may be desirable to have a pile that is either not driven, so that one does not get the residual compressive stresses shown in the Paper's Fig. 11, or to use a pile that is perhaps drilled and grouted. I would be interested in the Authors' comments.

A SPEAKER

In Paper 1, Dr Smith makes a comment about costly over-driving problems, and there is an inference that driving records might be analysed and on-the-spot decisions made as to whether pile driving could be curtailed. I would like to know if there has been any experience of this with the complex North Sea platforms.

109. A practical argument against it is that the position of the pile within the jacket may be fixed by other constraints, such as spiral bead welding, or the position of grout backers. There are also commercial considerations, related to problems of certification, and dealings with other oil company partners and with the contractor. The tendency for a lump sum type of contract means that the oil company operator and the partners will want to go for the desired plan, which is to achieve the pile to the full penetration.

110. Also in Paper 1 there is some reference to the use of curved conductors as being involved in a very limited way with the conductor setting programmes. Although I have no experience, I can think of two reasons why this could be a problem. One is that in conductor programmes, space on a platform is restricted and the tendency is to set the conductors as closely as possible to each other. Different consultants and contractors have different spacings, but generally about 8 ft centres has been adopted in the North Sea.

111. From conductor setting programmes, subsequent surveys of the setting of the 30 in conductors – usually about two joints are set, so one may get 80 ft penetration into the soil – have shown that this close spacing has caused a problem even in maintaining the spacing correctly between two adjacent wells. As I understand it, the curvature starts at some point further down, either with the 20 in casing or with the 13⅜ in casing, which is then further deviated. I would be interested to know of any experience of setting the 30 in casing with a curvature on it.

112. It seems to me that the Conference in general is concerned with driven piles; that is, with the calculations of pile capacity, whether axial, moment-wise or shear-wise, and with the problem of driving and installation of piling. Has the drilled-out insert pile gone out of fashion? Are there any new thoughts being generated over the problems of tension piles? A lot of work has been done, and I believe that Dr Matlock was involved in some preliminary design studies for tension piles for tension length structures. This is not the only form of structure in the North Sea where tension piles may be used. Permanent supply boat mooring systems alongside platforms may have tension piles. Although most offshore loading facilities have in the past had a gravity basis, it is not inconceivable that tension piles should be considered for this type of facility also.

Dr SMITH (Paper 1)

I am not aware of any work which has been done to predict directly the capacity of offshore piles from driving records. However, it was my understanding that Dr Goble's work is concerned with this. Presumably, there is nothing to stop him from having his computer on board and from doing all kinds of things in the future if experience improves and greater confidence is built up.

114. As far as curved conductors are concerned, in the discussion on Papers 1–7 Dr Mizikos mentions the driving of curved conductors. I think that the original work that I saw was done by Shell: certainly the idea was put forward within Shell, and one of the conceptual thoughts was that driving curved conductors would be like knocking in bent nails, and it would not be possible to knock them in at all. However, for the small curvatures that have been used – I think that Dr Mizikos confirms this – there is not any real problem if that is what one wants to do.

Mr TOOLAN & Mr HORSNELL (Paper 18)

With reference to the comments regarding the effect of sand inclusions in soil stratification on the lateral response of a pile (paragraphs 104–106), we are in total agreement. Paragraphs 11–19 of the Paper detail the methods adopted by us to overcome this problem.

116. Both Dr Banerjee and Mr Clausen draw attention to computer programs which have been developed to analyse pile group behaviour and which can account for coupling of response modes. Considerable time and effort have been put into developing these programs and Dr Banerjee and Mr Clausen are to be commended on their achievements.

117. As discussed in paragraphs 4–9 of the Paper, coupling of the response modes governing the interaction between pile groups and the structures they support is catered for within the structural analysis computer program. The structural designer requires as input data the response characteristics of each individual mode. The program developed by us was designed specifically to provide this data directly. It would be of interest to know how effective the more general programs of Dr Banerjee and Mr Clausen would be in providing this data.

118. Dr Semple re-emphasizes a point raised by us, that t–z curves were developed on the results of pile load tests carried out on relatively small piles in comparison with those used offshore. As discussed in paragraphs 26–31 of the Paper, the method proposed by us to predict the axial response of piles accounts for the effect of pile diameter.

119. In response to the latter part of Dr Semple's question, one must consider the phenomena contributing to the axial response of the pile. These are

(a) elastic compression of the pile;
(b) 'slip', or movement of the pile through the soil;
(c) elastic compression of the surrounding soil mass.

In the method described in the Paper, items (a) and (b) are modelled using conventional techniques and $t-z$ curves. Using the resulting load transfer distribution along the pile, the effect of item (c) can be determined.

120. Using this approach for single piles at relatively low load levels, we agree with Dr Semple that the soil at the tip of the pile does not participate in load transfer and that item (c) above may be neglected in comparison with items (a) and (b). However, for piles in groups, or heavily loaded single piles, the compressibility of the soil around the top of the pile plays a very important part in the computation of pile head settlement. This is in accordance with traditional formulation design techniques.

Dr MATLOCK (Paper 16)

We believe that there is a shear-affected zone very close to the pile, a discontinuity in the soil, which must be carefully examined to give an understanding under both cyclic and normal loading and certainly during pile driving what the ultimate shearing resistance felt by the pile is going to be. A great deal of attention must be given to the large strain, repeated loading, slip behaviour of the interface soil, and its effect not only on the ultimate drivability but on the ultimate capacity of the pile.

PROFESSOR DESAI (Paper 12)

I agree with the statement that the problem of consolidation after installation or driving a pile should be treated as a general coupled problem. It should be treated in such a way that, as the pile is progressively pushed into the ground, the time history of the behaviour of the surrounding soil is traced. Here the effects of driving should be considered as well as the remoulding of the soil. Also, the relative slip at the interface between the pile and soil can be important. If one wishes to be most general, it is necessary to compute changes in stresses and pore water pressures due to finite strains with geometric and material non-linearities.

123. I have performed an axisymmetric finite element analysis[45] in which the cavity expansion approach is used to simulate the process of driving and computation of the corresponding changes in stresses and pore water pressures. These changes are introduced in a consolidation code as initial conditions. The results from the code permit computation of time-dependent displacements, stresses and pore water pressures. It is also possible to compute values of distribution of shear stresses along the pile and axial stresses in the pile as the consolidation due to driving progresses. This procedure includes a linear elastic law, and a recent modification allows for a non-linear law based on a critical state plasticity model.

124. In addition to the consolidation due to driving, the procedure can also consider simultaneously consolidation due to external loads applied at given times after installation of the pile.

125. In contrast to the one-dimensional idealizations, this approach allows for both vertical and radial deformations and also for the coupling between displacements and pore water pressures.

PROFESSOR BUTTERFIELD

I would like to make a few points which are relevant to the Cambridge 'cavity-expansion' analyses of driven piles in clay. Such analyses necessarily incorporate an empirical relationship between a pseudo-elastic modulus, G say, and the undrained cohesion (C_u) of the soil in the form $G = \beta C_u$. The work reported in my paper with Ghosh (Paper 14) also generated a great deal of evidence supporting $\beta \approx 150$. Thus, if an undrained Poisson's ratio $\mu = 0.5$ is assumed, one obtains an equivalent Young's modulus $E = 2G(1 + \mu) = 450 C_u$ for use in elastic analyses. However, in many of our experiments we measured initial and final moisture contents

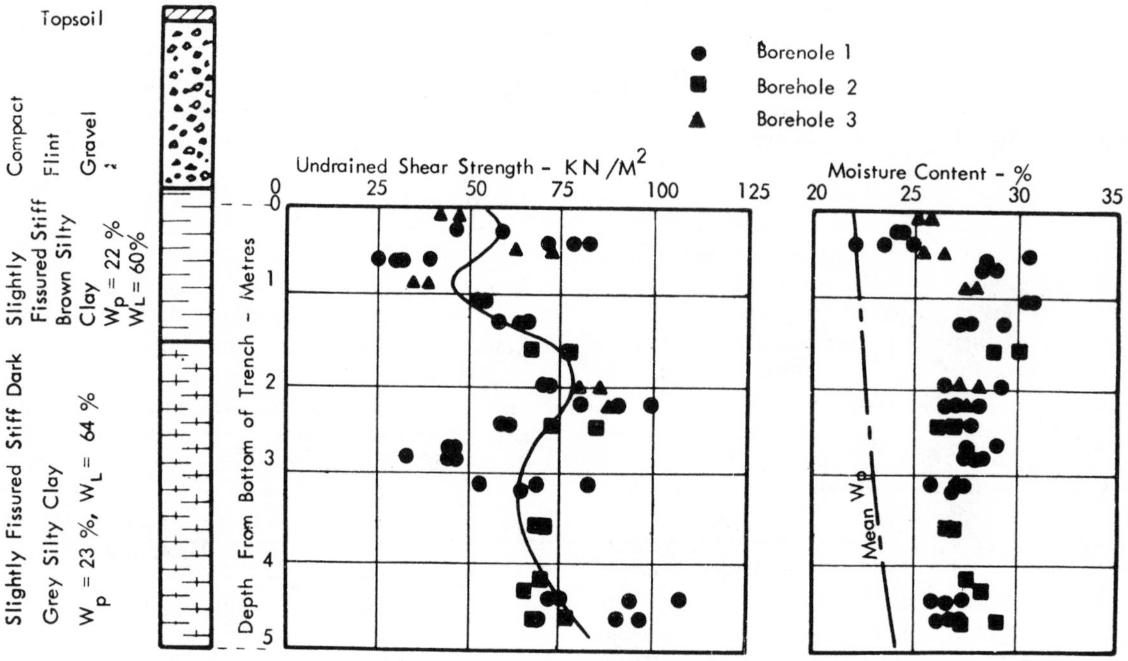

Fig. 24. Soil profile details

in the soil adjacent to the piles and found in London Clay a decrease of about 1½% and in Kaolin some 5–6%; the decrease in w being essentially zero at the pile toe and rather higher than average near the pile head. The piles were thereby stiffened and the C_u* value alongside them increased. From all such tests, which covered a range of (w, C_u*) values, we found that $G = \alpha + \beta C_u$* provided a more general interpretation of the results with $\alpha \approx 50$ kN/m^2 and $\beta \approx 70$. There is, of course, no reason why an empirical linear equation of this kind should pass through the origin ($G = 0$ when $C_u = 0$) and we may well be more successful in correlating our 'elastic' moduli with undrained strengths by using an expression of this form.

127. When Banerjee and I modelled driven piles as a cavity expansion problem[46] we only considered an 'elastic/Von Mises plastic' soil material but interesting, simple results were obtained which predicted the following: for a pile radius a, the radius (ρ) to the elastic–plastic interface was closely $\rho = a\beta^{1/2}$ (i.e., for $\beta = 36$ (a popular value at that time), $\rho \approx 6a$, whereas for $\beta = 144$, $\rho \approx 12a$); and initial ($t = 0$) pore pressures generated by an 'instantaneously' installed pile were around $6C_u$, as also were the total radial stresses (i.e., the 'instantaneous' radial effective stress changes were zero).

128. Figs 24 and 25, taken from published results[47] on a 100 mm dia. pile installed in situ by jacking, are also relevant to the discussion on radial total stresses and pore water pressures developed around preformed piles in clay. Local radial and shear stresses were measured by five pairs of diametrically opposed, Cambridge-style, load cells built into the pile as described in the original paper. Fig. 24 shows the rather inhomogeneous undrained strength profile determined from a large number of high quality 100 mm dia. triaxial tests. Plotted on Fig. 25 are the local total radial stress measurements from the leading pair of cells near the toe as they penetrated the ground. Superimposed on the figure are multiples of the mean C_u profile from Fig. 24. The results support both a total radial stress value of approximately $7C_u$ and, beyond about five diameters pile penetration, constant radial stress in line with the plane-strain cavity expansion analysis.

129. Since the simplest analysis of this kind[46] predicts zero initial effective radial stress change the ($t = 0$) pore pressures generated at the pile/soil interface were probably as high as 6–7 times C_u. The radial stresses measured on the cells following the initial ones into the ground were smaller,[47] typically about $5C_u$.

Dr BANERJEE

Leifer et al. (Paper 19) suggest that Kirby and Esrig[48] and Wroth et al.[49] found that expansion of a cylindrical cavity from zero radius to a finite radius (equal to the radius of the pile), under plain strain conditions, provides a reasonable model for pile installation. Nearly a decade ago Banerjee and Butterfield[46,50] first used the cavity expansion approach (from zero radius to a finite radius) in their effective stress analysis of the behaviour of a driven pile, and concluded that the increase in the total radial stress and the excess pore water pressure at the pile shaft surface are approximately $5C_u$ for a normally consolidated clay.

131. Although subsequent experimental studies of Butterfield and Johnston[47] confirmed these results, I had always felt that the cavity expansion approach could be regarded as a satisfactory idealization of the problem. Accordingly a comprehensive programme of theoretical and experimental studies of the behaviour of piles driven in saturated clay was undertaken at the University College, Cardiff, in 1974. The first stage of the work was completed in 1978[51] and the second stage is currently in progress. The following is a brief summary; further details can be found elsewhere.[51–53]

132. A theoretical analysis of the process of penetration of a pile within a soil mass which can be idealized as an

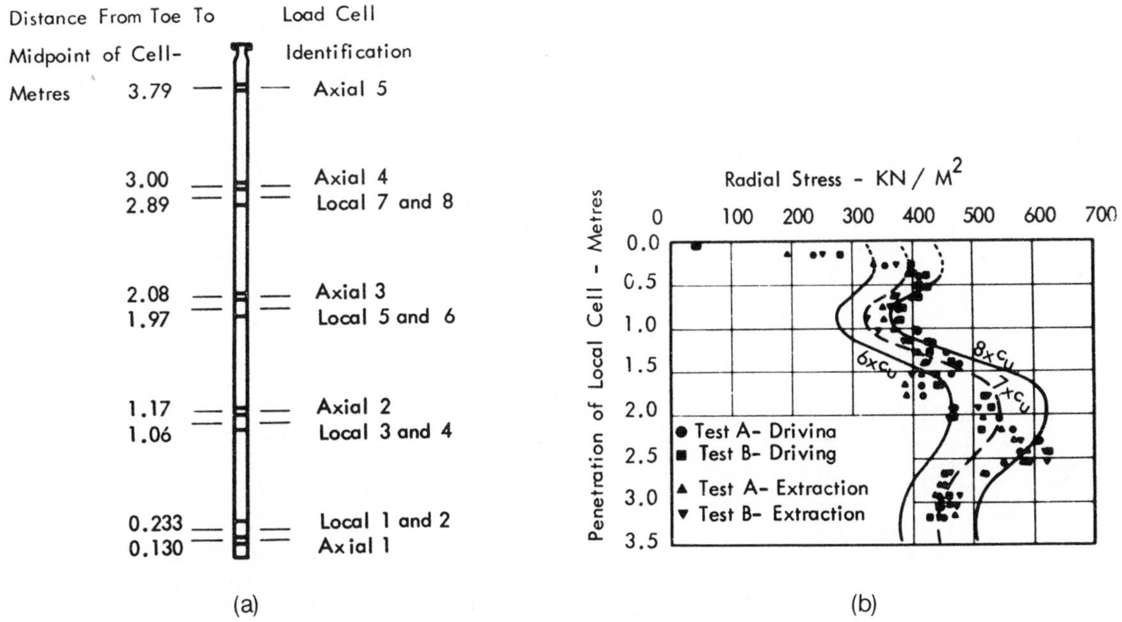

Fig. 25. (a) Identification and location of load cells; (b) radial stress on local cells 1 and 2

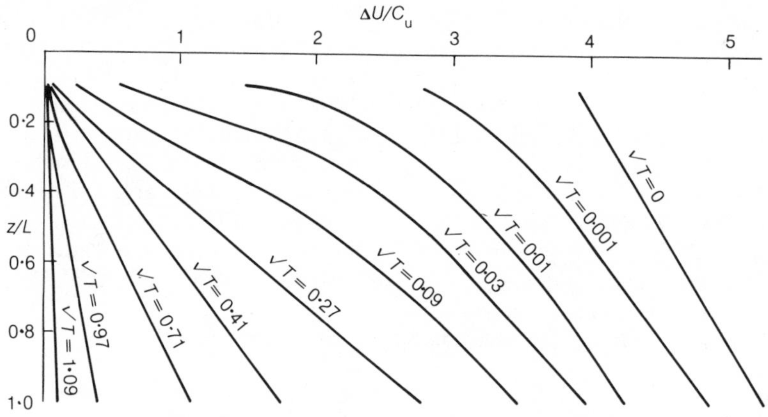

Fig. 26. Dissipation of excess pore water pressure near the pile shaft (impermeable pile)

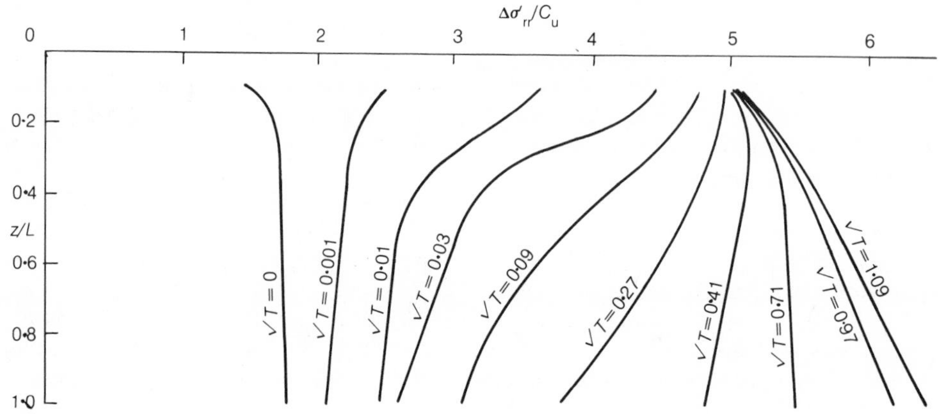

Fig. 27. Variation of the effective radial stress near the pile shaft due to consolidation (impermeable pile)

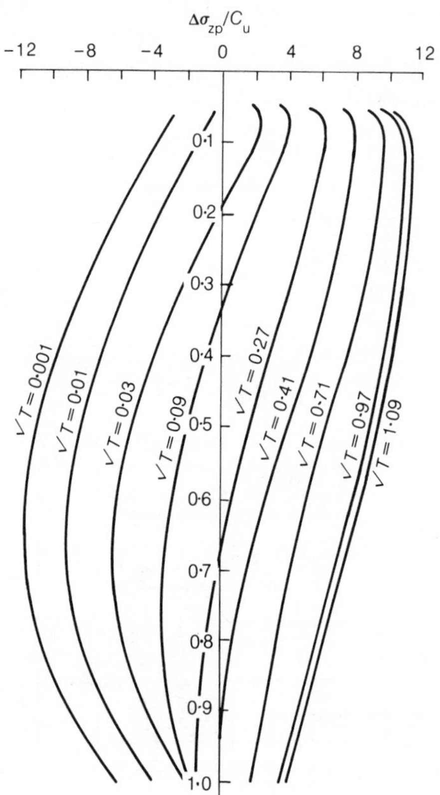

Fig. 28. Variation of vertical stresses in the pile shaft with time (impermeable pile)

elastic perfectly plastic or a compressible strain hardening plastic material (e.g., modified Cam clay) has been developed, using an updated Lagrangian formulation of the finite element method in an axisymmetric cylindrical co-ordinate system. The original analysis was developed using a linear shape function over a triangular element to represent the displacements and pore water pressure, and was later modified to have a quadratic shape function for displacements and a linear shape function for pore water pressure.

133. Figures 26–28 show typical results of the pile penetration and subsequent consolidation in a soil obeying Von Mises' yield criterion ($G = 36C_u$, $\nu' = 0.2$). The results for $\sqrt{T} = 0$ ($T = Ct/L^2$, where C is the coefficient of consolidation, t is the time and L is the embedded length of the pile) correspond to the end of pile penetration. Figs 26 and 27 show the distribution of the excess pore water pressure and the increase in the radial effective stresses at the pile face at different times after driving. The cavity expansion approach would show a uniform distribution with depth. Fig. 28 shows the distribution of the residual vertical stress at different times after the driving load has been released. The initial tensile stresses in the pile cross-section become totally compressive at later times.

134. In order to investigate whether such a sophisticated solution of the problem is really required, a carefully controlled experimental simulation of the problem was undertaken (Fig. 29). The soil sample (Kaolin) was consolidated from a de-aired slurry in a 48 in dia. × 60 in consolidation tank under a pressure of 16 lbf/in^2. A test pile (2½

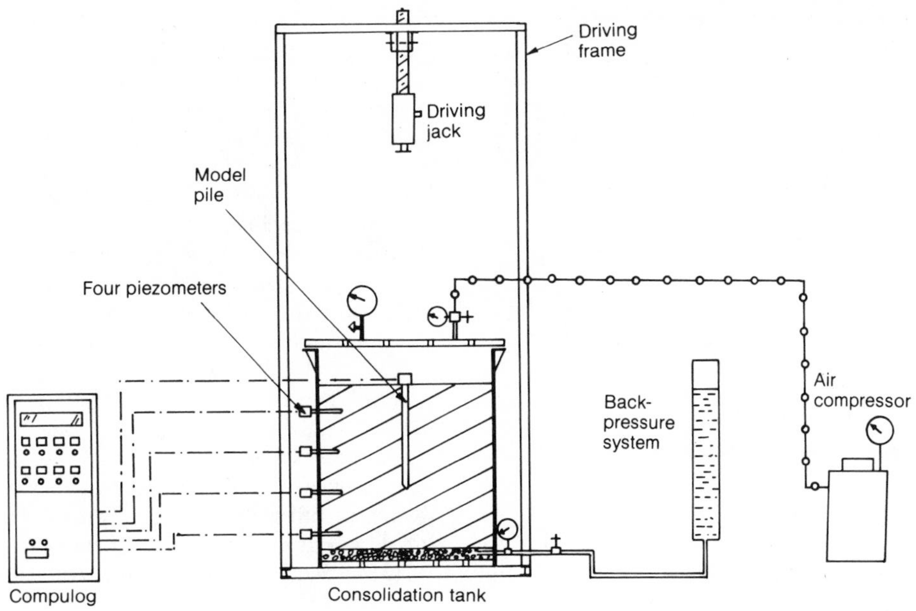

Fig. 29. Experimental apparatus

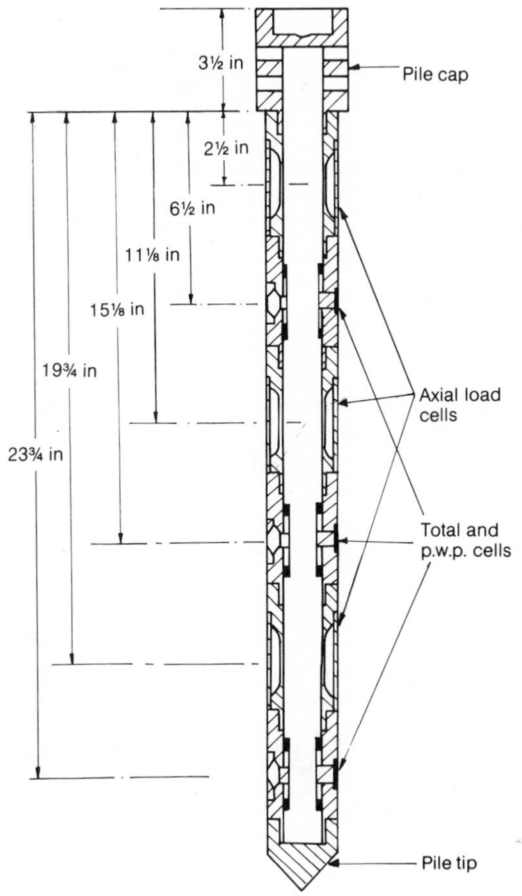

Fig. 30. Locations of the cells along the model pile shaft

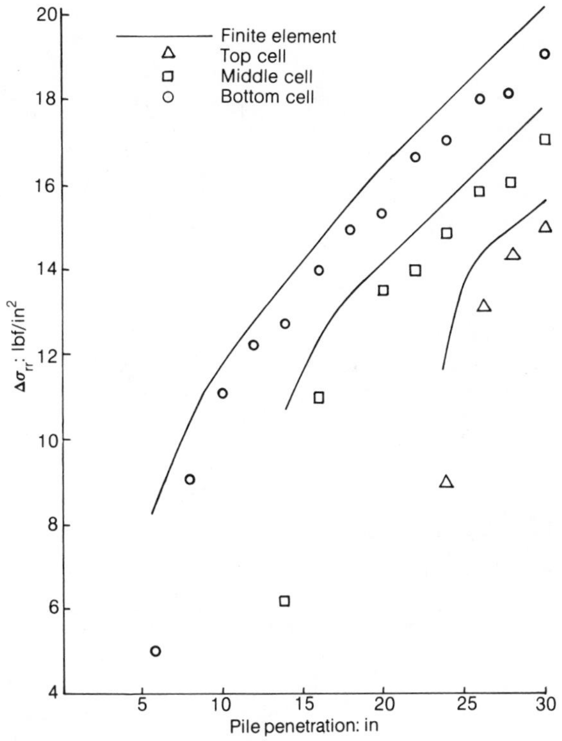

Fig. 31. Increase in total radial pressure on the pile shaft during pile penetration

in dia.) with axial load cells, total stress measuring cells and pore water pressure cells was constructed (Fig. 30). The soil parameters for the analysis using the modified Cam clay were measured to be $\lambda = 0.14$, $\kappa = 0.05$ and $M = 1.05$, and the permeability at an effective stress of 16 lbf/in^2 was found to be 0.2035×10^{-6} ft/day. In addition, values of $\nu' = 0.3$ and K_0 (the coefficient of earth pressure at rest) = 0.64 were assumed.

135. Figures 31–34 show typical comparisons between the theoretical and experimental results. Fig. 31 shows typical results of the increase in the total radial stress due to pile penetration at the positions of the total radial stress cells. The measured values of $\Delta\sigma_{rr}$ when each load cell enters the ground for the first time are consistently lower

than those obtained from the finite element analysis. This is probably due to the separation of the soil from the pile surface (which was clearly visible), resulting in stress relief. Figs 32 and 33 show the dissipation of excess pore water pressures and the variation of $\Delta\sigma_{rr}$ with time. The finite element results are generally in good agreement with the experimental data.

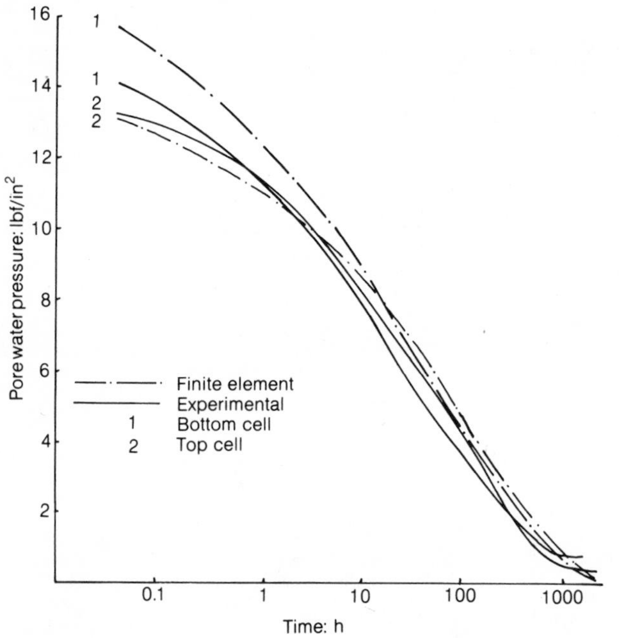

Fig. 32. Dissipation of excess pore water pressures near the pile shaft

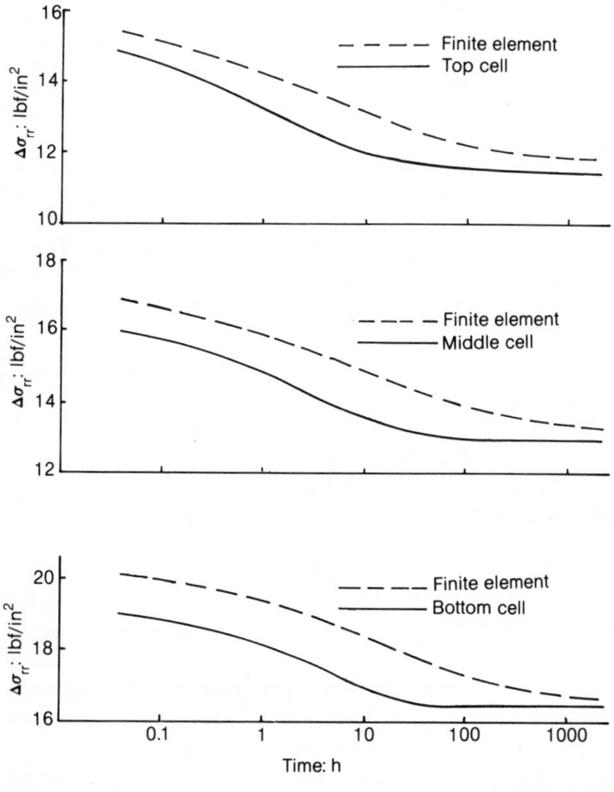

Fig. 33. Variations of $\Delta\sigma_{rr}$ at the pile–soil interface with time

136. At the end of the experiment a series of undrained triaxial tests was carried out on samples taken from various locations in the soil bed. Fig. 34 shows the measured shear strength profiles compared with those calculated using Skempton's empirical equation $C_u/p' = 0.11 + 0.37\text{PI}$ for normally consolidated clays. It is of interest to compare these results with the shear strength of the original homogeneous soil bed of 3.8 lbf/in². Typically, near the pile surface the shear strength has increased by about 50% near the ground surface to nearly 100% at the pile base.

137. Although this carefully controlled model simulation did help to calibrate our calculations and establish that the results are both relevant and meaningful, we wanted to see if these analyses could be applied equally well to full-scale problems. Unfortunately the reported results of the driving of full-scale piles are invariably incomplete because full details of the basic soil properties are not available.

138. Bakholdin and Bolshakov[54] present the results of a full-scale test on a pile, 177 in long and 7.87 in × 7.87 in in cross-section. They installed total pressure and the pore water pressure cells on the pile surface as well as in the ground. The site was described as fully saturated normally consolidated alluvial loamy sands and loams, with clay content ranging from 3.7% to 11.7% by weight. The average water content was 23.3% and the index properties were LL = 23.9% and PL = 17.25%.

139. Since the soil properties listed above are hopelessly inadequate to provide the necessary soil parameters, we had to look for a complete set of characteristic properties for a comparable soil documented elsewhere (e.g., Lambe and Whitman[55]). The estimated soil properties

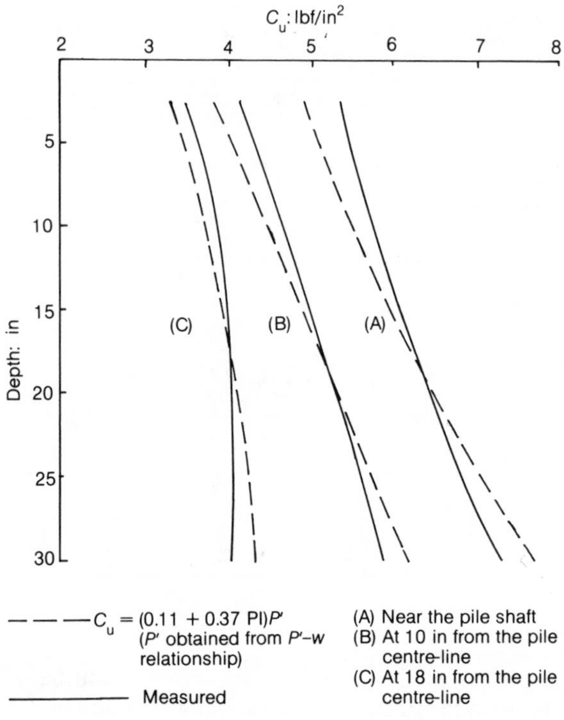

Fig. 34. Undrained shear strength (C_u) in the kaolin bed at the end of consolidation

were dry density = 96 lbf/ft^3, K_0 = 0.5, C_u/p' = 0.11 + 0.37PI, E' = 120C_u, ν' = 0.3, and permeability k = 2.32 × 10^{-4} ft/day.

140. Figures 35 and 36 show typical comparisons between the calculated (using an elastic ideally plastic idealization) and measured pore water pressures and total stresses at distance of 23.6 in from the pile centre during penetration of the pile. When the tip of the pile has reached the level of the cells ($L_z/L \approx 0.9$) the measured values reach a maximum and there is a rapid drop as the pile tip goes below this level. When the full length of the pile has been inserted in the ground (L_z/L = 1), ΔU and $\Delta\sigma_{rr}$ are approximately 3C_u and 4.4C_u respectively.

141. Clearly it can be concluded that although the problem of the expansion of a cylindrical cavity from zero radius to a finite radius[46,50] can help one to develop a qualitative understanding of this problem, it can never be regarded as a complete solution.

Mr E. P. HEEREMA (Paper 5)

I should like to comment on the cyclic degradation of horizontal stress during driving, and the question of whether it is better to use grouted piles or piles pushed in. Although it is considerable, the cyclic degradation of effective stresses is very short-lived. When a pile is load-tested several weeks after being driven, the friction has increased tremendously.

Mr LEIFER, Mr KIRBY & Mr ESRIG (Paper 19)

Dr Banerjee notes that plane strain cylindrical cavity expansion may not be a satisfactory model for estimating stresses and pore pressures around a pile in clay, and suggests that a finite element analysis which considers stress changes during pile penetration is required to adequately model the problem. We congratulate Dr Banerjee and his co-workers on the progress they have made in this complex area of engineering analysis. This work deserves careful study. Our initial impression is that, although in general the results appear reasonable, the pile stresses after installation and during reconsolidation are difficult to justify (Fig. 28). We would expect axial tension loads in the pile immediately after installation if only the outward displacements due to pile installation were considered. However, the rebound of a pile after removal of the jacking force, or after set under a hammer blow, should lead to axial compression loads immediately after pile installation. Our prediction is that the axial pile load must be compression immediately after pile installation. The change in residual pile load from compression to tension during consolidation (Fig. 28) is opposite to that predicted from the cavity expansion consolidation solution. It would be interesting to evaluate the predictions of pile stress in Fig. 28 to the measurements of the pile load from the laboratory experiment (Fig. 29).

144. We would also like to comment on the sophistication of analytical models. A useful analytical model must

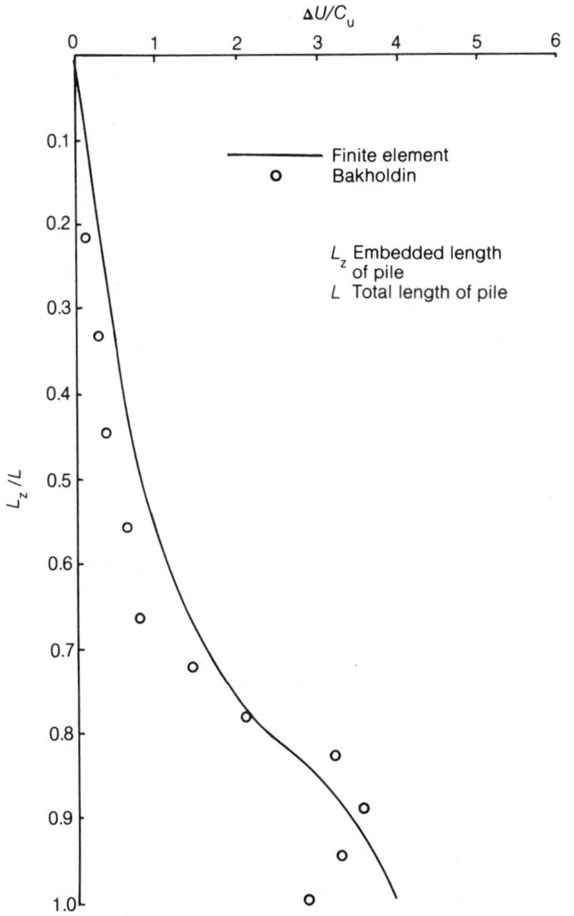

Fig. 35. Increase in pore water pressure at a distance of 23.6 in from the pile centre during driving

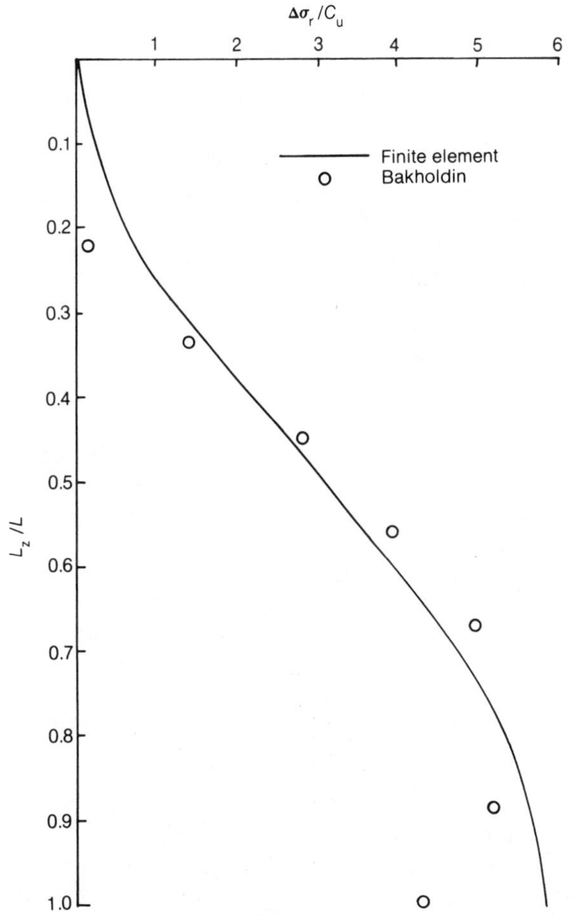

Fig. 36. Increase in total radial stress at a distance of 23.6 in from the pile centre during driving

capture the important characteristics of the real problem and allow solutions to be developed at a reasonable cost. The finite element model described by Dr Banerjee is a more complete idealization of pile installation than plane strain cylindrical cavity expansion. However, this increment in model completeness probably increases the cost for the computations by a factor of 100 or more, per problem. Therefore, we suggest that the pioneering work described by Dr Banerjee be directed toward evaluating the limitations of the expanding cavity model and developing quantitative estimates of the error associated with its use. We expect that the expanding cavity model provides very good estimates of radial stress and pore pressures for long piles (length > 25 diameters). The analysis of Dr Banerjee could be used to evaluate the range of relative pile lengths for which the expanding cavity model leads to small errors. All analytical models require estimates of engineering properties; use of an analytical model that errs by ±10% because of approximations is completely justified if the uncertainty in soil parameters is more than ±10%.

145. The concern about the analytical method used to predict pore pressure changes due to pile installation may not long be relevant. We have shown that a piezometer designed to simulate cavity expansion can be used as an exploration tool and can provide a direct measurement of pore pressures generated by pile driving, thereby eliminating the uncertainty of choosing soil properties for analysis. Such a piezometer has been designed, constructed and tested.

REFERENCES

1. BAGUELIN F. et al. Discussion contribution to Specialty Session 5: Determination of soil parameters from in-situ tests. Proc. 9th Int. Conf. Soil Mech., Tokyo, 1977, Vol. 3, 493–494. (Japanese Society of Soil Mech. Fdn Engng, Japan, 1978). Republished as: Comparison of results of tests carried out with various self-boring apparatus. Bull. Liaison Laboratoires Ponts et Chaussées, special issue VI E Soil mechanics, 1978, Apr., 88–90 (in English).
2. BAGUELIN F. et al. Le pressiomètre autoforeur et le calcul des fondations. In: Design parameters in geotechnical engineering: Proc. 7th European Conference Soil Mechanics and Foundation Engineering, Brighton, 1979. British Geotechnical Society, London, 1979, Vol. 2, 185–190.
3. SMITH I. M. and HOBBS R. Finite element analysis of centrifuged and built-up slopes. Géotechnique, 1974, 24, No. 4, 531–559.
4. HOBBS R. Finite element analyses of centrifuged soil slopes. PhD thesis, University of Manchester, 1975.
5. HOBBS R. et al. (ZIENKIEWICZ O. C. et al. (eds)). Some applications of numerical methods to the design of offshore gravity structure foundations. In: Numerical methods in offshore engineering. Wiley, London, 1978, chapter 14, 453–482.
6. FOX D. A. et al. Pile driving into North Sea boulder clay. Proc. Offshore Technology Conf. Houston, 1970, paper OTC 1200, 535–548.
7. AMERICAN PETROLEUM INSTITUTE. Recommended practice for planning, designing, and constructing fixed offshore platforms. API, 1979, RP2A, 10th edn.
8. TOMLINSON M. J. Pile design and construction practice. Cement and Concrete Association, London, 1977.
9. APPENDINO M. et al. Pore pressure of NC soft silty clay around driven displacement piles. In: Recent developments in the design and construction of piles. Institution of Civil Engineers, London, 1980, 123–129, paper 14.
10. APPENDINO M. Discussion contribution. In: Recent developments in the design and construction of piles. Institution of Civil Engineers, London, 1980.
11. MEYERHOF G. G. Bearing capacity and settlement of pile foundations. J. Soil Mech. Fdns Div. Proc. Am. Soc. Civ. Engrs, 1976, Vol. 106, Mar., No. GT3, 197–228.
12. BCP COMMITTEE. Field tests on piles in sand. Soils and foundations, Jap. Soc. SMFE, 1971, Vol. 11, June, No. 2, 29–49.
13. PETRASOVITS D. Forming of densified zones around piles driven in sand and its effect on bearing capacity. Proc. 8th Int. Conf. Soil Mech., Moscow, 1973, Vol. 21, 187–192.
14. BANERJEE P. K. and BUTTERFIELD R. (eds). Developments in boundary element methods – I. Applied Science Publishers, London, 1979.
15. BANERJEE P. K. and BUTTERFIELD R. Boundary element methods in engineering science. McGraw-Hill, London, 1979.
16. BANERJEE P. K. and DAVIES T. G. The behaviour of axially and laterally loaded single piles embedded in non-homogeneous soils. Géotechnique, 1978, Vol. 28, No. 3, 309–326.
17. WILSON R. B. and CRUSE T. A. An efficient implementation of anisotropic three-dimensional boundary integral equation stress analysis. Int. J. Num. Meth. Engng, 1978, Vol. 12, 1383–1397.
18. RANDOLPH M. F. and WROTH C. P. An analysis of the deformation of vertically loaded piles. J. Geotech. Engng Div. Proc. Am. Soc. Civ. Engrs, 1978, Vol. 104, No. GT12.
19. RANDOLPH M. F. and WROTH C. P. An analysis of the vertical deformation of pile groups. Géotechnique, 1979, Vol. 29, Dec., No. 4, 423–439.
20. WHITAKER T. Experiments with model piles in groups. Géotechnique, 1957, Vol. 7, No. 4, 147–167.
21. SOWERS G. F. et al. The bearing capacity of friction pile groups in homogeneous clay from model studies. Proc. 5th Int. Conf. Soil Mech., 1961, Vol. 2, 155–159.
22. POULOS H. G. and MATTES N. S. Settlement and load distribution analysis of pile groups. Aust. Geomech. J., 1971, Vol. G1, No. 1, 18–28.
23. MASTERS F. H. Timber friction pile foundations. Trans. Am. Soc. Civ. Engrs, 1941, 115–140, Paper 2174.
24. AMERICAN RAILWAY ENGINEERS ASSOCIATION, COMMITTEE 8. Steel and timber pile tests; West Atchafalaya Floodway – New Orleans, Texas and Mexico Railway. Proc. 50th Annual Convention, American Railway Engineers Association, Chicago, 1951, Vol. 52, 149–202.
25. SCHLITT H. G. Group pile loads in plastic soils. Proc. 31st Annual Meeting Highway Research Board, USA, 1952, 62–80.
26. KOIZUMI Y. and ITO K. Field tests with regard to pile driving and bearing capacity of piled foundations. Soils Fdns, Japan, 1967, Vol. 7, No. 3, 30–53.
27. HETÉNYI M. Beams on elastic foundations. University of Michigan Press, 1946.
28. POULOS H. G. Behaviour of laterally loaded piles: I – Single piles. J. Soil Mech. Fdns Div. Proc. Am. Soc. Civ. Engrs, 1971, Vol. 97, No. SM5, 711–731.
29. RANDOLPH M. F. A theoretical study of the perfor-

mance of piles. PhD dissertation, University of Cambridge, 1977.
30. KUHLEMEYER R. L. Static and dynamic laterally loaded floating piles. J. Soil Mech. Fdns Div. Proc. Am. Soc. Civ. Engrs, 1979, Vol. 105, No. GT2, 289–304.
31. REECE L. C. and MATLOCK H. Non-dimensional solutions for laterally loaded piles with soil modulus assumed proportional to depth. Proc. 8th Texas Conf. Soil Mech. Fndn Engng, Austin, 1956, special pubn 29, 1–41.
32. POULOS H. G. Load–deflection prediction for laterally loaded piles. Australian Geom. J., 1973, 1–8.
33. BANERJEE P. K. and DAVIES T. G. Analysis of pile groups embedded in Gibson soil. Proc. 9th Int. Conf. Soil Mech., Tokyo, 1977, Vol. 2, 381–386.
34. DAVIES T. G. Linear and nonlinear analysis of piles and pile groups. PhD thesis, University of Wales, University College, Cardiff, 1979.
35. POULOS H. G. Analysis of the settlement of pile groups. Géotechnique, 1968, 18, No. 4, 449–471.
36. POULOS H. G. Development of an analysis for cyclic axial loading of piles. Proc. 3rd Int. Conf. Numerical Methods in Geomechanics, Aachen, 1979, Vol. 4.
37. MATLOCK H. Correlations for design of laterally loaded piles in soft clay. Proc. 2nd Annual Offshore Technology Conf., Houston, Texas, 1970, 1, 577–594, Paper OTC 1204.
38. VIJAYVERGIYA V. N. Load–movement characteristics of piles. Ports '77 Conference, California, 1977.
39. COYLE H. M. and REESE L. C. Load transfer for axially loaded piles in clay. J. Soil Mech. Fdns Div. Proc. Am. Soc. Civ. Engrs, 1966, Vol. 92, No. SM2, Mar.
40. BANERJEE P. K. Analysis of axially and laterally loaded pile groups. In: Developments in soil mechanics. Applied Science Publishers, London, 1978, Chapter 9.
41. POULOS H. G. and DAVIS E. H. Elastic solutions for soil and rock mechanics. John Wiley and Sons, Inc., 1974.
42. BUTTERFIELD R. and BANERJEE P. K. Analysis of axially loaded compressible piles and pile groups. Géotechnique, 1971, Vol. 21, No. 1, 43–60.
43. BANERJEE P. K. and DRISCOLL R. M. C. Three-dimensional analysis of raked pile groups. Proc. Instn Civ. Engrs, Part 2, 1976, Vol. 61, 653–671.
44. BANERJEE P. K. et al. Two and three-dimensional problems of elasto-plasticity. In: Developments in boundary element methods. Applied Science Publishers, London, 1979, 65–96, chapter 4.
45. DESAI C. S. Effects of driving and subsequent consolidation on behavior of driving piles. Int. J. Numer. Anal. Meth. Geomech., 1978, Vol. 2, July–Sept., No. 3.
46. BUTTERFIELD R. and BANERJEE P. K. The effect of pore water pressures on the ultimate bearing capacity of driven piles. Proc. 2nd SE Asian Conf. Soil Mechanics and Foundation Engineering, Singapore, 1970, 385–394.
47. BUTTERFIELD R. and JOHNSTON I. J. The stresses acting on a continuously penetrating pile. Proc. 8th Int. Conf. Soil Mech., Moscow, 1973, Sect. 3/7, 39–45.
48. KIRBY R. C. and ESRIG M. I. Further development of a general effective stress method for prediction of axial capacity for driven piles in clay. In: Recent developments in the design and construction of piles. Institution of Civil Engineers, London, 1980.
49. WROTH C. P. et al. Stress changes around a pile driven into cohesive soil. In: Recent developments in the design and construction of piles. Institution of Civil Engineers, London, 1980.
50. BANERJEE P. K. A contribution to the study of axially loaded pile foundations. PhD thesis, Southampton University, 1969.
51. FATHALLAH R. C. Theoretical and experimental investigations of the behaviour of axially loaded single piles driven in saturated clays. PhD thesis, University of Wales, University College, Cardiff, 1978.
52. BANERJEE P. K. and FATHALLAH R. C. An Eulerian formulation of the finite element method for predicting the stresses and pore water pressures around a driven pile. Proc. 3rd. Int. Conf. Numerical Methods in Geomechanics, Aachen, West Germany, 1979, 1053–1060.
53. BANERJEE P. K. and FATHALLAH R. C. Stresses and pore water pressures around a continuously penetrating pile. Civil Engng Dept, University College, Cardiff, 1979.
54. BAKHOLDIN B. V. and BOLSHAKOV N. M. Investigation of the state of stress of clays during pile driving. Soil Mech. Fdn Engng, 1973, Vol. 10, No. 5, 300–305.
55. LAMBE T. and WHITMAN R. Soil mechanics. Wiley Interscience, New York, 1969.